Water Resources Allocation and Agriculture: Transitioning from Open to Regulated Access

Water Resources Allocation and Agriculture: Transitioning from Open to Regulated Access

Edited by

Josselin Rouillard, Christina Babbitt, Edward Challies and Jean-Daniel Rinaudo

IWA PUBLISHING

Published by **IWA Publishing**
Unit 104–105, Export Building
1 Clove Crescent
London E14 2BA, UK
Telephone: +44 (0)20 7654 5500
Fax: +44 (0)20 7654 5555
Email: publications@iwap.co.uk
Web: www.iwapublishing.com

First published 2022

British Library Cataloguing in Publication Data
A CIP catalogue record for this book is available from the British Library

ISBN: 9781789062779 (Paperback)
ISBN: 9781789062786 (eBook)
ISBN: 9781789062793 (ePub)

This eBook was made Open Access in August 2022

Content

About the Editors

Josselin Rouillard is Senior Fellow at the Ecologic Institute in Germany. He works on water governance and economics, with a background spanning both academic research and consultancy services. Josselin provides support to European institutions on the implementation of the Water Framework Directive. Between 2018 and 2020, he was on secondment at the French Geological Survey (Brgm) as part of a Marie-Skłodowska Individual Fellowship on sustainable groundwater management. Previously, he was a research associate for the IHP-HELP Centre for Water Law, Policy and Science at the University of Dundee. He also worked for the Centre for Ecology and Hydrology and the European Environmental Bureau in Brussels. His thesis from the University of Dundee was on flood risk management in Scotland. He holds an MSc in Environmental Change and Management from the University of Oxford and a BSc in Environmental Sciences from the University of Reading.

Christina Babbitt is a Director, Climate Resilient Water Systems, at Environmental Defense Fund, where she works at the intersection of water, climate, and agriculture to advance and scale water sustainability policies and practices across the western United States. Christina's research focuses on sustainable groundwater management and policy, water accounting, open-source data and tools, and water trading. Christina currently serves on the California Water Data Consortium steering committee; the Rosenberg International Forum on Water Policy advisory committee, and the board of the Aquaya Institute, a non-profit research and consulting organization dedicated to advancing global health through universal access to safe water and sanitation. She has previously worked on water issues in Europe and eastern Africa. Christina earned a PhD in natural resources from the University of Nebraska, Lincoln, an MSc in environmental science from Florida International University, a BA in international relations from Rollins College, and is currently an MBA candidate at the University of California, Berkeley.

Edward Challies is Senior Lecturer with the Waterways Centre for Freshwater Management at the University of Canterbury in Christchurch, New Zealand. His work focuses on environmental policy and governance, with particular attention to collaborative governance and collective management of land and water. His research has been funded by the European Research Council (ERC), the German Research Foundation (DFG), the New Zealand Endeavour Fund, and Our Land and Water National Science Challenge. Edward was previously Senior Research Associate with the Institute of

Sustainability Governance at Leuphana University in Lüneburg, Germany. He holds a PhD in human geography from Victoria University of Wellington.

Jean-Daniel Rinaudo is a researcher at Brgm, Montpellier University, where he coordinates the scientific programme on environmental and risk economics. Initially trained as an agricultural engineer (Montpellier SupAgro, 1994), he specialized in agricultural and resource economics (PhD University of Auvergne, 2000). Prior to joining Brgm, he worked for the International Water Management Institute in Pakistan, where his research focused on the political economy of irrigation management reforms. His current research mainly focuses on the institutional economic dimension of groundwater management. Most of his research is conducted in France, but he also works in Morocco and Chile. He is currently developing research activities in the field of natural disaster economics, focusing on the methods for assessing economic vulnerability and resilience. Dr. Jean-Daniel Rinaudo is also a member of the Scientific Council of the Adour Garonne River Basin Agency.

Contributors

Ak, Yasir M. University of Exeter, Department of Politics and International Relations, Penryn, Cornwall, United Kingdom

Aleskar, Uma Advanced Center for Water Resources Development and Management, Pune, India

Benson, David University of Exeter, Department of Politics and International Relations, Penryn, Cornwall, United Kingdom

Blomquist, William Indiana University, Indianapolis, United States

Buono, Regina M. Center for Energy Studies, Baker Institute for Public Policy, Rice University, Houston, Texas; Principal, Aither, United States

Burbach, Mark University of Nebraska-Lincoln, School of Natural Resources, United States

Closas, Alvar Water group, Department of Planning and Environment, Parramatta, New South Wales, Australia

Cook, Hadrian Harnham Water Meadows Trust, Harnham, Salisbury, United Kingdom

Demirbilek, Burcin Faculty of Economics and Administrative Sciences, Department of Political Science and Public Administration, Çankırı Karatekin University, Çankırı, Turkey

Eckstein, Gabriel Texas A&M University School of Law, Fort Worth, Texas, United States

Fragaszy, Stephen Independent consultant; at the time of writing, Ministry for the Environment, Wellington, New Zealand

García-Mollá, Marta Universitat Politècnica de València, Centro Valenciano de Estudios del Riego (CVER), València, Spain

Grantham, Theodore E. University of California, Berkeley, United States

Guillaume, Joseph H. A. Institute for Water Futures and Fenner School of Environment & Society, The Australian National University, Canberra, Australia

Hurlbert, Margot A. Canada Research Chair, Climate Change, Energy and Sustainability Policy; Johnson Shoyama Graduate School of Public Policy; University of Regina, Saskatchewan, Canada

Jedd, Theresa Technical University of Munich, School of Governance, Germany

Joshi, Dhaval School of Geosciences, University of Edinburgh, United Kingdom

Kulkarni, Himanshu Advanced Center for Water Resources Development and Management, Pune, India

Marques, Guilherme F. Institute of Hydraulic Research, Federal University of Rio Grande do Sul, Porto Alegre, Brazil

McCallum, Andrew Water group, Department of Planning and Environment, Parramatta, New South Wales, Australia

McClain, Michael E. IHE Delft Institute for Water Education; Delft University of Technology, The Netherlands

Nelson, Rebecca L. Melbourne Law School, University of Melbourne, Melbourne, Australia

Pérez-Blanco, Dionisio C. Universidad de Salamanca, Spain & Centro Euro-Mediterraneo sui Cambiamenti Climatici, Italy

Pulido-Velazquez, Manuel Universitat Politècnica de València, Institute of Water Engineering and Environment (IIAMA), València, Spain

Running, Katrina Idaho State University, Pocatello, United States

Sanchis-Ibor, Carles Universitat Politècnica de València, Centro Valenciano de Estudios del Riego (CVER), València, Spain

Schutz, Anthony University of Nebraska-Lincoln, College of Law, United States

Sengupta, Ashmita CSIRO, Black Mountain, ACT, Australia

Stein, Eric D. Southern California Coastal Water Research Project, California, United States

Tremblay, Hugo Associate Professor, Faculty of Law, University of Montreal, Canada

Valero de Palma, Juan Acequia Real del Júcar, Valencia, Spain

Yarnell, Sarah M. University of California, Davis, United States

Ziganshina, Dinara Scientific Information Center of Interstate Commission for Water Coordination in Central Asia; Tashkent Institute of Irrigation and Agricultural Mechanization Engineers, Tashkent, Uzbekistan

Zimmerman, Julie K. H. The Nature Conservancy, Sacramento, CA, United States

Foreword

Lucia De Stefano[1] and Ariel Dinar[2]

[1]*Universidad Complutense de Madrid, Spain*
[2]*University of California, Riverside, USA*

The days of disparate water allocation arrangements and of free, unregulated access to groundwater are over. Past decades when water resources were appropriated by humans only with socioeconomic development goals in mind and when groundwater was available to any user with capacity to drill and pump, have led to significant water quantity and quality degradation in many locations around the world. This has had consequences not only on freshwater ecosystems, traditional water-dependent livelihoods, or spiritual water uses, but also on the viability of economic uses themselves. The increasing costs of ensuring good-quality drinkable water, and water tables plummeting to depths that make groundwater pumping uneconomical, are just two of many signals that things have to change in the water world. And such change has to be real.

This book is timely and appropriate since allocation is at the very heart of any water management system. If water allocation is not revisited and substantively revised to adjust to present goals and challenges, it is highly unlikely that the water management system will be able to reverse unsustainable trends. This book explores water allocation strategies covering a large array of water sources and also considers water users that for quite some time have rarely been incorporated into existing water allocation systems. It reflects on the theory of water allocation and presents a number of diverse actual experiences, thus constantly reminding the reader that in water allocation schemes the interaction between humans, nature and rules rarely – if ever – happens exactly as planned by the policymaker.

Over recent years we have seen several attempts by scholars from different disciplines to contribute to the challenges posed by water allocation. Among others, we have seen innovative institutional arrangements at various levels, including institutional frameworks for joint water management by user groups, moving to decentralized water management based on local allocation mechanisms rather than on central government-led ones. We have also seen the reconsideration and modification of indigenous water allocation arrangements to face emerging water problems and conflicts. Recent years have brought works highlighting the importance of environmental flows and the usefulness of looking at the water system as a pool of highly interconnected water resources–groundwater, surface water, reclaimed water, desalinated water. This book covers and illustrates both methodologically and empirically the role and usefulness of those approaches and several others, including economic instruments such as water pricing and quantity and quality regulations.

As water scarcity and deteriorated quality endanger sustainability of water resources and water-dependent ecosystems, a holistic approach, as advocated in the book, might be proven as the most adequate and effective way forward. Managing water in a scarcity context will necessitate the combination of several approaches, the inclusion of several types of water, and the consideration of interests of and impacts on different types of sectors and users.

The larger and more complex the web of water users, the more urgent is the need to approach water allocation with sound, data-based science and with conflict-management techniques that can help stakeholders move from entrenched positions to options that can be agreeable by all. With the current advances in data collection methods and the still partially unexplored potential of Artificial Intelligence in water management, practitioners and stakeholders have an unprecedented opportunity to better understand natural and human systems and make truly informed decisions. With the current improved calculation and modeling capacity, policy interventions and interactions among water sectors and water users can be better understood by applying methods from disciplines such as Experimental Economics, Game Theory, Computable General Equilibrium models, and Hydro Economic modeling, to mention a few.

The cumulative knowledge contributed by science, however, does not produce by itself any durable change on the ground. The awareness of the need for both hard, data-based science, and soft, human-centered approaches, is key to learn from the various perspectives discussed in this book and to transition to a better and more sustainable allocation of scarce water.

Acknowledgements

We would like to warmly thank Aleksandra Lempp (Ecologic Institute) for her support in editing the book. The Editors collectively acknowledge financial support for open access publication from: German Unlatched, Germany; Ecologic Institute, Germany; Environmental Defense Fund, USA; the University of Canterbury (Library Open Access Fund), New Zealand; and the French Research Agency grant to the Belmont Forum INCLUSIVE project « Stakeholder-supported decision making for sustainable conjunctive management of soil and groundwater » (ANR-21-SOIL-0004-01).

doi: 10.2166/9781789062786_0001

Chapter 1

Introduction

Josselin Rouillard[1], Christina Babbitt[2], Edward Challies[3] and Jean-Daniel Rinaudo[4]

[1]*Ecologic Institute, Berlin, Germany*
[2]*Environment Defense Fund, San Francisco, United States*
[3]*University of Canterbury, Christchurch, New Zealand*
[4]*Brgm, G-EAU, Montpellier, France*

ABSTRACT

This chapter introduces the book by exploring the role of water allocation policies in the transition from open to regulated access in the use of water resources. Various allocation approaches and frameworks have developed over time, crafted by water users and communities, and by governments and public authorities. It examines the specific challenges of regulating agricultural water use and implementing water allocation policies in agricultural basins. In this context, the chapter then presents the overall aim of the book and the key dimensions that should be considered when characterising and assessing allocation systems in the context of agricultural water use. The chapter concludes with an outline of the content of the book. In particular, the book is structured in two main sections providing (i) an overview of cross-cutting issues related to the establishment of water allocation systems and (ii) a compilation of 13 chapters presenting water allocation systems across the world.

Keywords: Agriculture, allocations, regulated access, transition, water resources

1.1 TRANSITIONING AWAY FROM OPEN ACCESS IN THE USE OF WATER RESOURCES

1.1.1 The need to regulate water use

Throughout history, societies have developed large infrastructure, such as canals, reservoirs, pumping stations, wells and boreholes, to secure water supplies, benefiting millions in food production, improved drinking water provision, energy supply, and flood risk reduction. Yet, many regions around the world are increasingly facing water shortages, which are disrupting livelihoods and damaging economies, while depleting aquifers, rivers and lakes, and degrading associated ecosystems. Many river basins are 'closing', that is the flows required to meet societal and environmental needs cannot be met for at least part of the year (Molle *et al.*, 2010; Falkenmark & Molden, 2008; Garrick, 2015). In river basins approaching closure, competition over water and conflicts between users become more intense, with societal actors advocating for competing values and uses of the resource. Hard decisions must be taken over how water resources are used, shared and/or preserved for future generations and the environment, implying complex trade-offs between economic performance, social justice and environmental protection.

Authorities have generally responded to water scarcity by mobilising new water resources, through surface water storage and transfers, groundwater development and desalination. However, as these supply options become exhausted, prohibitively expensive or contested, authorities are increasingly looking to reduce demand, and shift from open and unlimited access, to regulated access of water resources. Regulated access relies on setting a sustainable limit on total water extraction in a river basin, groundwater body or other water management zone (i.e. a cap), and adjusting authorised extractions to meet that cap. The core issue when restricting access to water resources becomes how to best (re)allocate scarce water supplies among competing users, and between users and the environment (Arthington *et al.*, 2018).

1.1.2 Water allocation as a strategy to regulate water use

Water allocation regimes are 'the combination of actions which enable water users and water uses to take or to receive water for beneficial use according to a recognised system of rights and priorities' (Taylor, 2002). They define who is allowed to access water, how much may be taken and when, how it must be returned, and the conditions attached to the use of the extracted water (OECD, 2015). Allocation regimes need not only to specify and distribute water use authorisations, they may also regulate the exchange and transfer of such authorisations. Their prescriptions can remain informal and embedded in customs and local practice; or constitute rights explicitly codified in written agreements, legislation and formal permits (Abernethy, 2005; Bruns *et al.*, 2005).

Since the late 1980s, with growing awareness of environmental problems, policymakers have increasingly addressed the issue of water allocation as a trade-off between consumptive use and environmental protection. Hence, water policies have increasingly sought to regulate extraction within whole river basins or aquifers, and therefore to regulate access to water resources. Many water allocation regimes are moving away from solely avoiding resource exhaustion (Blomquist, 1992) and towards better providing for environmental and community needs (OECD, 2015). The most sophisticated regimes now design allocations with a more integrated and dynamic view of the water cycle, incorporating the ecological health of surface water and groundwater, environmental flows to maintain a flow regime supportive of diverse and rich aquatic biodiversity, and surface–groundwater exchanges benefiting groundwater-dependent ecosystems.

These allocation regimes are typically implemented by public authorities, although they can involve users and communities in the design and implementation of allocation rules in different ways. Allocations are then issued as time-limited allowances, permits or long-enduring entitlements. Water use charging schemes may also be implemented to recover administrative costs of the regulatory framework or to encourage efficient water use. Trading mechanisms may allow the temporary or permanent exchange of water use rights (Dinar *et al.*, 1997). More informal (re)allocation strategies may co-exist with, or override formal ones (Bruns *et al.*, 2005). For instance, water may be shared within communities on the basis of local customs and local agreements between users.

1.1.3 Regulating agricultural water use through allocation policies

Agriculture is the largest net water use in many regions (UNESCO, 2020). Surface water allocation regimes in agricultural systems have existed for centuries and millennia (Ostrom, 1990), usually to facilitate the sharing of water supplied via collectively developed irrigation schemes that capture and distribute water into networks of canals. Access to water then depends on contributions to the original investment, on fulfilling continuing obligations for operation and maintenance, and on complying with agreed procedures for distributing water during periods of scarcity. In the 20th century, agricultural water supply also benefited from larger, state-led surface water storage schemes, which regulate river flows to secure beneficial use during dry periods. Allocation decisions in infrastructure projects (user-based or state-led) are generally made by the user community or infrastructure operators, complemented by pricing mechanisms to recover the cost of infrastructure and, in some cases, to encourage efficient water use.

The issue of water allocation in agriculture has thus long been addressed as a problem of apportioning water resources between competing consumptive users, often within hydraulic systems. As water extraction from rivers and aquifers increases and ecosystems become further degraded, users increasingly have to deal with allocation decisions over larger and more loosely connected areas (including unregulated rivers and groundwater systems). They are also increasingly forced to account for minimum environmental flows while designing and implementing allocation regimes.

Allocation regimes have received considerable attention in the last three decades, and research in the field has offered general guidance on the design of allocation systems (Bruns *et al.*, 2005; Speed *et al.*, 2013; OECD, 2015, 2017). Despite progress in the understanding of institutions underpinning allocation decisions, most research has focused on traditional, user-based irrigation systems, and few studies have examined how integrated water resource allocation regimes manage extraction across whole river basins and aquifers with consideration for environmental, community and agricultural user needs. In addition, relatively few allocation systems integrate groundwater and surface water. In many places, these two resources are allocated through separate institutional arrangements, adding a layer of complexity to the challenge of meeting community, environmental, and agricultural needs through reform of allocation regimes. It is in integrating these dimensions that this edited collection makes a significant and novel contribution to the literature.

1.1.4 Challenges of establishing allocation policies in agricultural basins

Implementing water allocation regimes is particularly challenging in rural areas where water resources have been progressively developed and used outside any regulatory framework, generating a feeling of appropriation by thousands of historical agricultural users. Policies aiming at capping and reducing water use then face strong acceptability problems, as they have severe consequences for agricultural businesses and rural livelihoods. In such contexts, acceptability problems may emerge at different sensitive stages of the establishment of allocation regimes.

The first sensitive stage corresponds to the development of a registry of water users, that is identifying who is currently using water, where, when and how much. This initial inventory of users may trigger political debate and opposition for different reasons. Some users may resist, as they fear this first step will lead to greater control on extraction, announcing future restriction on use and possibly the implementation of an extraction fee. Opposition may also come from stakeholders fearing that historical users will be given (unfair) advantage in future allocation decisions, if the grandfathering principle applies. Overall, initiating an inventory of users inevitably triggers intense debates on which users are legitimate and which criteria should be used to perform future allocation. Thus, this first stage must be carried out with attention to participation, transparency, and accountability.

The second sensitive stage corresponds to the definition of a global extraction limit that will constrain allocation to users. Due to the complexity of water resources, insufficient knowledge of interactions between surface, groundwater resources and dependent ecosystems, but also to the variability of climatic and environmental conditions, there are huge uncertainties associated with the assessment of extraction limits. This fuels controversies among stakeholders who contest scientific assumptions when this can serve their own interests. Transparency and participation are here again the keywords to ensure that the extraction limit imposed on users is perceived as technically and scientifically sound, in spite of remaining uncertainties.

The third sensitive step is when the new public, river basin-wide allocation regime superimposes onto pre-existing localised, user-based or customary arrangements. This may create institutional complexities, synergistic or conflictual, disrupting established practices and norms (Bruns *et al.*, 2005). Replacing or augmenting historically derived institutions with new ones is likely to face resistance, especially when older rules favour particular local appropriative issues rather than tackling provisioning ones (Schlager and López-Gunn, 2006). For instance, allocations derived in irrigation systems are more likely to integrate irrigators' concerns and requirements regarding the timing and intensity of irrigation, while river basin-wide allocations are designed to protect environmental flows and make fair allocations. These overlapping definitions of allocations can be seen as problematic and a source of confusion and conflict.

Different countries, states and regions have made unique choices on how to deal with the socio-political sensitivity of these issues. There is an urgent need to take stock of recent institutional developments and present alternative strategies and options for designing robust allocation rules in agriculture.

1.2 OBJECTIVE AND SCOPE OF THE BOOK

The main objective of this book is to present and evaluate integrated water resource allocation regimes that aim to reduce and adapt agricultural water demand to available resources, taking into account environmental, community and other needs. This book aims to contribute to the literature on water governance, by drawing lessons on alternative allocation mechanisms and providing insights into the design of more robust allocation regimes for agricultural water use.

The originality of the book is two-fold. First, at a conceptual level, it examines governance frameworks on allocations along the full groundwater–surface water continuum, rather than considering them separately. In addition, it considers how diverse allocation regimes integrate environmental and community needs, instead of focusing on allocation regimes in the context of supply infrastructure development or complete resource exhaustion.

Second, this book intends to highlight the range of institutions (e.g. regulations, formal and informal rules, incentives, organisations, etc.) that have been developed to control agricultural extraction based on detailed analysis of different advanced cases of water allocation regimes in selected countries. Allocation systems in the reviewed countries and states exhibit a wide diversity of design parameters regarding the institutional framework guiding allocation decisions, the approach for defining the available resource pool, the rules underpinning allocations and reallocations, monitoring and enforcement mechanisms, and the wider policy mix within which the allocation regime is embedded.

1.3 KEY THEMATIC AREAS OF THE BOOK

Allocation systems exhibit a wide diversity of design options regarding:

- The institutional framework guiding allocation decisions,
- The basis for defining allocation limits,
- The rules underpinning allocations and reallocations,
- Compliance and enforcement mechanisms.

1.3.1 The institutional framework

Allocation regimes work within a legal and policy context characterised by governmental priorities for water management and procedures for integrated water resources management. Regulating water use involves establishing rules on the management of available water resources (i.e. defining spatial and temporal conditions to access and extract water) and on the rights to access and use the resource by excluding specific types of water uses, and controlling how rights to access and use the resource can be transferred between users. Who will be empowered with the right to manage, alienate or transfer water is a key question which will influence the effectiveness of any allocation regime (Ostrom, 1990; Rouillard *et al.*, 2021).

The nature and characteristics of water use rights is important to consider, as this affects the level of institutional 'rigidity' faced when implementing reallocations. Rights to water will vary in character between countries and states. Allocations may be issued as formal permits, concessions, or full property rights, or, more informally, via decisions among users or the community (Rinaudo *et al.*, 2020). The legal status of water rights varies widely, as does the level of oversight afforded to authorities, user groups and communities, and the degree of flexibility that exists in adjusting allocations.

Of particular interest is the relationship between rules established at national or state level and those set locally, by users and/or communities. Some authors warn of the inherent limitations of state-driven controls on water extraction and allocations, for instance due to the lack of acceptability amongst users of state-set rules, or the lack of capacity of the state to monitor compliance by users (Ostrom, 1990; Molle & Closas, 2020). They emphasise that the role of higher-level authorities is to empower water users and community groups in making allocation decisions and in implementing these decisions. Other authors warn of the fundamental risk in self-regulatory systems of 'capture' by specific users, resulting in poor environmental performance and unfair allocation outcomes (López-Gunn, 2006).

Finally, allocations cannot be examined in isolation. They need to be viewed as part of a wider policy instrument mix, which supplements controls on water use with for example incentives to promote behavioural change or mitigate the negative social impacts of reducing allocations (Rey *et al.*, 2019; Rouillard, 2020). Molle and Closas (2020) insist on the importance of 'carrots', through for example compensations, to secure commitments by water users when implementing 'sticks', such as regulated water extraction and use.

1.3.2 Setting allocation limits

Allocation regimes issue allowances to extract from specified water resource pools for specific periods of time. How authorities and users deal with these characteristics to issue functional allocations is of interest, including the use of particular water accounting frameworks and assessments of water balances. This must provide for user needs, but also environmental needs through for example recognition of environmental flows, impacts on protected habitats and species, and key groundwater-dependent ecosystems. Allocation must respond to variability between years and set out how the cap might be modulated accordingly. Hence, several factors may be taken into account when setting a cap on allocations: the temporal and spatial variability of water resources, such as periods of low or high flows, storage capacity of dams and aquifers, the role of groundwater in provision of sustainable baseflow in rivers, groundwater recharge rates, and so on.

Overall, allocations should ideally be consistent with the way water is stored and how it flows, accounting for return flows and connectivity between water bodies, and taking into account the impact of transferring water from one extraction point to another. For instance, regulating surface water but leaving groundwater to landowner appropriation could encourage shifting use from surface to groundwater. Also, establishing a limit on individual withdrawals at farm level may encourage increased water use efficiency as farmers strive to maximise the production value of their allocated water. However, reduced return flows to the natural environment and downstream contributes to the well-known rebound effect, changing the overall water budget at the basin scale. Adequate monitoring and return-flow accounting can mitigate this issue.

1.3.3 Allocation rules

Allocation relies on a set of rules defining:

- How individual allowances are defined (flow rates, volumes or proportional shares) and over which period they are valid (seasons, months, weeks, days, hours).
- Who has the right to use water, how much, how and when.
- How individual allocations will be ramped down in cases where the total amount of water allocated exceeds the extraction cap (which is likely in the early stages).
- How water can be reallocated between users over time to provide flexibility, and account for new users and uses and changing conditions (e.g. climate change).
- How to prioritise among agricultural uses (if different farm types for instance), and between agricultural uses and other uses such as communities and environmental needs.

- How to account for interactions between surface and groundwater and implement conjunctive use of surface and groundwater.
- How to deal with fluctuation of resource availability: in times of extreme scarcity (drought), allocations may not follow the agreed fixed time slots or proportional shares, but rather may be distributed on an agreed priority ladder between users, for instance when drinking water supply is prioritised over irrigation.

The design of those rules raises issues of social justice (i.e. what is a fair allocation of a scarce natural resource?) and of economic efficiency (i.e. how should water be allocated to maximise economic production and social welfare?). How to integrate the needs of agricultural water users and communities (social resilience) while protecting environmental flows (ecological resilience) is an often overlooked but central question in the elaboration and implementation of allocation rules. Different countries, states and regions have made unique choices on trade-offs between social, environmental and economic priorities. There is an urgent need to take stock of recent institutional developments and present alternative strategies and options for designing robust allocation rules in agriculture.

1.3.4 Compliance and enforcement

Ostrom (1990) has shown how important monitoring and enforcement procedures were in irrigation systems to increase compliance with allocation rules. Designing and implementing an effective compliance and enforcement strategy raises three main issues:

- The first one relates to the role played by the State and users' communities. While some countries have opted for a fully decentralised approach where users are given legal powers to monitor water use and impose sanctions in need, others rely on a hybrid approach where powers are shared or strictly keep enforcement as a duty of public administrations.
- Organising effective water use monitoring is a second key issue. Technology can now help with controlling use, for instance through satellite images or smart volumetric meters. Encouraging social control by users themselves is a complementary strategy which can also be supported by ICT (information and communications technology; smartphone applications to report extraction points for instance).
- The cost of monitoring and enforcement is the third issue: human and financial resources invested in compliance and enforcement must be proportionate to the level of water scarcity in the basin, with potential conflicts between agriculture and the environment.

1.3.5 Performance of allocation regimes

When crafting allocation rules, stakeholders are (implicitly or explicitly) making trade-offs between four main competing water management objectives: effectiveness, economic efficiency, social justice and resilience. However, when implemented in practice, allocation rules may not exactly match initial expectations and they may prove to perform less well than anticipated at the design stage. Evaluating the performance of rules in use along the aforementioned criteria is an exercise that should help improve allocation regimes.

Effectiveness corresponds to the ability of the allocation regime to ensure predictability of supply to water users (including domestic as well as agricultural and industrial water users) and environmental sustainability. Effectiveness depends on how actors integrate the complexity of hydrological systems into the allocation rules in order to enhance their environmental effectiveness. Economic efficiency is achieved when the apportionment of water among users maximises social welfare, with minimum transaction costs. Economic efficiency is dependent on the capacity to transfer allocations between uses, and to do this in such a way that water ideally moves towards the highest use value. Equity or fairness refers to two distinct dimensions: distributive justice which refers to fairness in the allocation itself (who gets how much water) and procedural justice which refers to the way allocation rules have been crafted (users' participation in the decision process) (Syme *et al.*, 1990). Resilience refers to the ability of allocation rules to maintain effective, fair and economically optimal outcomes during multi-year droughts and over time, taking into account the impacts of climate change.

1.4 STRUCTURE OF THE BOOK

This is an edited book with 20 chapters, including the present introduction and a conclusion. It is divided into two sections.

The first section deals with five cross-cutting issues in the transition from open to regulated access through allocation regimes. Blomquist and Babbitt (Chapter 2) focus on the political nature of the transition to regulated access, focusing on groundwater. Based on several experiences across the world, they provide nine concrete recommendations to support the transition process.

Nelson (Chapter 3) then describes how the 21st century has seen a geographic broadening of arrangements for allocating water sources. She argues that allocation regimes across the world increasingly include non-traditional water sources, interactions between sources, environmental needs, and cultural purposes. There is a broadening beyond current water rights holders to include a wider range of values in decision-making, and to recognise human rights to water. Similarly, Hurlbert (Chapter 4) reviews how previously ignored Indigenous rights to water are now increasingly recognised, drawing from examples in Canada, the United States, Central and South America, and New Zealand. She argues that recognising these rights and worldviews, such as respecting Mother Earth and the concept of Buen Vivir, move law, practice and water governance closer to a fairer and more socially just sharing of water resources.

Stein and colleagues (Chapter 5) provide insights into the increasing integration of environmental needs in water allocation regimes. They argue that a holistic environmental water allocation approach focuses on protecting overall ecological structure and functions, including preserving environmental flows at broad spatial and temporal scales, and consideration of surface–ground water interactions and the relationships between flow, sediment, temperature, and water quality. At the same time, they emphasise that environmental flow programmes will only be successful if they are sensitive to social issues and concerns, and integrate traditional values and perspectives.

Finally, Perez-Blanco (Chapter 6) explores the major economic challenges in implementing allocation regimes, and proposes key design features for an optimal water allocation framework, which achieves sustainable, equitable and robust economic growth. The chapter also provides examples of economic instruments that can facilitate the transition to regulated access through allocations where agriculture is a major water use.

The second section of the book presents 13 examples of transitions away from open access through the development of water allocations. These examples were selected to cover a wide range of geographical, environmental, social, economic and political contexts, while all addressing allocations to major agricultural irrigation water uses (Figure 1.1). Most cases present allocation systems applied to surface water and groundwater resources, although some focus more specifically on surface water or groundwater, especially where they represent the more dominant resource.

The first four cases experiences are linked to the implementation of the EU Water Framework Directive (WFD). Benson and colleagues (Chapter 7) present how authorities in England and Wales have established a catchment-based approach to regulating water abstraction, and reformed licensing arrangements to better take into account environmental sustainability. Sanchis-Ibor and colleagues (Chapter 8) examine the case of Spain and the consequences of a historically permissive policy in issuing water use rights in a context where scarcity is more pronounced. To tackle widespread overallocation and institutional rigidities of the concessional regime, new economic instruments have been sought to induce more flexibility in allocation decisions. In France, Rouillard and Rinaudo (Chapter 9) describe how authorities have started to devolve allocation decisions to catchment groups

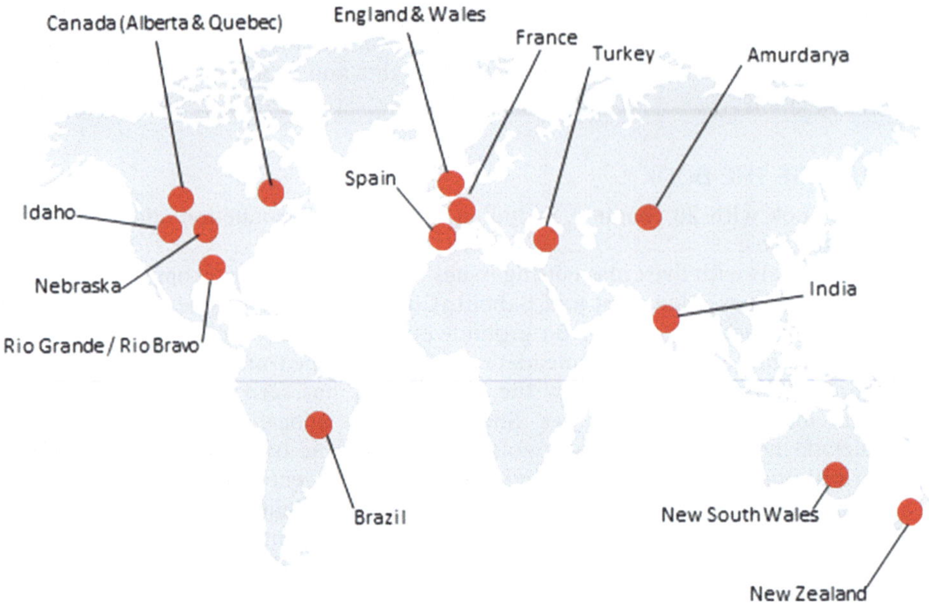

Figure 1.1 Cases included in the book as individual chapters.

and agricultural user associations in a move towards collective management of water use rights. Ak and colleagues (Chapter 10) present the Turkish case of national governance of allocations in a context shaped by the adoption of several principles set out in the WFD.

Two contrasting chapters from the Pacific region and with vastly different climates (i.e. New Zealand and Australia) are then presented. Challies and colleagues (Chapter 11) present the institutional framework in New Zealand which combines a decentralised approach to allocating water in a context where Aboriginal Maori rights over water resources are increasingly recognised nationally. Guillaume *et al.* (Chapter 12) describe the experience of New South Wales (Australia) in regulating groundwater use. A robust framework of extraction limits and a shares approach to allocations has been implemented, together with flexibility built-in thanks to a regulated water market.

The following three chapters present insights into approaches carried out in three large, federal states. First, Marques (Chapter 13) characterises key water allocation strategies followed in multiple Brazilian states, showcasing innovative solutions crafted often collaboratively between users and authorities. Aleskar and colleagues (Chapter 14) report on the overtly technical India's experience in groundwater allocations, and propose an alternative socio-hydrogeological approach that promotes participatory mapping of aquifers and decentralised groundwater allocation for agricultural decisions. Tremblay (Chapter 15) contrasts the legal framework for water allocation of two Canadian provinces (i.e. Québec and Alberta) to show the implications of different legal histories over water use rights (i.e. riparian and prior-appropriation).

There follow two cases from the United States which illustrate the potential for successful local governance of water allocations. Running (Chapter 16) describes how a cooperative five-year agreement between ground- and surface-water farmers in Idaho's Eastern Snake Plain Aquifer has contributed to the recovery of groundwater levels. The agreement was the result of multi-decadal legal, regulatory, and policy disputes, and had to operate within a rigid institutional setting comprising overlapping existing water use rights. In contrast, Jedd and colleagues (Chapter 17) present Nebraska's transition towards conjunctive use of surface water and groundwater, and provide insights into the benefits and limitations of the polycentric water governance model based on strong local control over water resources in the form of Natural Resources Districts.

The last two case chapters present transboundary cases of water allocation. Ziganshina (Chapter 18) presents the key principles and rules of water allocation in the Amudarya basin shared by Afghanistan, Kyrgyzstan, Tajikistan, Turkmenistan, and Uzbekistan. Starting in Soviet times and validated by the Almaty Agreement in 1992, the basin now has a long experience of international cooperation for sharing water. This longevity can be explained partly by the treaty's flexible modalities and specific water allocation formula. Major improvements are nevertheless needed to deal with the dual pressures of increasing water demand and diminishing water supply in the near future.

Similarly, Buono and Eckstein (Chapter 19) report on the experience of the Rio Grande basin, shared by several US states and Mexico. Its innovative and collaborative cross-border governance model has come under intense pressure in recent years. The authors explore, in particular, three major challenges regarding consideration of groundwater and ground–surface water interactions; processes for greater participatory governance; and the recent crisis linked to Mexico's water debt.

The conclusion (Chapter 20) provides a comparative assessment of the 13 cases together with insights from the cross-cutting chapters. Critical reflections on the key design features, implementation processes, and performance of water allocation regimes are made, concluding with recommendations for future research.

REFERENCES

Abernethy C. L. (2005). Constructing new institutions for sharing water. Lessons for Institutional Design. In: Water Rights Reform: Lessons for Institutional Design. B. R. Bruns, C. Ringler and R. S. Meinzen-Dick (eds.), Intl Food Policy Res Inst, Washington, USA, pp. 55–86.

Arthington A. H., Kennen J. G., Stein E. D. and Webb J. A. (2018). Recent advances in environmental flows science and water management–innovation in the anthropocene. *Freshwater Biology*, **63**(8), 1022–1034. https://doi.org/10.1111/fwb.13108

Blomquist W. (1992). Dividing the Waters: Governing Groundwater in Southern California. ICS Press, San Francisco.

Bruns B. R., Ringler, C. and Meinzen-Dick, R. S. (eds.). (2005). Water Rights Reform: Lessons for Institutional Design. Intl Food Policy Res Inst, Washington, USA.

Dinar A., Rosegrant M. W. and Meinzen-Dick R. (1997). Water Allocation Mechanisms: Principles and Examples. The World Bank, Washington, DC, USA.

Falkenmark M. and Molden D. (2008). Wake up to realities of river basin closure. *International Journal of Water Resources Development*, **24**(2), 201–215. https://doi.org/10.1080/07900620701723570

Garrick D. (2015). Water Allocation in Rivers under Pressure. Edward Elgar, Cheltenham, UK.

López-Gunn E. and Cortina L. M. (2006). Is self-regulation a myth? Case study on Spanish groundwater user associations and the role of higher-level authorities. *Hydrogeology Journal*, **14**(3), 361–379. https://doi.org/10.1007/s10040-005-0014-z

Molle F. and Closas A. (2020). Comanagement of groundwater: a review. *Wiley Interdisciplinary Review Water*, **7**(1), e1394.

Molle F., Wester P. and Hirsch P. (2010). River basin closure: processes, implications and responses. *Agricultural Water Management*, **97**(4), 569–577. https://doi.org/10.1016/j.agwat.2009.01.004

OECD. (2015). Water Resources Allocation: Sharing Risks and Opportunities, OECD Studies on Water. Éditions OECD, Paris. https://doi.org/10.1787/9789264229631-en

OECD. (2017). Groundwater Allocation: Managing Growing Pressures on Quantity and Quality, OECD Studies on Water. Éditions OECD, Paris. https://doi.org/10.1787/9789264281554-en

Ostrom E. (1990). Governing the Commons: The Evolution of Institutions for Collective Action. Cambridge University Press, Cambridge, UK.

Rey D., Pérez-Blanco C. D., Escriva-Bou A., Girard C. and Veldkamp T. I. (2019). Role of economic instruments in water allocation reform: lessons from Europe. *International Journal of Water Resources Development*, **35**(2), 206–239. https://doi.org/10.1080/07900627.2017.1422702

Rinaudo J. D., Holley C., Barnett S. and Montginoul M. (eds.). (2020). Sustainable Groundwater Management. Global Issues in Water Policy. Springer, Cham, Switzerland, Vol. **24**, pp. 47–65.

Rouillard J. (2020). Tracing the impact of agricultural policies on irrigation water demand and groundwater extraction in France. In: Sustainable Groundwater Management: A Comparative Analysis of French and Australian Policy and Implications to Other Countries, J. D. Rinaudo, C. Holley, M. Montginoul and S. Barnett (eds.), Springer, Cham, Switzerland, pp. 461–479.

Rouillard J., Babbitt C., Pulido-Velazquez M. and Rinaudo J. D. (2021). Transitioning out of Open Access: A Closer Look at Institutions for Management of Groundwater Rights in France. Water resources research: e2020WR028951, California, and Spain.

Schlager E. and López-Gunn E. (2006). Collective systems for water management: is the tragedy of the commons a myth. In: Water Crisis: Myth or Reality, P. P. Rogers, M., Ramón Llamas L. M. and Cortina (eds.), Taylor & Francis, Leiden, pp. 43–59.

Speed R., Yuanyuan L., Zhiwei Z., Le Quesne T. and Pegram G. (2013). Basin water allocation planning: Principles, procedures and approaches for basin allocation planning.

Syme G. J., Nancarrow B. E. and McCreddin J. A. (1999). Defining the components of fairness in the allocation of water to environmental and human uses. *Journal of Environmental Management*, **57**(1), 51–70. https://doi.org/10.1006/jema.1999.0282

Taylor P. (2002). Some principles for development and implementation of water-allocation schemes. ACIAR PROCEEDINGS, pp. 62–74. ACIAR; 1998.

UNESCO UN-Water. (2020). United Nations World Water Development Report 2020: Water and Climate Change. UNESCO, Paris.

doi: 10.2166/9781789062786_0011

Chapter 2

The politics of groundwater allocation and the transition from open access

William Blomquist[1] *and Christina Babbitt*[2]

[1]*Indiana University, Indianapolis, US*
[2]*Environmental Defense Fund, San Francisco, US*

ABSTRACT

Groundwater allocation is a key institutional instrument for restoring sustainability in overdrawn aquifers, especially those that have been used on an open-access basis in the past. Many suggestions for improving groundwater management – pumping reductions, fees on overuse, transferability of pumping rights or shares – are based on the establishment of groundwater allocations. Therefore, how groundwater allocations can be created and maintained is a critical foundational issue in sustainable water resource management. In this chapter, we review the institutional and policy issues associated with establishing groundwater allocations: how they differ from and relate to allocations of surface water; the various values to be taken into account when developing allocations; and why the allocation of groundwater is an inescapably political process. We also review examples of groundwater allocation efforts and identify some patterns among those examples. Combining the review and the examples, we present and explain a set of considerations regarding the process of establishing groundwater allocations. Our considerations are intended to be useful to practitioners as well as researchers interested in this subject, and therefore potentially beneficial in practical ways in the advancement of water sustainability.

Keywords: Agricultural use, allocation, groundwater management, open access, politics, stakeholder involvement, transition, water policy

2.1 INTRODUCTION: THE IMPORTANCE OF ESTABLISHING GROUNDWATER ALLOCATIONS

Groundwater is a largely hidden resource, filling the porous rock and soil layers that exist beneath the places we live, work, and recreate. While less visible than the water flowing in our rivers, streams and reservoirs, groundwater is critical to sustaining people, economies, and natural systems. Globally, groundwater is the primary water source for more than 2 billion people and makes up half of drinking water and 40% or more of irrigation water worldwide (Lall *et al.*, 2020). Increasing groundwater withdrawals have led to significant depletion, posing risks to drinking water reliability, groundwater quality, ecosystem health, agricultural productivity, and water, food, and livelihood security, and raising dangers from land subsidence and sea water intrusion (Bierkens & Wada, 2019; Foster & Chilton, 2003; Gorelick & Zheng, 2015).

The extent and severity of the groundwater challenge is daunting. Satellite observations show major aquifers are being depleted on every continent except Antarctica (Lall *et al.*, 2020). Famiglietti (2014) warned, 'Because the gap between supply and demand is routinely bridged with non-renewable groundwater, even more so during drought, groundwater supplies in some major aquifers will be depleted in a matter of decades.' Compounding the challenge, many impacts of over pumping are effectively irreversible, either physically (i.e., land subsidence) or economically because remedies are cost prohibitive (e.g., saltwater intrusion, anthropogenic contamination, mobilization of geogenic contaminants) (Lall *et al.*, 2020). Climate change is exacerbating these problems in many locations by reducing soil moisture and increasing evapotranspiration, both of which affect the vulnerability of groundwater-dependent species (Condon *et al.*, 2020).

Consequently, in areas where people, economies, and natural systems are at risk from groundwater depletion or degradation, unregulated groundwater use is no longer tenable. The severity and urgency of the situation varies by location but increasing competition for water overall intensifies the need for improved water resource management, including systems to improve how water is allocated between and within user groups (Meinzen-Dick & Bruns, 2000).

Groundwater allocation is a key institutional instrument for restoring sustainability in overdrawn aquifers, especially those that have been used on an open-access basis in the past. How groundwater allocations can be created and maintained is a critical foundational issue in sustainable water resource management. Many suggestions for improving groundwater management – pumping reductions, fees on overuse, transferability of pumping rights or shares – are based on the establishment of groundwater allocations.

In this chapter, we focus on the institutional and policy issues associated with establishing groundwater allocations, including: how they differ from and relate to allocations of surface water; the values to be considered when developing allocations; and why the establishment of allocations is an inescapably political process. We also review a small but diverse set of examples where groundwater allocations have been developed, with particular attention to allocation development in agricultural and mixed-use settings. Combining the review and the examples, we present and explain a set of recommended considerations regarding the process of establishing groundwater allocations. There is no proven formula or linear path for devising an allocation system to manage groundwater resources. Instead, we offer a set of considerations intended to be useful to practitioners in the advancement of water sustainability.

2.2 OPPORTUNITIES AND DIFFICULTIES IN ALLOCATING GROUNDWATER

The primary strategy for arresting groundwater depletion in overdrawn aquifers involves setting limits on the quantity of water that can be pumped and assigning enforceable allocations defining how much users can pump per time period (Garner *et al.*, 2020). This concept is well accepted and widely used in surface water management. While there are similarities in managing surface water and groundwater, establishing rules to allocate groundwater entails distinct opportunities and challenges.

2.2.1 Opportunities

Groundwater's slower movement through the subsurface layers of the earth and the ability of aquifers to receive and store water partially dampen the variability and vulnerability that are characteristic of surface water. Consequently, groundwater can serve as a strategic reserve, or savings account, to buffer against water scarcity and prolonged drought. Moreover, the coordinated management of groundwater and surface water can enhance the overall resilience of water systems in ways that benefit multiple uses and users of water (Blomquist *et al.*, 2004). In most places groundwater and surface water systems are closely linked (OECD, 2017) and in many circumstances, groundwater can be recharged either naturally or through designed groundwater replenishment operations.

Groundwater also allows for distributed access without massive infrastructure (unless large-scale treatment is needed), which can complement surface water systems' reliance on reservoirs, canals, and other conveyance facilities.

These characteristics of groundwater and their complementarities with surface resources open up prospects for improved water management. As we will discuss further in Section 2.3, the process of developing groundwater allocations can be pursued in ways that capitalize on these advantages.

2.2.2 Difficulties

Other characteristics of groundwater present challenges to allocation development. Many arise from the relative invisibility of the resource. Unlike surface water, groundwater is nearly impossible to observe directly, making its use and conditions difficult to monitor and manage (Moench, 2004). Further, compared with the data available for surface water, in most countries, groundwater quality, quantity, and use data are poor and incomplete (OECD, 2017; Theesfeld, 2010), partly due to insufficient investment in data collection but also because groundwater's movement creates considerable lag times that complicate monitoring and assessment. These challenges make it harder to determine what amount of groundwater use can be sustained over time without harm, and a groundwater resource can be in substantial difficulty by the time it is realized. These challenges do more than increase the technical difficulties associated with groundwater management. They also make allocating and managing groundwater more prone to conflict.

To start, groundwater's relative invisibility blurs governance boundaries. Uncertainties about recharge areas, flow and discharge characteristics, and connectivity with surface water make it difficult to identify hydrogeological boundaries precisely (Theesfeld, 2010). Uncertain boundaries complicate the identification of users and stakeholders (Blomquist & Schlager, 2005; OECD, 2017) and diminish users' awareness of their interdependent reliance on a shared aquifer system – especially compared to the more recognizable upstream/downstream dynamic among surface water users. Furthermore, aquifer boundaries often do not neatly align with surface water watersheds or river basins (OECD, 2017), so jurisdictional boundaries established to manage surface water do not necessarily match those appropriate for managing a groundwater resource even where groundwater and surface water are interconnected.

Surface water allocation and management have usually developed first, and groundwater rules have lagged or been absent altogether (Mechlem, 2012). Where it has been managed at all, groundwater has been governed and managed differently from surface water. Surface water is often regulated by a state authority through a standardized permitting system, whereas groundwater governance has generally evolved toward local planning and management (Peck *et al.*, 2019). Whether one governance mode is better than the other is largely beside the point; what matters more is that the two modes exist and reconciling them in any way – including any effort to fit the governance of surface water use into the management of groundwater or vice versa – can be expected to spark battles over control.

In addition, the physical nature of aquifer structures often renders overlying groundwater users to be differently situated (Blomquist, 2020). Two groundwater users may extract water from the same aquifer, but the aquifer's underlying geological structure can render one groundwater user more or less vulnerable than their neighbor. Users' relative vulnerability also depends on their financial or technical capability to access the resource (e.g., dig a deeper well). Situational disparities such as these complicate reaching an agreement on allocations.

Other differences among groundwater users – such as wealth, family size and influence, history in the area, ethnicity or tribal identity, and extent and location of land holdings – may also contribute to making allocation assignment conflictual. When we model or theorize about groundwater allocations, we typically disregard those differences and treat users as identical. In actual settings, any effort to transition from open access to allocations will occur among users who are not abstractly homogeneous, and their differences mean that distributional effects of any allocation system will be layered onto these distinctions as well.

Groundwater use also varies. Even where agricultural use predominates, groundwater will also be used for purposes such as a village's water supply and sanitation. Drinking water and other domestic uses clearly must be accounted for and protected, even if the predominant use of groundwater is for agriculture. Groundwater also supports surface water flows and habitat. Allocating groundwater therefore becomes more than a question of how much each user gets; it becomes a matter of how much is allocated to one kind of use compared with another (Endo, 2019; Jarvis, 2014).

Taken together, these differences among users and uses make choosing any basis for groundwater allocations prone to conflict. Groundwater allocations can be, and have been, based on various criteria, including property size, current land and water use, historical use, access to other water sources, and more (Babbitt *et al.*, 2018; Bruns & Meinzen-Dick, 2005; Ostrom, 2005; Theesfeld, 2010). These options will have location-specific effects on groundwater management – ease or difficulty of monitoring and enforcement, incentives and disincentives for greater conservation, and so on.

Further, choosing and implementing any of these allocation bases raises equity questions. If property size is used, should all land count or only arable land? If land use is the basis, should allocations be adjusted by soil type, crop type, irrigation needs? If historical use is chosen, what time period will be selected for determining users' histories, and will users inflate their use to secure a larger entitlement? If access to other water sources is used, should groundwater allocations take account of the variability or quality of those other sources and, if so, how?

The selection of any allocation basis will be contestable. No allocation basis is objectively or neutrally best, and each choice will leave some groundwater users comparatively better off and others worse off. The perceived fairness or unfairness will affect how users respond to or resist an allocation arrangement (Daigneault *et al.*, 2017; Hammond Wagner & Niles, 2020; Rinaudo *et al.*, 2016; Syme *et al.*, 1999).[1] As Babbitt *et al.* (2018) observed:

> *Groundwater management often requires asking people to change what they do in a way that has an actual or perceived financial impact. This requires establishing trust within that group of people – acceptance of a fair system that will allow them to use a sustainable amount of groundwater that supports their livelihood over the long-term.*

Because of these unavoidable aspects of developing groundwater allocations, adopting an allocation system is ultimately a political decision. It is political in the fundamental sense of who gets to decide and how. The decision might be made by users themselves through some locally situated process or by some governing body on which their interests and opinions are represented, mandated and imposed on them by some external authority without their voice, or some variation or combination along this spectrum. In the end, a decision to adopt and implement an allocation system will be made by someone through some means, meaning that at the core, developing groundwater allocations is inescapably political. Deferring to whatever decision-making body allocates surface water is also a political decision – one that should not be expected to be received well by groundwater users as already noted. Relying on larger jurisdictions also does not erase the politics of groundwater allocation – it just means that the array and relative influence of the interests being heard and attended to will differ from what they would have been at a smaller scale (Lebel *et al.*, 2005).

One more important consideration is that people usually do not attempt to solve problems until they are aware they have them. A logical and amply verified extension is that by the time a group of people is having a problem, deciding upon and implementing a solution is inherently difficult because people will have to change something they have been doing that brought about the problem. Molle and Closas (2020c) characterize the resulting situation as a dilemma. There is little incentive (and

[1] Arguing that perceived unfairness can be mitigated later through side payments or transfers of water entitlements is logically compelling but temporally backwards; in transitioning from open access to groundwater allocations, what matters is getting an allocation system adopted in the first place. Even if there are potential *ex post* remedies, users can be expected to resist an allocation system they perceive *ex ante* as unfair.

substantial disincentive) to incur the costs of adopting and implementing a groundwater allocation or licensing program *before* negative effects on the aquifer occur, but it is harder to adopt and implement one *after* conditions have deteriorated. By the time people contemplate allocations, the resource will be showing signs of overuse and degradation and developing allocations will occur in the context of trying to remedy problems rather than prevent them.

2.3 MANAGING THE TRANSITION TO GROUNDWATER ALLOCATIONS

Viewed in terms of the groundwater allocation dilemma, the pressing challenge is to move away from the already problematic open-access status quo. Since that transition will almost certainly be conflictual and political, it is not necessarily a quest for an optimal allocation system. We do not propose a particular allocation system, let alone an optimal one. The water allocation literature is rich with optimization models. In practice, however, diversity of allocation rules is the norm and optimality the exception. Instead of an optimization scenario, users of an open-access groundwater resource that is already overexploited or degraded may need suggestions for proceeding from the status quo toward something different. Our recommendations therefore focus on the transition itself.

2.3.1 Examples of transitions away from open access

Thanks to the work of many researchers in recent years, there are numerous examples of the transition from open access to groundwater allocation. Some appear in subsequent chapters of this book. In this section we draw upon six previously published case studies for most of our illustrations:

- the Gakunen Council for Coordinated Groundwater Pumping (CCGP), in Japan (Endo, 2019; Jinno & Sato, 2011);
- the Pioneer Valley groundwater basin, in Australia (Queensland 2016, 2019; Thomann *et al.*, 2020); and,
- in the US, the Main San Gabriel groundwater basin in California (Main San Gabriel Basin Watermaster 2020; Steed, 2010), the Sheridan 6 Local Enhanced Management Area (LEMA) in Kansas (Peck *et al.*, 2019), and the Central Platte and Upper Republican Natural Resources Districts (NRDs) in Nebraska (Hiatt & Zellmer, 2018; Hoffman & Zellmer, 2013).

Table 2.1 relates these cases to our recommendations regarding the transition process.

Table 2.1 Six groundwater allocation cases.

	Central Platte NRD	Gakunen CCGP	Main San Gabriel Basin	Pioneer Valley	Sheridan 6 LEMA	Upper Republican NRD
Measure and report extractions	×	×	×	×	×	×
Buyouts	×	×			×	×
Establish and maintain incentives	×	×		×	×	×
Involve as many users as possible	×	×	×		×	×
Allow carry-over			×		×	×
Make production tradeable			×		×	×
Allocations as shares			×	×		×
Set an initial cap	×	×	×	×	×	×
Data on basin conditions updated regularly and transparently and cap adjusted as needed	×	×	×	×	×	×

2.3.2 Recommended considerations for the transition process

From these examples and other contributions to the natural resource management literature, we have identified several factors that may contribute positively to transitions from open access to groundwater allocations. Our recommendations are intended as helpful observations organized in a sequence that we believe to be beneficial. We emphasize that what follows is not a recipe and the recommendations should be considered and applied according to the specific conditions and dynamics of each basin.

2.3.2.1 Measure and report extractions

There may be exceptional circumstances where users are measuring and reporting their withdrawals from an open-access groundwater resource, but typically where there are no allocations or limits there is also no reporting of groundwater use. Transitioning away from open access may begin with developing some means of measuring and reporting pumping (Bruns & Meinzen-Dick, 2005; Evans & Dillon, 2018; Hoffman & Zellmer, 2013; Thomann *et al.*, 2020). The information generated by monitoring and reporting is essential for subsequent steps in allocation development and for monitoring compliance with whatever future allocation system is worked out (Babbitt *et al.*, 2018).

In addition, the act of measuring and reporting raises users' own awareness of their individual and collective use of the resource and this alone can have beneficial impacts (Syme *et al.*, 1999). Sharing the information – being aware of their own use and knowing others will see it – is sometimes associated with a drop in withdrawals; in other words, the reporting may be as important as the measurement (Ostrom, 2005).

In the six cases we listed above, measurement and reporting were early features of the transition process. In some instances, this step began with users simply reporting their own estimated use; subsequently, provisions for the metering of wells were added to assure more consistent and accurate measurement. All six cases now feature measurement and reporting of extractions (see Table 2.1). Other cases reported in the literature reinforce the importance of this step in the transition to allocations (Rittenhouse, 2018; Singh & Zaragosa-Watkins, 2018).

Larger jurisdictions can play a catalytic role (requiring or incentivizing the initiation of measurement and reporting), a reinforcing role (making measurement and reporting a condition for state recognition of users' allocations), or both. As remote-sensing technology has become more familiar and accurate, larger jurisdictions may support its adoption as an alternative to metering large numbers of spatially dispersed agricultural wells.[2] For a transition from open access to allocations, beginning to measure and report is more important than what method is used – methods can be refined over time but getting started is the key consideration.

2.3.2.2 Developing allocations takes time; in an emergency, consider buyouts

The transition away from open access will entail tensions and dilemmas. Transition is unlikely to begin until a groundwater basin is showing signs of trouble and developing and implementing an allocation system can take a long time, during which degradation will continue. That time is necessary because building cooperation and trust 'can be the difference between successful and unsuccessful groundwater management' (Babbitt *et al.*, 2018). Patience is required and typically a system of allocations will be assembled in stages rather than all at once (Bruns, 2005).

Fortunately, due to characteristics noted in Section 2.2, changes in groundwater conditions tend to be slower than for surface water. If open-access overexploitation has occurred at high rates or for long periods, however, conditions may have worsened to the point of an emergency (e.g., failing wells, water quality impairment) by the time the allocation development process gets underway. Even in

[2] See https://blogs.edf.org/growingreturns/2021/10/21/measuring-water-california-delta-openet/ for a description of such a system, OpenET. Starting in January 2022, California will allow farmers to use OpenET to report their annual water use in the San Francisco-San Joaquin Delta, which supplies water to 25 million people and 3 million acres of Central Valley farmland in that US state.

these circumstances there is an alternative to rushing an allocation system into place. We recommend that users and other policymakers consider buyouts as a short-term measure to address emergency conditions and provide the needed breathing space for the allocation development process to continue in a deliberate fashion.

In agricultural settings, buyouts would typically take the form of compensation for temporarily fallowing or permanently retiring irrigated land, which may mean shifting to dryland farming. Buyouts can produce prompt reductions in pumping. There may be constructive roles for larger jurisdictions, such as assisting with technical analyses to target lands for buyouts or providing funds for the buyouts (EDF, 2021). Smart implementation of buyouts should target irrigated land with comparatively lower-value production relative to water use and consider whether alternative land use benefits are possible. We do not endorse buyouts as a panacea but they may be used selectively and constructively to remediate dire groundwater problems quickly and allow a considered transition to allocations to proceed.

Buyouts have been used in most of the six basins summarized in Table 2.1. We highlight here the Gakunen case in Japan where a buyout scheme was devised to target seawater intrusion. Pumpers away from the intruded area subsidized an alternative water source for the pumpers overlying the intruded area so they could cease pumping, thereby allowing groundwater levels to recover and arrest further inland movement of seawater (Endo, 2019). There are many other examples in the literature of buyouts being used in overdrafted groundwater basins (e.g., Rittenhouse, 2018; Rosenberg, 2020a,b; Ross and Martinez-Santos, 2010; Singh and Zaragosa-Watkins, 2018).

2.3.2.3 Establish and maintain incentives to complete the transition

Given the time necessary to develop groundwater allocations, it is helpful to establish some motivation to complete the process, that is, some negative consequence users will experience if the transition is not made. Otherwise, the open-access status quo may persist despite deteriorating conditions, and users may abandon or undermine the transition process once it becomes difficult, as it inevitably will.

Here too, constructive roles may be played by larger jurisdictions. A threat of intervention if users and others at the local level do not develop an allocation system or fail to show progress within a specified time frame can be highly motivating (Molle and Closas, 2020a; Rouillard *et al.* 2021). Groundwater users might prefer no governance to local governance, but they will normally prefer local to external governance. Another source of motivation can come from neighboring users or jurisdictions. It is rare for a groundwater basin to be completely isolated: commonly there is inflow from or outflow to adjacent water sources. To the extent that groundwater overuse in one basin affects adjacent ones, the threat of intervention initiated by neighbors can provide incentives to develop an allocation system. Droughts, especially severe or multi-year ones, sometimes provide impetus to initiate or complete a transition away from open access and may trigger threats of intervention from larger or neighboring jurisdictions.

Although the details vary, all six of our comparative cases involved some motivation or incentive to keep the transition process moving. In both the Central Platte and Upper Republican cases, interstate compacts govern rivers that are either fed or depleted by the groundwater supply conditions. In Pioneer Valley there is the threat of intervention by a larger jurisdiction: the State of Queensland possesses authority to take over if basin-level efforts fail and the State audits basin-level performance every five years. In the Main San Gabriel Basin there was no threat of state intervention, but there was a prospect of litigation from downstream users who depended on the basin's outflow.

2.3.2.4 Involve as many users as possible

As stated in the Organization for Economic Co-operation and Development (OECD) report: 'A clear and transparent process should be in place to facilitate stakeholder engagement in the determination of a sustainable exploitation strategy and other key allocation decisions' (OECD, 2017). There are understandable temptations to 'streamline' the process in order to reduce the transaction costs and time associated with allocation development. In practice, however, those measures often entail excluding subsets of users or other stakeholders – for example, just negotiating allocations among the largest users or just negotiating allocations for one sector.

As mentioned earlier, the allocation development process need not and should not be rushed. An allocation system should be designed for duration; shortcuts to the transition process run the risk of neglecting issues or interests in ways that generate problems during or after implementation. Even common shortcuts such as excluding small users should be reconsidered. Involvement is a form of recognition, a way of being taken seriously (Bruns, 2005). In many groundwater basins, small users have the most at stake – their reliance on groundwater may be nearly total. Excluding them from the transition process may equate to excluding the voices, concerns, and ideas of users for whom the groundwater resource is most essential.

Although an inclusive process cannot guarantee that an allocation system will achieve consensus acceptance, trouble-free implementation, and smooth adaptation over time, excluding users or other stakeholders raises the chances of opposition, resistance, and rigidity. Involving as many users as possible is a means of incorporating the most information into the allocation system design in the present and allowing some flexibility in the future. Leaving individuals or interests out of the process may appear expedient in the short term but prove a poor bargain later.

Of course, involvement can occur in a variety of ways. Large numbers of users may have to be represented (Evans & Dillon, 2018), and local or regional governmental or non-governmental bodies may have to represent some interests or groups (as occurred in the Main San Gabriel Basin case). In the cases we compared (Table 2.1), involvement took myriad forms as one would expect: agricultural water users were directly involved in the design and implementation of allocations in Sheridan 6, for example, and users' representatives negotiated the allocation arrangements in the Main San Gabriel Basin.

2.3.2.5 Allow carry-over pumping or multi-year allocations

For many and probably most users, any proposed movement away from the status quo, even in the name of reform or sustainability, prompts worry. Most users will expect to end up worse off and will weigh prospective losses more heavily than gains. Accordingly, we see setting a cap – which some readers might have expected to come next – as better tackled after users are assured of some control over their post-cap circumstances and we therefore discuss carry-over and transferability options ahead of cap setting.

One way to provide some assurance of control and flexibility under a cap is through provisions for carrying over unused pumping from one period to the next. Although users will be concerned that their new allocation under the cap will be insufficient, there is some advantage in knowing that if one is able to make do with the assigned allocation – or even with a little less in some years – under-utilization in period one can leave a little cushion in period two. Multi-year allocations are another means to this end. Users who receive an allocation that can be used over a period of years are similarly assured that reduced usage in one year can be available if needed in a subsequent year.

One of the most destructive dynamics in the use of any common-pool resource is the 'use it or lose it' calculus by which users strive to extract whatever they can in the present for fear that consumption foregone today will be foreclosed tomorrow. The prospects for successful adoption and continuation of an allocation system are enhanced by replacing the use-it-or-lose-it mindset with an assurance that forbearance does not equate to forfeit. In agricultural settings in particular, users may value knowing that pumping less in a year when precipitation or surface water is good allows one to save that unused pumping in case the next year offers less precipitation.

Reasonable constraints may be placed on carry-over allowances, for example, limiting how much unused pumping can be carried over from one period to the next or how much stored carryover water can be withdrawn in any subsequent period. Such precautions can mitigate negative effects from harmful rises or drops in groundwater levels (Hiatt & Zellmer, 2018). In practice, carry-over provisions and multi-year allocations are often accompanied by such limits but still provide users some control over their use and some incentive to exercise restraint in the present without jeopardizing the future. In most of the cases we have compared (see Table 2.1), either carry-over or multi-year allocations are in place subject to some rules. In the Sheridan 6 Local Enhanced Management Area (LEMA) in Kansas, for instance, with the assistance and support of the state's Chief Engineer, pumpers negotiated a multi-year allocation arrangement in which carry-over of unused pumping was both allowed and limited (Peck *et al.*, 2019). In the first five years of operation under those allocations, the target reduction of groundwater use was 20% and irrigators exceeded that goal by achieving an overall reduction of 23% and therefore had carry-over water available for the next five-year period.

2.3.2.6 Make production tradeable

Transferability of water allocations is recommended widely in the water resource management literature. Most such recommendations identify transferability as something to be added on to an allocation system after a cap on pumping has been put in place and allocations have been assigned to users. For example, the OECD states: 'Once the elements of a robust allocation scheme are in place, allowing water entitlement holders to trade, lease or transfer water entitlements can improve efficiency in allocation and resource use' (OECD, 2017).

As with carry-over provisions, and for similar reasons, we recommend that users be assured of transferability even before a cap is set and allocations made. Putting transferability assurances into place ahead of the assignment of a cap and allocations can facilitate the transition process: users will know that if they need more water than they were allocated they may be able to acquire some from another user and if they are able to use less than their allocation, they may be able to monetize the savings or exchange water transferred in one period for access to a comparable amount later. In combination with carry-over provisions, transferability gives groundwater users more control over their situation in the future which may reduce their anxiety about and resistance to the adoption and implementation of an allocation system.

Incorporating transferability into the transition process does not have to mean full development of a water market. That can come later if desired. The key is to get from open access to allocation by lowering water users' apprehension. Actual transfers may be relatively rare, especially at first, and 'large investments in developing formal registries, building capacity to scrutinize potential third-party impacts, and other costly measures, may not be justified by the potential level of transfers' (Bruns, 2005).

Our rationale for making allocations transferable is therefore different from the rationales found in the water resource literature of the past half-century or so. Those arguments were offered mainly on the grounds of economic efficiency (allowing those with higher-valued uses of water to bid some away from those with lower-valued uses) and adaptive management (allowing the initial assignment of allocations to be adjusted marginally through transfers as water resource conditions and/or valuations change). Both rationales are sound and important, but our argument for transferability is as an aid to getting an allocation system into place.

As with carry-over, transfers may be subject to limitations. In most of the cases we reviewed where transfers of allocations are allowed (see Table 2.1), there are constraints to account for potential impacts of shifting the location and intensity of pumping within a groundwater basin or for other purposes. In the Upper Republican Natural Resources District in Nebraska, for example, transfers are subject to approval by the district's board of directors and are limited to nearby lands (less than 6 miles or approximately 8 km away) (Hiatt & Zellmer, 2018).

2.3.2.7 Develop allocations as shares rather than fixed quantities

Providing assurance that no current users will be shut out can address users' anticipation about an allocation system's effects. This can be achieved by assigning users shares in the total amount of allowed pumping rather than assigning them fixed quantities of allowed pumping. Many users would prefer the certainty of a fixed quantity, but in an already overdrawn basin they also rationally anticipate there will be a cap someday and it will mean limitations on pumping. In such a scenario, fixed-quantity allocations for all users and reductions in total pumping necessarily collide. Those who anticipate negative impacts can be expected to resist and try to undermine the allocation arrangement. Young (2014) and Young and McColl (2003, 2009) have advocated allocating pumping shares for precisely this reason.

Share assignments are also more easily adjustable when a cap on total pumping is adjusted downward or upward based on groundwater conditions. This latter advantage is emphasized by Young (2014) and others, for example, Bruns (2005), Burchi (2018), Evans and Dillon (2018), and OECD (2017). Share allocations may therefore benefit both the transition process itself and the implementation of the resulting allocation system. Assigning shares is no panacea, however, and location-specific factors may require adjustment if some water users or uses are more flexible than others (Meinzen-Dick & Bruns, 2000).

Among our cases (Table 2.1), shares have not been as common as fixed quantities thus far. We draw attention here in particular to Pioneer Valley in Australia, a country that is a leader in the shift toward share-based allocations (Burchi, 2018). Each groundwater user in Pioneer Valley possesses a pumping license that nominally assigns a volumetric quantity of allowed pumping, but each year the basin administrator announces an adjustment percentage to be applied to each user's licensed quantity based on groundwater conditions in the basin. The application of an annual percentage to each user's allocation effectively transforms the volumetric allocations into relative shares. In Nebraska's Upper Republican Natural Resource District, each irrigated acre is allocated a specific number of inches of groundwater that can be pumped, and this allocation is adjusted every five years based on water availability, Republican River Compact compliance, and how much water is reasonably needed (Hiatt & Zellmer, 2018). The adjustable five-year allocation essentially mimics a shares-based system with allocations adjusted through time based on changing conditions. In California, the Main San Gabriel Basin was the first adjudicated groundwater basin to assign shares (Blomquist, 1992) and other basins followed that lead in subsequent allocation transitions.

2.3.2.8 Set an initial cap

At this point we turn to establishing an initial limit on total pumping, that is, the initial cap. Presumably the cap will be set below the aggregate amount of current extractions; otherwise there is little point in transitioning away from the status quo. The initial cap (and subsequent revisions to it) should also account for non-agricultural uses that may need to be protected, such as groundwater availability for drinking water and sanitation and for ecological needs.

On the other hand, there are reasons why the initial cap does not have to be at an ideal level for permanent sustainability. First, that level may not be known yet, and there are sound arguments for managing adaptively as more becomes known about groundwater conditions and how they (and the users) respond to the initial limitation. Second, groundwater resources generally have more buffering capacity than surface water resources and are more amenable to a phased approach. Third, as noted, we presume and recommend that if groundwater conditions are in crisis some emergency measures should be taken; in the absence of that, there should be at least some time for adjustment toward a sustainable path.

Our view is that (a) getting an initial cap in place is an important element of developing and implementing allocations, but (b) the initial cap need not be the final cap, and therefore (c) establishing the initial cap does not have to wait until everything is known and everyone has agreed on what the final cap should be. In some of the cases we have compared for this chapter, initial pumping limits have been put in place and then adjusted – typically downward – once groundwater users have been assigned their allocations and the allocation system has begun operation (see Table 2.1). In Kansas, the agricultural pumpers who established the Sheridan 6 LEMA agreed in 2012 to set their initial cap at 20 percent below their estimated aggregate pumping, with a multi-year allocation of 55 inches of irrigation water per acre for the initial five-year period. This was a substantial reduction from the status quo, but the farmers also were aware that it was still probably too high and would have to be adjusted (Peck *et al.*, 2019). It was nonetheless an important starting place.

2.3.2.9 Update data on groundwater use and basin conditions frequently and transparently and use this data to adapt production limits periodically as needed, until what is perceived to be sustainable is reached

Once an initial cap is established and allocations assigned, it is vital that groundwater users and basin managers understand how the allocation system is operating and what effects it is having. Credible monitoring and reporting undergird sustainable management arrangements (Evans & Dillon, 2018). Data on groundwater use are essential to establish confidence that users are complying – which reinforces users' future compliance – or to identify non-compliance so it can be addressed quickly before it erodes trust. Data on groundwater conditions are equally essential to determine the allocation system's effects, although conditions may take a while to show results because groundwater can respond more slowly. These data can be used to adjust production limits over time until desired conditions are maintained.

There is no uniform recommendation regarding how frequently updates should occur. Technological advancement has certainly made it possible to communicate data faster than in the past, when annual reports might have been regarded as frequent enough. Annual updates on usage may still be sufficient and are the norm among the cases we have compared for this chapter. The Natural Resources Districts in Nebraska and the court-appointed watermaster in the Main San Gabriel Basin issue publicly available annual reports, which include data not only for the current year but also showing how usage and conditions have changed over time.

In regard to adjustments, production limits have been adapted over time in the cases we have compared for this chapter (see Table 2.1). We draw particular attention to the Upper Republican Natural Resources District in Nebraska, where pumpers' initial allocations in 1978–1979 were 20 inches of groundwater per year for irrigation (100 inches over five years). By 2013–2017, the five-year allocation had been adjusted downward to 65 inches or an average of 13 inches per year – a drop of more than one-third compared with the initial allocation (Hiatt & Zellmer, 2018). The politics of groundwater allocation has not prevented even this substantial adjustment; the allocations are made by the District's elected board of directors, most of whose constituents are the agricultural users receiving those allocations. Transparent, credible data on groundwater use and basin conditions, combined with the external pressure of an interstate agreement governing the river to which the groundwater resource is connected, have made reduced allocations acceptable even if they are unwelcome.

2.4 CONCLUSIONS: CONTEXT AND VARIATION IN GROUNDWATER ALLOCATION DEVELOPMENT

In any location, allocations will reflect basin characteristics and conditions, uses, preferences and priorities, and the historical, cultural, and political contexts of land and water use. Groundwater allocations also will and should reflect how groundwater and its use in that location interact with surface water and surface water use. The configuration of these conditions, characteristics, and contexts is unlikely to be the same from one location to another. Accordingly, we fully expect the transition from open access to groundwater allocations to proceed differently and produce different results in each groundwater basin.

The importance of contextual influence also means that there is and can be no blueprint or recipe for the transition to allocations or the resulting allocation system. The recommended considerations in Section 2.3 are intended only as suggested ways that a transition process may be facilitated, based upon our recognition of (a) the opportunities and challenges associated with groundwater and its allocation,

(b) the inherently and unavoidably conflictual and political nature of the transition to groundwater allocations, and (c) some lessons and illustrations drawn from places where the transition has been made.

Observing that the transition process and resulting allocation system will vary from place to place is, in our view, an empirical statement. It is not a normative argument that local allocation systems will automatically be efficient, equitable, and effective. There are no panaceas, and that includes local management itself (Boelens *et al.*, 2005; Molle & Closas 2020b; Ostrom, 2005). The importance of local variation derives from the significance of context rather than from a faith that local arrangements will always or inherently be best. Sustainable groundwater management is a complex endeavor, and on balance the development of allocations is better undertaken with careful regard for contextual factors (Boelens *et al.*, 2005; Daigneault *et al.*, 2017), which include not only physical conditions but social and historical ones, such as what users perceive to be fair and legitimate (Bruns & Meinzen-Dick, 2005; Hammond Wagner & Niles, 2020). Water 'scarcity and competition are not standard problems for which universally valid solutions can be formulated' (Boelens *et al.*, 2005; see also Ostrom, 2005).

Although allocation development is context-specific, larger-scale jurisdictions such as national and regional governments can fulfill valuable roles to enable and encourage the transition process. They can aid in the development and availability of technical information about water resources, provide financial or other assistance for emergency actions such as buyouts if needed, and adjust water resource policies, property law, or other rules to remove impediments to flexibility-enhancing practices such as carry-over or multi-year allocations, transferability, and allocations of shares. Within such a policy environment, efforts at the groundwater basin level to transition to allocations will face better prospects for success. Larger jurisdictions can also support transition processes by providing institutional arrangements that facilitate conflict resolution and equitable access to participation.

There is no reason to expect that transitions from open access to allocations will be easy, quick, or inexpensive, or will be successful upon first attempt. Diversity in allocation systems is neither unexpected nor undesirable, and policymakers may need to resist temptations to impose or induce uniformity in the name of harmonization. Similarly, there is no reason to expect transitions away from open access will result in optimal allocation systems as defined by modelers or other researchers. The transition process itself is more important and offers water users a more sustainable, enduring pathway than the open-access alternative. Further, a transition away from open access sets the stage for other water management improvements (transfers, conjunctive management, etc.) to follow.

REFERENCES

Babbitt C., Gibson K., Sellers S., Brozovic N., Saracino A., Hayden A., Hall M. and Zellmer S. (2018). Summary of lessons learned with implications for SGMA implementation. In: The Future of Groundwater in California: Lessons in Sustainable Management from Across the West, Report. Environmental Defense Fund, New York, USA, 15–18.

Bierkens M. F. P. and Wada Y. (2019). Non-renewable groundwater Use and groundwater depletion: a review. *Environmental Research Letters*, **14**, 1–44.

Blomquist W. (1992). Dividing the Waters: Governing Groundwater in Southern California. ICS Press, San Francisco, USA.

Blomquist W. (2020). Beneath the surface: complexities and groundwater policy-making, *Oxford Review of Economic Policy*, **36**(1), 154–170. https://doi.org/10.1093/oxrep/grz033

Blomquist W. and Schlager E. (2005). Political pitfalls of integrated watershed management. *Society and Natural Resources*, **18**(2), 101–117. https://doi.org/10.1080/08941920590894435

Blomquist W., Schlager E. and Heikkila T. (2004). Common Waters, Diverging Streams: Linking Institutions to Water Management in Arizona, California, and Colorado. Resources for the Future, Washington, DC.

Boelens R., Zwarteveen M. and Roth D. (2005). Legal complexity in the analysis of water rights and water resources management. In: Liquid Relations: Contested Water Rights and Legal Complexity, D. Roth, R. Boelens and M. Zwarteveen (eds.), Rutgers University Press, New Brunswick, USA, pp. 1–20.

Bruns B. (2005). Routes to water rights. In: Liquid Relations: Contested Water Rights and Legal Complexity, D. Roth, R. Boelens and M. Zwartveen (eds.), Rutgers University Press, New Brunswick, NJ, USA, pp. 215–236.

Bruns B. R. and Meinzen-Dick R. (2005). Frameworks for water rights: an overview of institutional options. In: Water Rights Reform: Lessons for Institutional Design, B. R. Bruns, C. Ringler and R. Meinzen-Dick (eds.), International Food Policy Research Institute, Washington DC, USA, pp. 3–25.

Burchi S. (2018). Legal principles and legal frameworks related to groundwater. In: Advances in Groundwater Governance, K. G. Villholth, E. Lopez-Gunn, K. I. Conti, A. Garrido and J. Van der Gun (eds.), CRC Press, Boca Raton, USA, pp. 119–136.

Condon L. E., Atchley A. L. and Maxwell R. M. (2020). Evapotranspiration depletes groundwater under warming over the contiguous United States. *Nature Communications*, **11**, 1–8 https://doi.org/10.1038/s41467-020-14688-0

Daigneault A., Greenhalgh S. and Samarasinghe O. (2017). Equitably slicing the pie: water policy and allocation. *Ecological Economics*, **131**, 449–459. https://doi.org/10.1016/j.ecolecon.2016.09.020

Endo T. (2019). A fifty-year experience of groundwater governance: the case study of Gakunan council for coordinated groundwater pumping, Fuji city, Shizuoko prefecture, Japan. *Water*, **11**(12), 2479. https://doi.org/10.3390/w11122479

Environmental Defense Fund (EDF). (2021). Advancing Strategic Land Repurposing and Groundwater Sustainability in California: A Guide for Developing Regional Strategies to Create Multiple Benefits, EDF, San Francisco, USA.

Evans R. S. and Dillon P. (2018). Linking groundwater and surface water: conjunctive water management. In: Advances in Groundwater Governance, K. G. Villholth, E. Lopez-Gunn, K. I. Conti, A. Garrido and J. Van der Gun (eds.), CRC Press, Boca Raton, USA, pp. 329–351.

Famiglietti J. S. (2014). The global groundwater crisis. *Nature Climate Change*, **4**, pp. 945–948. https://doi.org/10.1038/nclimate2425

Foster S. S. D. and Chilton P. J. (2003). Groundwater: The processes and global significance of aquifer degradation. *The Royal Society*, **358**, 1957–1972.

Garner E., McGlothlin R., Szeptycki L., Babbitt C. and Kincaid V. (2020). The sustainable groundwater management Act and the common law of groundwater rights–finding a consistent path forward for groundwater allocation. *UCLA Journal of Environmental Law & Policy*, **38**(2), 163–215. https://doi.org/10.5070/L5382050109

Gorelick S. M. and Zheng C. (2015). Global change and the groundwater management challenge. *Water Resources Research*, **51**, 3031–3051. https://doi.org/10.1002/2014WR016825

Government of Queensland, Department of Natural Resources and Mines. (2016). Pioneer Valley Water Resource Operations Plan.

Government of Queensland, Department of Natural Resources, Mines and Energy. (2019). Minister's Performance Assessment Report, Water Plan (Pioneer Valley) 2002.

Hammond Wagner C. R. and Niles M. T. (2020). What is fair in groundwater allocation? Distributive and procedural fairness perceptions of California's sustainable groundwater management Act. *Society and Natural Resources*, **33**(12), 1508–1529. https://doi.org/10.1080/08941920.2020.1752339

Hiatt J. and Zellmer S. (2018). Case study 6/Nebraska – upper republican natural resources district. In: The Future of Groundwater in California: Lessons in Sustainable Management from Across the West, C. Babbitt (ed.), Environmental Defense Fund, New York, USA, pp. 72–80.

Hoffman C. and Zellmer S. (2013). Assessing institutional ability to support adaptive, integrated water resources management. *Nebraska Law Review*, **91**(4), 805–863.

Jarvis W. T. (2014). Contesting Hidden Waters: Conflict Resolution for Groundwater and Aquifers. Routledge, London, UK.

Jinno K. and Sato K. (2011). Groundwater resources management in Japan. In: Groundwater Management Practices, A. N. Findikakis and K. Sato (eds.) CRC Press, Leiden, The Netherlands, pp. 17–31.

Lall U., Josset L. and Russo T. (2020). A snapshot of the world's groundwater challenges. *Annual Review of Environment and Resources*, **45**, 171–194. https://doi.org/10.1146/annurev-environ-102017-025800

Lebel L., Garden P. and Imamura M. (2005). The politics of scale, position, and place in the governance of water resources in the Mekong region. *Ecology and Society*, **10**(2), 18. https://doi.org/10.5751/ES-01543-100218

Main San Gabriel Basin Watermaster. (2020). 2019–2020 Annual Report. Azusa, USA.

Mechlem K. (2012). Legal and Institutional Frameworks. Groundwater Governance: A Global Framework for Action, Thematic Paper 6, GEF, FAO, UNESCO-IHP, IAH, and the World Bank, Rome, Italy.

Meinzen-Dick R. S. and Bruns B. R. (2000). Negotiating water rights: introduction. In: Negotiating Water Rights, B. R. Bruns and R. S. Meinzen-Dick (eds.), Vistaar Publications, New Delhi, India, pp. 23–55.

Moench M. (2004). Ground water: the challenge of monitoring and management. In: The World's Water, 2004–2005, P. Gleick (ed.), Island Press, Washington, DC, USA, pp. 79–100. Washington, DC.

Molle F. and Closas A. (2020a). Co-management of groundwater: a review. *WIREs Water*, **7**(1), e1394.

Molle F. and Closas A. (2020b). Why is state-centered groundwater governance largely ineffective? *WIREs Water*, **7**(1), e1395.

Molle F. and Closas A. (2020c). Groundwater licensing and its challenges. *Hydrogeology Journal*, **28**(6), 1961–1974. https://doi.org/10.1007/s10040-020-02179-x

OECD. (2017). Groundwater Allocation: Managing Growing Pressures on Quantity and Quality. OECD Studies on Water. OECD Publishing, Paris, France.

Ostrom E. (2005). Understanding Institutional Diversity. Princeton University Press, Princeton, USA.

Peck J. C., Illgner R., Wiley J. and Crittenden Owen C. (2019). Groundwater management: the movement toward local, community-based, voluntary programs. *Kansas Journal of Law and Public Policy*, **29**(1), 1–49.

Rinaudo J.-D., Moreau C. and Garin P. (2016). Social justice and groundwater allocation in agriculture: a case study. In: Integrated Groundwater Management: Concepts, Approaches and Challenges, A. J. Jakeman, O. Barreteau, R. J. Hunt, J.-D. Rinaudo and A. Ross (eds.), Springer, Cham, Switzerland, pp. 273–293.

Rittenhouse K. (2018). Case study 5/Colorado – Rio Grande water conservation district (subdistrict No. 1). In: The Future of Groundwater in California: Lessons in Sustainable Management from Across the West, C. Babbitt (ed.), Environmental Defense Fund, New York, USA, pp. 62–71.

Rosenberg A. (2020a). Incentives to Retire Water Rights Have Reduced Stress on the High Plains Aquifer. United States Department of Agriculture, Economic Research Service, Washington, DC.

Rosenberg A. B. (2020b). Targeting of water right retirement programs: evidence from Kansas. *American Journal of Agricultural Economics*, **102**(5), 1425–1447. https://doi.org/10.1111/ajae.12102

Ross A. and Martinez-Santos P. (2010). The challenge of groundwater governance: case studies from Spain and Australia. *Regional Environmental Change*, **10**, 299–310. https://doi.org/10.1007/s10113-009-0086-8

Rouillard J., Babbitt C., Pulido-Velazquez M. and Rinaudo J.-D. (2021). Transitioning out of open access: a closer look at institutions for management of groundwater rights in France, California, and Spain. *Water Resources Research*, **57**(4), e2020WR028951. https://doi.org/10.1029/2020WR028951

Singh K. and Zaragosa-Watkins M. (2018). Case study 8/Texas – Edwards aquifer authority. In: The Future of Groundwater in California: Lessons in Sustainable Management from Across the West, C. Babbitt (ed.), Environmental Defense Fund, New York, USA, pp. 93–105.

Steed B. C. (2010). Natural Forces, Human Choices: An Over Time Study of Responses to Biophysical and Human Induced Disturbance in Los Angeles, California Groundwater Governance. Ph.D. thesis, Indiana University, Bloomington, USA.

Syme G. J., Nancarrow B. E. and McCreddin J. A. (1999). Defining the components of fairness in the allocation of water to environmental and human uses. *Journal of Environmental Management*, **57**(1), 51–70. https://doi.org/10.1006/jema.1999.0282

Theesfeld I. (2010). Institutional challenges for national groundwater governance: policies and issues. *Ground Water*, **48**(1), 131–142. https://doi.org/10.1111/j.1745-6584.2009.00624.x

Thomann J. A., Werner A. D., Irvine D. J. and Currell M. J. (2020). Adaptive management in groundwater management: a review of theory and application. *Journal of Hydrology*, **586**, 124871. https://doi.org/10.1016/j.jhydrol.2020.124871

Young M. D. (2014). Designing water abstraction regimes for an ever-changing and ever-varying future. *Agricultural Water Management*, **145**, 32–38.

Young M. D. and McColl J. C. (2003). Robust reform: the case for a new water entitlement system for Australia. *Australian Economic Review*, **36**(2), 225–234.

Young M. D. and McColl J. C. (2009). Double trouble: the importance of accounting for and defining water entitlements consistent with hydrological realities. *Australian Journal of Agricultural and Resource Economics*, **53**(1), 19–35.

doi: 10.2166/9781789062786_0025

Chapter 3

Allocations and legal trends in the 21st century

Rebecca L. Nelson

Melbourne Law School, University of Melbourne, Melbourne, Australia

ABSTRACT

Around the world, allocating water is a major task of national and sub-national water laws. Although each jurisdiction develops its laws uniquely in response to local conditions, common trends emerge across multiple jurisdictions. The 21st century has seen a geographic broadening of administrative planning and permitting or licensing arrangements for allocating traditional water sources. Allocation regimes also increasingly cover non-traditional water sources that previously fell outside their bounds, including brackish groundwater, rainwater and recycled water. Recognizing interactions between sources is of increasing concern to allocation regimes. Legal developments also provide for an increasing array of water users and other participants in allocation processes. Traditionally, allocation arrangements have centered on significant agricultural, municipal, industrial and other commercial uses. More recently, water law increasingly contemplates granting allocations to, or controlling, uses for environmental purposes and cultural purposes, and broadening access beyond current water rights holders under water market rules. Both constitutional and water laws increasingly recognize a human right to water, and increasingly inclusive processes apply to formulating water plans that guide or influence allocation regimes. These developments are not uniformly present, or present to the same degree, in all jurisdictions. However, encouragingly, they demonstrate increased recognition of the need to respond to water scarcity and give greater attention to equity and inclusion among water users and participants in allocation regimes.

Keywords: Allocation, environmental flows, groundwater, human right to water, law, participation, water licensing, water planning, water transfers, water user institutions

3.1 INTRODUCTION

Around the world, allocating water is a major task of national and sub-national water laws. Although each jurisdiction develops its laws uniquely in response to local conditions, overarching global legal trends have emerged over time, carrying important implications for agricultural water uses. First, water law increasingly uses administrative planning and permitting systems to cover more water sources, including those that traditionally have been less regulated, and it does so in more complex ways. Second, it increasingly both accommodates and controls more diverse social and environmental, as well as economic, stakeholders, such as water users and participants, in administrative regimes.

While both categories of trends are notable, they are not without difficulties and challenges. These point to varying gaps between law on paper and law in practice, and, in some cases, more fundamental questions about the appropriateness of reform directions.

This chapter synthesizes insights into these overarching legal trends and accompanying difficulties and challenges in 21st century reforms to allocating water between water users, with a focus on implications for the agricultural context. More detailed, jurisdiction-specific discussion follows in subsequent chapters. As befits these introductory purposes, the chapter draws on secondary literature,[1] attempting to represent as many countries as possible, especially those that are less frequently discussed in the literature. A number of important matters fall outside its scope, including allocations between nations or sub-national jurisdictions; detailed institutional aspects of water allocation regimes; water charging arrangements; and, except as they relate to 'formal' allocation laws, customary regimes for water allocation, which are in particularly wide use in Africa (Schreiner & van Koppen, 2020).

3.2 THE GLOBAL GROWTH AND DEVELOPMENT OF ADMINISTRATIVE REGIMES FOR ALLOCATING WATER

3.2.1 Adoption of permitting and planning systems across more nations

Across diverse jurisdictions, a common starting point for formal legal water allocation systems is the vesting of water resources in the state (e.g. Indonesia, Kenya: Dai *et al.*, 2017; Israel, Nicaragua: Global Legal Research Center, 2013; Bangladesh, Bhutan: Hirji *et al.*, 2018).[2] Water is still capable of being privately owned in some jurisdictions, and this occurs much more commonly for groundwater than for surface water (e.g., Austria, Japan, Portugal, some areas of the USA: OECD, 2015), but these are best considered 'isolated pockets' of water allocation regimes (Burchi, 2019).

Public control of water is associated with the increasingly common use of administrative regimes to allocate water, often by using statutory water plans and state permitting systems (China, Kenya, South Africa: Dai *et al.*, 2017; Tanzania, Kenya, Burkina Faso, Swaziland: van Koppen, 2017). These administrative regimes advantageously 'allow[] prospective consideration of the consequences of allocation … [and] a systemic approach to consideration of the secondary impacts of management actions' (Cosens, 2018). They are also able clearly to define the relevant entitlements, which is considered vital to an allocation system (Hirji *et al.*, 2018; OECD, 2017). However, these advantages of administrative regimes can only reliably be secured if the management plans are binding on water allocation decisions. Otherwise, they may act as mere plans 'on the shelf', with reduced on-ground influence on allocations. In a recent World Bank survey of 101 countries, 65 required water management plans that were binding on water allocation decisions, and 63 required a permit to abstract water accompanied by a public notice and comment process (World Bank, 2019). In other jurisdictions, water plans are a mere consideration (Finland: Soininen, 2014) and do not bind water permit decisions. In some cases, water planning and permitting regimes only apply to some parts of jurisdictions that are considered particularly water-stressed (Australia: Cosens, 2018) or where environmental problems are significant (New Zealand: Daya-Winterbottom, 2014). There is also great diversity in the scale at which water planning occurs, from the national level (Nepal: Hirji *et al.*, 2018) to the state level (states of the western USA: Cosens, 2018), to the local watershed level (some areas of Australia, California, USA: Cosens, 2018; South Africa: Pejan *et al.*, 2014). Though it appears

[1]For reasons of feasibility and accessibility to a broad readership, where possible, the materials cited rely preferentially on readily available books, reports and multidisciplinary journal articles that discuss multiple national jurisdictions, rather than law journal articles, which usually discuss a single jurisdiction and are often only accessible through legal databases.

[2]For clarity and brevity, where a reference describes particular national legislative systems, these are stated in parentheses with the reference.

relatively rare, some jurisdictions are attempting to link statutory water plans with land use plans (e.g. Vietnam, Ecuador, California, USA, Japan, Zambia, Tanzania: Burchi, 2019).

A key task of administrative allocation regimes is adapting allocations to scarcity. At the level of individual allocations, a variety of mechanisms for adapting to reduced water availability have emerged in different water allocation regimes over the last two decades. Some statutes expressly allow regulators to reduce licensed abstraction volumes in response to reduced availability, with or without compensation (Ecuador, Tanzania, Namibia, Zambia: Burchi, 2019). Others now specify allocations as 'shares' of the available resource, rather than as absolute volumes (Australia, also considered in England and Wales: Burchi, 2019). At the extreme, reduced water availability may lead water regulators to restrict the granting of permits that draw from certain sources for agricultural uses in favor of drinking water, as is occurring in the Netherlands in relation to dwindling fresh groundwater (Dai *et al.*, 2017).

At the level of water planning instruments, climate change and associated reductions in water availability are increasingly considered (India; Rufiji Basin, Tanzania: Hirji *et al.*, 2018), but this is far from the case in all jurisdictions (Afghanistan, Bangladesh, Bhutan, Nepal, Pakistan, Sri Lanka: Hirji *et al.*, 2018). More generally, growing use of groundwater is frequently suggested as a climate change adaptation measure (Hirji *et al.*, 2018). This highlights the increasing need to ensure sustainable groundwater allocation regimes are in place, particularly given that millions of wells globally are at risk of running dry with even modest reductions in water table levels (Jasechko & Perrone, 2021).

3.2.2 Application of permitting and planning regimes to more water sources

As administrative permitting and planning regimes are covering water sources in more nations, these regimes also appear to be applied to an increasing range of water sources within nations. This is consistent with formal allocation regimes having increasing benefits when water sources are used more intensively (OECD, 2017).

From a traditional focus on surface water, water allocation laws have increasingly broadened to allocate access to groundwater (Mechlem, 2016), even in comparatively water-rich regions (British Columbia, Canada: Mechlem, 2016). Indeed, permitting arrangements for groundwater allocations are now considered to be 'the central element' of groundwater laws to control demand (Mechlem, 2016). In some cases, significant recent changes to groundwater allocation have been triggered by changes to surface water allocation. In Australia, where sub-national states have historically controlled water allocation, the federal Parliament's first legislative foray into this area was prompted by concerns about over-allocation of surface water to agriculture, but the resulting legislation introduces comprehensive caps on allocation of both groundwater and surface water for consumptive purposes. This federal intervention instituted the first such controls on groundwater in some areas, which had not previously been subject to state-level caps on allocation (Nelson, 2018). By contrast, India's national government has encouraged, but not required, states to adopt groundwater-specific legislation that often includes permitting regimes in notified areas. However, rather than adopting this permitting approach, some irrigation-dependent states have favored incentive-based approaches to crop diversification and micro-irrigation to address sustainability concerns (Cullet, 2009). A still-common alternative to direct groundwater allocation and permitting systems is land-based restrictions, such as restricting the construction of new wells or regulating the allowable expansion of irrigated land (OECD, 2015).

Water management regimes also increasingly cover sources that have not traditionally been regulated, including brackish groundwater, rainwater and recycled water, as well as water that is a by-product of industrial processes (OECD, 2015). However, there are diverse approaches to formalizing access to non-traditional sources, and many commentators indicate the need to create improved and legally certain allocation frameworks for non-traditional sources (Hirji *et al.*, 2018; OECD, 2017). In Africa, commentators have noted a largely unmet need for law reform to facilitate irrigated agriculture using wastewater (African Union, 2020).

A different perceived solution to scarcity has attracted significant attention, and the development of legal frameworks, in some western jurisdictions: managed aquifer recharge, or aquifer storage and recovery (e.g. California, Texas, Florida, USA: Bray, 2020). These practices involve intentionally placing water into aquifers for later use using injection wells or infiltration ponds, which raises initial permitting considerations related to water quality impacts (Bray, 2020). Associated permitting regimes for the 'recovered' water must deal with many complex issues, including the percentage of the stored water that is recoverable and where it may be recovered, preventing others extracting it, and allowable rates of recovery (Nelson & Casey, 2013). There is some evidence of over-reliance on managed aquifer recharge as a solution to water scarcity problems in some jurisdictions, and perhaps under-preparedness for its potential legal complexity (e.g. California, USA: Ulibarri *et al.*, 2021). Managed aquifer recharge occurs in jurisdictions outside the USA (e.g. Kitui District, Kenya: Clifton *et al.*, 2010; Kumamoto, Japan: OECD, 2017; India: Sakthivel *et al.*, 2015; European jurisdictions: Sprenger *et al.*, 2017). However, significant analysis of the accompanying legal arrangements, including how formal water allocation regimes accommodate these practices, has been rarer (e.g. Clifton *et al.*, 2010; OECD, 2017).

3.2.3 Increasing complexity of permitting and planning

Allocation regimes are also increasing in their complexity. The issue of recognizing interactions between water sources, particularly groundwater and surface water, has characterized important legal developments across several jurisdictions (OECD, 2015; states of the western USA: Trout Unlimited, 2007; states of the eastern USA: Weston, 2008). For example, water allocation reforms in the laws of the Australian states have been driven by an intergovernmental National Water Initiative policy agreed in 2004, which highlighted the need to manage interconnected surface water and groundwater in an integrated manner (Cosens, 2018). In some places, innovative legal mechanisms provide for groundwater users to 'offset' or mitigate their indirect use of connected surface water, for example by purchasing or leasing (and not using) rights to connected surface water, or 'pumping and dumping' water from unconnected sources into streams that would be depleted by a groundwater pumping proposal (states of the western USA: Nelson, 2015). In these contexts, integrating groundwater and surface water refers to considering connections between surface water and groundwater to ensure that the use of one does not unintentionally impact the other. Integration can also refer to the 'complementary use of surface water and groundwater' to increase productivity (World Bank, 2006). The latter approach can be facilitated by allocation regimes that allow users to switch between surface water and groundwater (e.g., some western states of the USA: Thompson, 2011).

3.2.4 Implementation challenges to permitting and planning systems

Putting to one side trends in law 'on paper', as time passes, commentators increasingly recognize that rather than being a panacea for water problems, allocation permitting and planning regimes carry their own challenges. This is the case even in jurisdictions in which they are well-established, as in New Zealand, where regional plans define the quantum of water available for irrigation, industry and the environment, and set out rules for water allocation (Daya-Winterbottom, 2014). There, uncertainties have emerged where regional plans lack rules about how to consider competing applications to use the same water resource, and the statute itself is silent on the matter (Daya-Winterbottom, 2014). South Africa initially sought to implement a system of 21 catchment management strategies, which relate to water management decisions, based on catchment boundaries, but subsequently revised this proposal to 9 due to 'operational challenges' (Pejan *et al.*, 2014).

In some jurisdictions, rollouts of area-based permitting systems, or their implementation, have stalled, partially because they require greater administrative resources than are available in lower income countries (South Africa, Tanzania, Malawi, Ghana: van Koppen, 2017). Some African jurisdictions lack sufficiently secure agricultural water rights in general to facilitate private investment in irrigation, for example, where government departments lack the administrative capacity to issue small water user permits (African Union, 2020). Social equity concerns pose a further challenge to permitting systems in some jurisdictional settings, discussed further below.

3.3 ACCESS TO WATER FOR MORE WATER USERS, AND PARTICIPATION BY MORE STAKEHOLDERS

The past two decades have also seen legal developments that recognize an increasing array of water users and other participants in allocation and re-allocation processes, achieved through both market and non-market mechanisms. Traditionally, allocation arrangements have centered on significant agricultural, municipal, industrial and other commercial uses. More recently, environmental and social equity concerns are prompting some jurisdictions to recognize more diverse water users either within water allocation systems, or in a way that constrains the grant or exercise of formal allocations.

3.3.1 Human right to water

International recognition of the human right to water emerged in the early 21st century, notably with the United Nations Committee on Economic, Social and Cultural Rights issuing a General Comment on the right to water, in 2002, and a United Nations General Assembly resolution in 2010 (Winkler, 2017). The right is considered to encompass both distributive and procedural components (Hey, 2009), both of which are relevant to water allocation laws. Conceptually, the distributive aspect either supports directly allocating water to beneficiaries of the right, or constraining allocations to others in order to protect access for beneficiaries of the right. The procedural aspect intersects with a greater focus on the participation of more diverse stakeholders in water allocation and management processes.

International right-to-water developments have been accompanied by diverse jurisdictional approaches to the issue as, some have argued, is entirely appropriate (Lugaresi, 2014). An increasingly common trend in less wealthy nations is to constitutionalize the right to water in itself (e.g., Ecuador, Bolivia, Gambia, Uruguay) or jointly with another right (Ethiopia, Kenya, Uganda, Panama, South Africa, Nicaragua, Democratic Republic of Congo), such as a right to health or food (Lugaresi, 2014). In other nations, courts have recently interpreted older constitutional rights provisions as encompassing the human right to water where the constitutional text does not explicitly refer to water (e.g., Argentina: Onestini, 2017). Other, weaker, approaches include legislatively requiring agencies to consider the human right to water when taking action in relation to policies, regulations and grant criteria (Nelson & Quevauviller, 2016).

While the diversity of approaches to providing for a human right to water makes it difficult to generalize about its possible effects on allocating water for agricultural purposes, several possibilities emerge. Agricultural uses of water may be constrained if they threaten the human right to water (Soininen, 2014), even if the right is not accompanied by allocations that appear in formal permitting or planning systems. This could conceivably include threats to quantity as well as quality, although the latter appear most prominently in the literature (e.g. Argentina: Onestini, 2017). Though international conceptions of the human right to water appear restricted to domestic and sanitation uses, and have been criticized for this (van Koppen, 2017), national conceptions of the right may extend to subsistence agriculture, thereby placing water allocations for subsistence agriculture (covered by the right) in tension with those for large-scale agriculture (not covered by the right). Comprehensive comparative analysis of how constitutional courts interpret the potential for competing rights in this context remains relatively limited.

3.3.2 Water for environmental purposes

Environmental uses of water traditionally have been under-represented in water allocation regimes, but this is now changing, with implications for the allocation of water for agricultural purposes. Water laws increasingly contemplate securing water for environmental purposes in one of two broad ways: 'rules-based' approaches protect water for the environment by controlling the way that water is allocated for consumptive purposes, including agriculture; and 'rights-based' approaches allocate water directly to environmental purposes (Horne *et al.*, 2017). The latter, which may allocate water to statutory holders of environmental water (O'Donnell, 2018), appear relatively rare in a global sense. These latter approaches are also currently largely restricted to the surface water context, though there is potential, and arguably value, in extending them to groundwater (Nelson, 2022), as has occurred with rules-based approaches (Australia: Pierce & Cook, 2020; western USA: Saito *et al.*, 2021).

Under rules-based approaches, water for environmental purposes is not formally conceived as a use that is subject to allocation permit systems, but environmental purposes nonetheless constrain access to water for uses that hold or require formal allocations. This is the case, for example, where water for basic needs, ecosystems or both, is conceived of as a 'reserve' that is not available for allocation (Kenya, South Africa: Dai *et al.*, 2017), or where the judicially developed public trust doctrine obliges government trustees to supervise and protect water as a trust resource (USA as a matter of state law, India: Scanlan, 2017). Other nations subject consumptive uses to temporal or spatial rules on withdrawing water, with overall caps on extraction calculated by reference to ecological requirements, or require environmental matters to be considered in water allocation permitting processes on a case-by-case basis (western USA, Australia: Horne *et al.*, 2017; Nelson, 2013).

Within permitting processes, environmental issues may be considered under formal environmental impact assessment-like processes (e.g. Peru, Honduras: Burchi, 2019); or using broader and vaguer 'principles' (western USA, Australia: Nelson, 2013; western USA: Squillace, 2020). Increasing attention to the environmental implications of permitting water uses in Finland's 2012 Water Act has led to a statutory test that requires balancing of harms and benefits, though its practical application is made difficult by the lack of a hierarchy or weighting criteria (Soininen, 2014). Similarly, New Zealand's Court of Appeal has determined that the legislated sustainable management principles considered in the context of water allocation decisions require taking a 'balanced judgment' in relation to competing considerations (Daya-Winterbottom, 2014). While facilitating contextualized decision-making, this approach raises questions about the scope of administrative discretion, and the possibility of widely divergent local approaches in the absence of specific guidance material (Daya-Winterbottom, 2014).

These developments may pose challenges to new and established agricultural water uses in a variety of ways. Rules-based approaches constrain the amount of water available for agricultural allocations, and rights-based approaches place environmental water holders in more direct competition with agricultural users. Some legal approaches may allow for re-allocating agricultural water to environmental purposes. This may occur where rules are newly established or adapted in response to new information, such as revised caps on total consumptive allocations to reflect new information about greater environmental water needs. New environmental concerns may prompt pressure on governments to decline the renewal of time-limited consumptive allocation permits (India: Scanlan, 2017). In some contexts, environmental water holders may purchase agricultural water to dedicate to environmental purposes (see Section 3.3.3 below). In others, the public trust may require the government to revise previously issued permits to secure ecological outcomes (California, USA: Scanlan, 2017). In other cases, this is less clear in relation to agriculture, specifically: India's adoption of the public trust occurs alongside preferences for both domestic and agricultural requirements over commercial uses (Scanlan, 2017), making its implications for agricultural water uses less clear.

The practical implications for agriculture of these developments may also be limited by design, or by incomplete implementation. In some places, legal mechanisms that secure water for ecological needs are only active during drought (Netherlands, China: Dai *et al.*, 2017). In others, legislation or policy may recognize the need to allocate water for the environment, but not yet systematically do so (Bangladesh, Bhutan, India, Nepal, Pakistan: Hirji *et al.*, 2018). Similarly, growing statutory and judicial recognition of the legal personality of rivers and other water bodies around the world (e.g., Aotearoa New Zealand, India, Bangladesh, Colombia: O'Donnell, 2021) certainly carries strong rhetorical weight. However, these developments universally appear to lack water rights for rivers (O'Donnell, 2021), so impacts on agriculture are likely to be less direct than competition for allocations. Formal pro-ecology policy declarations in the context of water allocations in other jurisdictions (e.g., 'ecological civilization' in China: Jia & Zhu, 2020) also carry uncertainties about the extent to which, and how, they might be implemented in practice, and affect new or existing agricultural water allocations on the ground.

3.3.3 Transferring allocated water

As well as recognizing the legitimacy of new classes of water use, such as ecological uses, water law reforms increasingly provide for transferring water between established categories of uses. In some cases, it may be more palatable to establish a new allocation system if the system affords users flexibility through a water transfer or water market system (e.g. California: Nylen *et al.*, 2017).

Re-allocating water from agricultural to urban purposes is increasing around the world, with a systematic review having identified 103 major rural-to-urban re-allocation projects undertaken from 2000 to 2018, with most occurring in Asia and North America (Garrick *et al.*, 2019b). Whether re-allocations are voluntary or involuntary, they occur in the context of legal and institutional frameworks dealing with water rights (Garrick *et al.*, 2019a), and can take a variety of administrative, negotiated and judicial guises, of which market-based transactions are one. Interestingly, most rural-to-urban re-allocations around the world do not occur under formal market and trading rules, but as administrative decisions to re-allocate water, and under formal negotiated agreements (Garrick *et al.*, 2019b). Despite a general trend of recognizing the economic value of water (Dai *et al.*, 2017), including through the development of markets, the practical prominence of this concept in the context of rural-to-urban re-allocations appears muted. Rural-to-urban re-allocations also tend to concern surface water more than groundwater (Garrick *et al.*, 2019b), but more generally, markets are also developing in relation to groundwater allocations (western USA, Australia, New Zealand, Chile, Mexico, Spain: OECD, 2015). The development of water markets, and facilitative laws and regulations, have enabled the transfer of allocations of agricultural water between users to varying degrees in different jurisdictions. Some jurisdictions allow only intra-sectoral transfers (e.g. farmer to farmer) (e.g. Honduras: Burchi, 2019); and others expressly ban transfers or implicitly do so by tying water rights to land (e.g. Peru, Zambia: Burchi, 2019).

Where transfers are facilitated by law, permits and operational rules for re-allocations are considered critical for making re-allocations effective and equitable; and conversely, uncertainty about who owns water is a common cause of controversy in re-allocations (Garrick *et al.*, 2019a). It is possible to establish water markets for diverse types of rights, though trading is more complex and time-consuming in some situations. In Australia's much-discussed Murray-Darling Basin, the high activity of agricultural water markets is facilitated by laws that require a transfer of a typically perpetual water right to be consistent with a watershed plan, avoiding the need to assess transaction-specific third party impacts, including to 'donor' agricultural communities, and thereby reducing transaction costs (Cosens, 2018). Trade of older-style, time-limited rights that are associated with specific parcels of land (termed 'take and use licenses' in Victoria, e.g.) is also possible, though more time-consuming. In the western USA, the heavy regulation of water transfers more closely analyses third party impacts than occurs in Australia, but dramatically increases transaction costs (Cosens, 2018).

3.3.4 Water for other consumptive users

In some ways, the converse of allocation regimes allowing for, and legitimating, more diverse water uses, is the legal trend of requiring formal allocations for more types of water uses–in other words, seeking to bring more uses under closer legal control. While not yet common, some water allocation regimes are seeking to encompass indirectly used water, which was formerly exempt from formal allocation permit requirements. This has occurred, for example, in relation to water use by trees in dryland tree farms (South Australia, Australia: Burchi, 2019) and water that is a by-product of mining and oil and gas production (New South Wales, Australia: Nelson, 2018; western states of the USA: Thorne & Caile, 2013). This trend may protect agricultural water uses from the effects of what might otherwise be unregulated and uncontrolled withdrawals, and reflects the concern of water allocation regimes to better regulate the cumulative effects of water withdrawals.

3.3.5 Implementation challenges to facilitating access

While there is increasing legislative recognition of the importance of water for both environmental and other public benefit uses, barriers remain to the wide availability of water allocations for more diverse water users. In some cases, even relatively recently introduced water allocation permit systems may tie water permits to land ownership, which marginalizes the landless, especially women (Kenya: Dai *et al.*, 2017; sub-Saharan Africa: van Koppen, 2017). This is especially a concern in relation to groundwater (OECD, 2015), though there is an increasing trend of separating even groundwater rights from land ownership (Mechlem, 2016). Many allocation systems in former colonies have only recently begun addressing water rights for Indigenous peoples based on historical uses or in other ways (Macpherson *et al.*, 2018). Some jurisdictions focus on providing rights in the form of spiritual and cultural rights rather than for commercial uses (Australia: Macpherson *et al.*, 2018) that would enable Indigenous peoples to access water for agriculture or other forms of economic development. The latter approach is more common in Latin America (Macpherson *et al.*, 2018). Concerns can even arise with legally strong, constitutionally based forms of protection: despite South Africa's strong constitutional focus on social equity, and a constitutional commitment of government to bring about equitable access to water resources (Pejan *et al.*, 2014, citing s25 South African Constitution), little of the intended redistribution of rights to water has occurred in practice (South Africa: van Koppen, 2017).

Some commentators suggest that the growing use of state permit systems to allocate water is fundamentally in tension with the human right to water. This may be particularly the case in agrarian low- and middle-income countries where permit regimes 'annul' customary law and introduce a 'bias towards single water uses', rather than domestic water and subsistence agriculture, ultimately 'finishing the unfinished business of colonialism' (van Koppen, 2017). Some recent water allocation legislation deals with this by formally recognizing uses established under customary law, accompanied by mechanisms such as local accreditation measures (Bhutan), a requirement to consider traditional uses when deciding on new abstraction applications (Zambia, Namibia), or according customary rights a generic priority or equal status to 'modern' permits (Peru, Tanzania) (Burchi, 2019). If well implemented, such measures would presumably go some way to protecting traditional, small-scale agriculture from the adverse effects of newer, large-scale agricultural withdrawals.

3.3.6 More diverse participants in processes that influence allocations

As well as substantive changes to the scope of water allocation regimes, procedural reforms that relate to stakeholder participation may indirectly influence allocations. The 21st century has seen a reduced role for central governments in the irrigation sector globally, away from a bureaucratic, top-down approach, and towards a paradigm of participatory irrigation management and water user associations (India: Cullet, 2009; World Bank, 2006), and frequent decentralization of control to the river-basin level (Dinar, 2005). Influenced at least in part by developing international norms, national-level allocation regimes have broadened stakeholder processes to involve water users in water resources management processes (Brantas River Basin, Indonesia: Dai *et al.*, 2017; Guatemala, Lao PDR, Sierra Leone, reforms from 2016–2018: World Bank, 2019).[3] In some places, First Nations peoples, who traditionally have been disempowered or disadvantaged within these regimes, are increasingly considered in reforms or reflected in developments in rights claims (Macpherson, 2019; Womble *et al.*, 2018).

However, these reforms have not escaped critique. Establishing new water users' institutions, rather than using existing local institutions, can threaten equity: new institutions may represent only landowner farmers, often privileging high-status men, when 'irrigation is an issue which concerns not only landowners but everyone else as well', including through impacts on domestic users and women who may not be perceived to be water users (Egypt: Barnes, 2015; India: Cullet, 2009). Sometimes, 'participation' may amount to simple notification of decisions or other shallow mechanisms, rather

[3] For information for 101 countries on whether a country requires water users to be represented in water resource management institutions, see https://eba.worldbank.org/en/data/exploretopics/water.

than influence on decision-making processes (Kenya, China: Dai *et al.*, 2017). Even in jurisdictions that have long traditions of involving stakeholders in allocation-related processes like water planning and permit decision-making, participation may not extend to the ready availability of court challenges to water plans or already granted permits (Netherlands: Dai *et al.*, 2017). The practical effectiveness of participation provisions may also be affected by a lack of interest or awareness on the part of stakeholders (Netherlands: Dai *et al.*, 2017). In other regions, central agencies may be reluctant to give up power, or state-supported decentralization may be perceived to erode local allocation traditions in favor of new arrangements that are inadequately resourced (Middle East/North Africa region: Closas & Villholth, 2020). Even when political trends support increased stakeholder representation, and environmental trends lead to legal restrictions on access to water (see above Section 3.3.2), corruption, existing power disparities, and 'local tyrannies' may frustrate the achievement of these aims on the ground (Middle East/North Africa region: Andersson & Ostrom, 2008; Closas & Villholth, 2020). Further investigation of the effects on equity of the push to decentralize institutions with power over water allocations is warranted.

3.4 REFLECTIONS AND CONCLUSION

These developments are not uniformly present, or present to the same degree, in all jurisdictions. Apparently similar developments may also mask surprising legal differences, like different definitions of 'groundwater' (Nelson & Quevauviller, 2016). However, the generalized emergence of these overarching categories of developments – allocation regimes that facilitate access to more water sources, for more water users – is perhaps unsurprising against larger biophysical and socio-political trends.

Allocation rules that broaden the water sources contemplated by allocation rules and recognize greater complexity in physical water sources respond to increasing competition for water and water scarcity and the need to steward it more carefully. Allocation systems are becoming more elaborate, in line with increasing use. However, it is unclear to what extent increasingly complex allocation systems are being introduced preventatively, ensuring that 'the basic building blocks of a robust regime [are] put into place at an early stage to avoid lock-in to unsustainable use and allow for adjustment at least cost, as needed, over time' (OECD, 2017). Future empirical research on the triggers for these legal changes would help resolve this issue.

Allocation rules that expand the range of legitimate water uses and participants in allocation regimes mirror the more diverse voices and values increasingly safeguarded – or at least debated – across other areas of law. The interactions and relationship between substantively providing for new water users/uses to improve distributional equity, and broadening procedural stakeholder involvement processes, provide a fertile ground for further investigation.

Explicitly recognizing these trends is important for several reasons. First, it underscores that further legal change may be needed to fully operationalize these developments. This includes better monitoring and compliance; reliable, trusted and readily accessible information about water sources and water allocations; and coordination between jurisdictional allocation regimes (Mechlem, 2016). Enforcement problems associated with permit systems are noted in a wide range of jurisdictions (Kenya: Dai *et al.*, 2017), especially in relation to agricultural water use and use of groundwater (e.g. illegal or unregulated wells) (e.g. southern Europe, Mexico: OECD, 2015). Similar concerns may arise about the legal clarity and certainty of the governing water allocation legislation (Kenya: Dai *et al.*, 2017; New Zealand: Daya-Winterbottom, 2014; states of the western USA: Squillace, 2020). On the other hand, some perceive that the practical difficulty of implementing a legal system of water allocations relates to tensions with underlying views of water as a social good (Kenya: Dai *et al.*, 2017), which raises questions about its appropriateness in the first place. Second, recognizing these trends raises questions about ways of more fully reflecting the values underlying them through better integrating water law and other areas of law to which these trends speak, for example biodiversity-focused environmental laws, broader natural resources laws, and cultural heritage laws (Nelson, 2020; Nelson *et al.*, 2018). Finally, recognizing these trends facilitates the achievement of a key goal of this book: cross-jurisdictional learning about responses to common challenges.

ACKNOWLEDGEMENTS

I gratefully acknowledge the support of the Australian Research Council (#DE180101154), feedback from Amanda Cravens, A. R. Siders, Nicola Ulibarri, Christina Babbitt and Josselin Rouillard, and research assistance by Lucas Volfneuk and the Melbourne Law School's Academic Research Service.

REFERENCES

African Union. (2020). Framework for Irrigation Development and Agricultural Water Management in Africa. African Union, Addis Ababa, Ethiopia.

Andersson K. P. and Ostrom E. (2008). Analyzing decentralized resource regimes from a polycentric perspective. *Policy Sciences*, **41**(1), 71–93. https://doi.org/10.1007/s11077-007-9055-6

Barnes J. (2015). Who Is a water user?: The politics of gender in Egypt's water user associations. In: Contemporary Water Governance in the Global South, L. M. Harris, J. A. Goldin and C. Sneddon (eds). Routledge, Abington, Oxon, pp. 185–198.

Bray Z. (2020). The fragile future of aquifer storage and recovery. *San Diego Law Review*, **57**(1), 1–60.

Burchi S. (2019). The future of domestic water law: trends and developments revisited, and where reform is headed. *Water International*, **44**(3), 258–277. https://doi.org/10.1080/02508060.2019.1575999

Clifton C., Evans R., Hayes S., Hirji R., Puz G. and Pizarro C. (2010). Water and Climate Change: Impacts on Groundwater Resources and Adaptation Options. World Bank, Washington, DC.

Closas A. and Villholth K. G. (2020). Groundwater Governance: Addressing core concepts and challenges. *WIREs Water*, **7**(1), 1–16. https://doi.org/10.1002/wat2.1392.

Cosens B. (2018). Governing the freshwater commons: lessons from application of the trilogy of governance tools in Australia and the Western United States. In: Reforming Water Law and Governance: From Stagnation to Innovation in Australia, C. Holley and D. Sinclair (eds.), Springer, Singapore, pp. 281–298.

Cullet P. (2009). Water Law, Poverty, and Development: Water Sector Reforms in India. Oxford University Press, Oxford.

Dai L., Van Rijswick M. and Schmidt B. (2017). Towards a sustainable, balanced and equitable allocation of water use rights. In: Water Resource Management and the Law, E. J. Hollo (ed.), Edward Elgar Publishing, Northampton, Massachusetts, pp. 151–195.

Daya-Winterbottom T. (2014). Sustainability, governance and water management in New Zealand. In: Water and the Law: Towards Sustainability, M. Kidd, L. Feris and T. Murombo (eds.), Edward Elgar Publishing, Cheltenham, UK, p. 107.

Dinar A. (2005). Decentralization of River Basin Management: A Global Analysis. World Bank Publications, Washington, DC, Vol. **3637**.

Garrick D., De Stefano L., Turley L., Jorgensen I., Aguilar-Barajas I., Schriener B.... and Horne, A. (2019a). Dividing the Water, Sharing the Benefits: Lessons from Rural to Urban Water Reallocation.

Garrick D., De Stefano L., Yu W., Jorgensen I., O'Donnell E., Turley L.... and Wight, C. (2019b). Rural water for thirsty cities: a systematic review of water reallocation from rural to urban regions. *Environmental Research Letters*, **14**(4), 043003. https://doi.org/10.1088/1748-9326/ab0db7

Global Legal Research Center. (2013). Legislation on Use of Water in Agriculture: Afghanistan, Argentina, Brazil, Chile, Egypt, Iran, Iraq, Israel, Kyrgyzstan, Lebanon, Libya, Mexico, Nicaragua, Saudi Arabia, Tajikistan, Turkey, Uzbekistan, Venezuela, Yemen. The Law Library of Congress, Global Legal Research Center, Washington DC.

Hey E. (2009). Distributive justice and procedural fairness in global water law. In: Environmental Law and Justice in Context, J. Ebbesson, P. Okowa and E. Hey (eds.), Cambridge University Press, Cambridge, pp. 351–370.

Hirji R., Nicol A. and Davis R. (2018). South Asia Climate Change Risks in Water Management. World Bank; International Water Management Institute, Washington, DC, USA.

Horne A. C., O'Donnell E. L. and Tharme R. E. (2017). Mechanisms to allocate environmental water. In: Water for the Environment: From Policy and Science to Implementation and Management, A. C. Horne, J. A. Webb, M. J. Stewardson, B. Richter and M. Acreman (eds.), Elsevier, London, pp. 361–398.

Jasechko S. and Perrone D. (2021). Global groundwater wells at risk of running dry. *Science*, **372**(6540), 418–421. https://doi.org/10.1126/science.abc2755

Jia S. and Zhu W. (2020). China's achievements of water governance over the past seven decades. *International Journal of Water Resources Development*, **36**(2–3), 292–310. https://doi.org/10.1080/07900627.2019.1709422

Lugaresi N. (2014). The right to water and its misconceptions, between developed and developing countries. In: Water and the Law, M. Kidd, L. Feris, T. Murombo and A. Iza (eds.), Edward Elgar Publishing, Cheltenham, UK, pp. 331–348.

Macpherson E. J. (2019). Indigenous Water Rights in Law and Regulation. Cambridge University Press, Cambridge, UK.

Macpherson E., O'Donnell E., Godden L. and O'Neill L. (2018). Lessons from Australian Water Reforms: Indigenous and Environmental Values in Market-Based Water Regulation. Springer, Singapore, pp. 213–234.

Mechlem K. (2016). Groundwater governance: the role of legal frameworks at the local and national level—established practice and emerging trends. *Water*, **8**(8), 347. https://doi.org/10.3390/w8080347

Nelson R. (2013). Groundwater, rivers and ecosystems: comparative insights into law and policy for making the links. *Australian Environment Review*, **28**, 558–566.

Nelson R. (2015). Paying back the river: a first analysis of western groundwater offset rules and lessons for other natural resources. *Stan. Envtl. LJ*, **34**, 129.

Nelson R. (2018). Regulating cumulative impacts in groundwater systems: global lessons from the Australian experience. In: Reforming Water Law and Governance: From Stagnation to Innovation in Australia, C. Holley and D. Sinclair (eds.), Springer, Singapore, pp. 237–256.

Nelson R. (2020). Challenges to improved integrated management of the Murray-darling basin. In: The Murray-Darling River System: Its Future Management from Catchment to Coast, B. Hart, N. Bond, N. Byron, C. Pollino and M. Stewardson (eds.), Elsevier, Amsterdam, pp. 339–362.

Nelson R. (2022). Water rights for groundwater environments as an enabling condition for adaptive water governance. *Ecology and Society* **27**(2), 1–22, Article 28. https://doi.org/10.5751/ES-13123-270228

Nelson R. and Casey M. (2013). Taking Policy from Paper to the Pump: Lessons on Effective and Flexible Groundwater Policy and Management from the Western U.S. And Australia. Stanford Woods Institute for the Environment, Palo Alto, California.

Nelson R. and Quevauviller P. (2016). Groundwater law. In: Integrated Groundwater Management, T. Jakeman, O. Barreteau, R. Hunt and J.-D. Rinaudo (eds.), Springer Publishing, Switzerland, pp. 173–196.

Nelson R., Godden L. and Lindsay B. (2018). Cultural Flows: A Multi-Layer Plan for Cultural Flows in Australia: Legal and Policy Design, National Native Title Council, North Melbourne, Victoria.

Nylen N. G., Kiparsky M., Archer K., Schnier K. and Doremus H. (2017). Trading Sustainably: Critical Considerations for Local Groundwater Markets under the Sustainable Groundwater Management Act. Center for Law, Energy & the Environment, UC Berkeley School of Law, Berkeley, CA.

O'Donnell E. (2018). Legal Rights for Rivers, Routledge, Abington, Oxon.

O'Donnell E. (2021). Rivers as living beings: rights in law, but no rights to water? *Griffith Law Review*, **29**(4), 1–26.

OECD. (2015). Drying Wells, Rising Stakes: Towards Sustainable Agricultural Groundwater Use. OECD, Paris.

OECD. (2017). Groundwater Allocation: Managing Growing Pressures on Quantity and Quality. IWA Publishing, Ashland, USA.

Onestini M. (2017). Human right to water: Argentine cases, human rights – are they enforceable? In: Water Resources Management and the Law, E. J. Hollo (ed.), Edward Elgar Pub. Ltd, Massachusetts, p. 118.

Pejan R., Toit D. D. and Pollard S. (2014). Using progressive realization and reasonableness to evaluate implementation lags in the South African water management reform process. In: Water and the Law: Towards Sustainability, M. Kidd, L. Feris and T. Murombo (eds.), Edward Elgar Publishing, Cheltenham, UK, p. 305.

Pierce D. and Cook P. (2020). Conceptual approaches, methods and models use to assess extraction limits in Australia: from sustainable to acceptable yield. In: Sustainable Groundwater Management: A Comparative Analysis of French and Australian Policies and Implications to Other Countries, J.-D. Rinaudo, C. Holley, S. Barnett and M. Montginoul (Eds.), Springer, Cham, Switzerland, pp. 275–290.

Saito L., Christian B., Diffley J., Richter H., Rohde M. M. and Morrison S. A. (2021). Managing Groundwater to Ensure Ecosystem Function. Groundwater, (early view).

Sakthivel P., Elango L., Amirthalingam S., Pratap C. E., Brunner N., Starkl M. and Thirunavukkarasu M. (2015). Managed aquifer recharge: the widening gap between law and policy in India. *Water Supply*, **15**(6), 1159–1165. https://doi.org/10.2166/ws.2015.074

Scanlan M. (2017). A comparative analysis of the public trust doctrine for managing water in the United States and India. In: Routledge Handbook of Water Law and Policy, A. Rieu-Clarke, A. Allan and S. Hendry (eds.), Taylor & Francis Group, Florence, UK, p. 23.

Schreiner B. and van Koppen B. (2020). Hybrid water rights systems for pro-poor water governance in Africa. *Water*, **12**(1), 155. https://doi.org/10.3390/w12010155

Soininen N. (2014). Water and the law. In: Weighing of Interests in the Finnish Water Law – from Financial Evaluation to Normative Weight of Interests, M. Kidd, L. Feris, T. Murombo and A. Iza (eds.), Edward Elgar Publishing, Cheltenham, UK, pp. 227–244.

Sprenger C., Hartog N., Hernández M., Vilanova E., Grützmacher G., Scheibler F. and Hannappel S. (2017). Inventory of managed aquifer recharge sites in Europe: historical development, current situation and perspectives. *Hydrogeology Journal*, **25**, 1909–1922. https://doi.org/10.1007/s10040-017-1554-8

Squillace M. (2020). Restoring the public interest in western water law. *Utah Law Review* **2020**, 627–683.

Thompson B. H., Jr. (2011). Beyond connections: pursuing multidimensional conjunctive management. *Idaho Law Review*, **47**, 273–323.

Thorne C. L. and Caile W. H. (2013). Produced water extraction from oil and gas wells: implications for western water rights. *Natural Resources & Environment*, **27**(3), 16–19.

Trout Unlimited. (2007). Gone to the Well Once Too Often: The Importance of Ground Water to Rivers in the West: A Report. Trout Unlimited, Boulder, CO.

Ulibarri N., Escobedo Garcia N., Nelson R. L., Cravens A. E. and McCarty R. J. (2021). Assessing the feasibility of managed aquifer recharge in California. *Water Resources Research*, **57**(3), e2020WR029292, 1–18. https://doi.org/10.1029/2020WR029292

van Koppen B. (2017). Water allocation, customary practice and the right to water: rethinking the regulatory model. In: The Human Right to Water: Theory, Practice and Prospects, M. Langford and A. F. S. Russell (eds.), Cambridge University Press, Cambridge, UK, pp. 57–83.

Weston R. T. (2008). Harmonizing management of ground and surface water use under Eastern water law regimes. *University of Denver Water Law Review*, **11**, 239–292.

Winkler I. T. (2017). The human right to water. In: Routledge Handbook of Water Law and Policy, A. Rieu-Clarke, A. Allan and S. Hendry (eds.), Taylor & Francis Group, Florence, UK, p. 109.

Womble P., Perrone, D., Jasechko, S., Nelson, R. L., Szeptycki, L. F., Anderson, R. T. and Gorelick, S. M. (2018). Indigenous communities, groundwater opportunities. *Science*, **361**(6401), 453–455.

World Bank. (2006). Reengaging in Agricultural Water Management: Challenges and Options. World Bank, Washington DC, USA.

World Bank. (2019). Enabling the Business of Agriculture 2019. The World Bank, Washington, DC.

doi: 10.2166/9781789062786_0037

Chapter 4

Indigenous water and Mother Earth

Margot A. Hurlbert

Canada Research Chair, Climate Change, Energy and Sustainability Policy, Johnson Shoyama Graduate School of Public Policy, University of Regina, Saskatchewan, Canada

ABSTRACT

This chapter reviews Indigenous water law in Canada and the United States, highlighting the protection of Indigenous rights and their priority in law, even though the historical practice has often excluded or denied them. Through treaties, court cases, and advancing laws respecting Mother Earth, the worldviews and values of Indigenous relations and Buen Vivir inform and mediate water management solutions and competing claims, overexploitation, and water conflict. New legal developments including the rights of nature, protecting them in constitutions, and recognizing river rights in laws in Central and South America, the United States and New Zealand move law, practice and water governance closer to a fairer and more socially just sharing of water resources between Indigenous communities and competing water claims including irrigated agriculture. Addressing water challenges of the 21st century will require innovative solutions and approaches to water management. Including and giving full voice to Indigenous people and their water laws and practices is a necessity both in law and water management practice.

Keywords: Indigenous groundwater rights, Indigenous rights, Indigenous water rights, *Sui Generis*, United Nation Declaration of the Rights of Indigenous Peoples

4.1 INTRODUCTION

Water has diverse sacred meanings for Indigenous peoples. Water is the artery of Mother Earth (Paul, 2020). A few examples from the language of North American Indigenous people illustrate this idea. In Cree, 'Nipiy' is the word for water joining 'n'niya' or 'I' or 'I am', together with 'iy' or 'pimatisiwin' or 'life'. In this way Niipiy is equivalent to 'I am Life' (Littlechild, 2014). For the Tlingit, water is characterized by respect and relations of responsibility between people, and water is seen as 'more-than-a-human person' (Wilson & Inkster, 2018). Indigenous water meanings are just the beginning. The diverse Indigenous thought, practices and governance (patterns of decision making) approaches destabilize modernism and expose colonial structures (including colonial water law constructs) and there are benefits of recognizing multiple values in enhancing water management and governance and ensuring just and socially accepted water allocation decisions through decolonization. Acknowledging Indigenous water meanings and rejecting settler views of water that subsume Indigenous water laws within settler-colonial society allows for pluralism. In this process, water can be decolonized and the ontological violence authorized by Eurocentric epistemologies in scholarship and everyday life can be exposed (Wilson & Inkster, 2018).

Agriculture and Indigenous water rights have an interesting history. Colonizing practices did not negate pre-existing Indigenous water rights. However, the state of Indigenous drinking water across the world reflects the legacy of colonialism and the widespread and intensive development of dryland agriculture by settlers. More recently, Indigenous laws and practices are increasingly being acknowledged in modern treaties. Their exercise and practice are now advancing in Canada and the United States, and elsewhere through, for instance, the rights of Mother Earth, rivers, and water. A space exists for the emergence of Indigenous water law, rooted in treaty protection, Indigenous sovereignty, and the United Nations Declaration of the Rights of Indigenous Peoples. This emergent space requires nurturing and the seeds lie in Indigenous traditions and practices, and ultimately Indigenous sovereignty.

This chapter will focus on Indigenous water law in Canada and the United States in Sections 4.2 and 4.3, but expand the discussion globally with considerations of Mother Earth in Section 4.4 and the Rights of nature in Section 4.5 concluding with the United Nations Declaration of the Rights of Indigenous Peoples in Section 4.6.

As a white settler, I write this chapter acknowledging its inherent weakness in recounting Indigenous water law within the confines of colonial structures, rules and practices that have confined it. I will provide an overview of Indigenous water rights (as determined by courts and written by governments) and I will also try to reference and recognize Indigenous water scholarship that is not constricted and confined by the former. However, a comprehensive review is beyond the scope of this chapter.

4.2 ENSURING THE PLACE OF INDIGENOUS WATER RIGHTS: PARALLELISM AND PLURALISM

Defining Indigenous law, or Indigenous water law, is a subject of debate. For some scholars, Indigenous law includes customary, common, statutory, and government law that specifically and only affects Indigenous people and the relationship between Indigenous people (University of Melbourne, 2021). For others, Indigenous law is characterized by legal pluralism which recognizes more than one legal system in the same social field (Merry, 1988), an Indigenous legal system and a Canadian, civil or common law, municipal, provincial or federal law system (Borrows, 2010). For Borrows *et al.* (2019) a vision of Indigenous law that is braided is preferred:

> *The braiding of Indigenous law with international and national law is thus a unique undertaking that helps us to reconceive the very idea of law. As suggested, Indigenous Peoples' law questions the claims of both international and national laws to universality and supremacy. Law can be multidirectional in sources and applications. It might be created by clans, flow from experiences with glaciers or rivers, or be sourced in custom and grassroots practices, as well as being created by legislatures, courts and executive authorities.*

For many Indigenous scholars, Indigenous laws move beyond pluralism to parallel expressions of self-determination overlapping with claims often incommensurable to those of the Canadian State (Day, 2001). In the words of James Sakej Youngblood Henderson (2002):

> *The task of Indigenous peoples is to encourage diversity as the prime assumption of legal systems, and to resist any false universality, despite the consequences of existing legal theory.*

In Canada, Indigenous law could be defined as 'a source of law apart from the common and civil legal traditions in Canada' (White, 2021) or Indigenous people's own legal systems. White's (2021) view of Indigenous law as parallel and separate is preferred by the author and recognized in Section 4.2.1. However, due to space, time, and the positionality of the author, this chapter mostly recounts the Indigenous law braided into the fabric of law through scholarship, court decision, and statute. But first, inherent Indigenous water law will be outlined followed by the law of *Sui Generis* that allows for the uniqueness of Indigenous rights and also builds a bridge back to Settler legal systems.

4.2.1 Inherent Indigenous water law

For Indigenous peoples, Indigenous rights, including Indigenous water rights, arise based on Indigenous peoples' existence as sovereign nations residing and governing throughout their territories and are 'inherent' rights (Phare, 2009). These laws are both given and limited by the Creator and include laws of stewardship and reciprocity with nature that Indigenous people cannot alter or narrow, and cannot be altered or narrowed by other humans, governments or their laws. These inherent Indigenous water rights are separate and apart from Indigenous water rights claimed through treaty or constitutional rights. While a discussion of these rights is beyond the scope of this chapter, these water rights are as unique and diverse as Indigenous people and their communities, and not restricted to those recognized by colonial governments and legal systems (Gullason, 2018).

4.2.2 *Sui Generis* Indigenous rights

Indigenous rights are *Sui Generis*, having a unique meaning reflecting the uniqueness and diversity of Indigenous people and their laws. They exist in a unique class by themselves (Battiste & Barman, 1995). In law generally, *Sui Generis* is a legal classification existing as a singularity or independent of other categorizations. In North American law, it reflects aspects of community, both human and ecological, interconnected human and Earth relations, and teachings of 'Honoring Earth'. Indigenous rights are separate and different from human rights. Unlike human 'rights', they are not advanced from a place of victimhood or disadvantage that requires redress through creation of a right, but they represent long-enduring rights that arose from the prior occupation of land by Indigenous peoples and the prior social organization and distinctive cultures of Indigenous peoples on that land (McNeil, 2013).

Indigenous rights are recognized as far back as the Royal Proclamation of 1763, wherein King George III received lands surrendered by France in the Treaty of Paris and specifically reserved lands to the Indigenous people as their traditional hunting grounds in the British colonies. This Proclamation established a nation-to-nation relationship between Indigenous people and the English monarch (McNeil, 2013). While human rights claimants are often regarded as 'victims', less powerful, downtrodden, and in need of protection (Ball, 2013), Indigenous rights holders are not.

Several features of Indigenous rights are unique and provide enhanced protection beyond notions of human rights. The unique Crown-First Nations relationship is described as *Sui Generis* because it is derived from both Indigenous and English legal regimes. It is neither exclusively public nor private, but flows from a unique creation of legally enforceable duties of the Crown (R. v. Sioui, 1990). The relation is between those of sovereign states and states with their own citizens (Delgamuukw v. British Columbia, 1997). Indigenous title is a *Sui Generis* right arising from prior occupation of Canada by Indigenous people. Chief Justice Lamer in Delgamuukw stated if occupation was also pursuant to Indigenous law, there would be a second source of Indigenous title arising from the relationship between common law and pre-existing systems of Indigenous law. He stated in his judgement, that this title can't be sold, conveyed to private persons or corporations or held by individual Indigenous persons. It is held by all members of the nation and decisions are made communally. Indigenous land cannot be used in a manner irreconcilable with the nature of the attachment to the land forming the basis of the group's claim of Indigenous title. Indigenous land rights are tied to a unique common interest, fiercely different from private property and its assumed individuality of decision making. It is this legal protection that allows for the Indigenous tradition of honoring Mother Earth.

Human rights are often regarded by Indigenous scholars as assimilative and contrary to the diverse Indigenous people and their rights (Howard-Hassmann, 2014). This is because human rights arise as a construct of Euro-centric structures. Their very conception, their structures of recognition and protection, occur within institutions and cultural practices of European origin. Universal human rights are regarded as 'totalizing' rights, erasing Indigenous cultural differences, and thus may imply assimilation. Far different from Indigenous collectivities, human rights, including water rights, assimilate rights into liberal, individualist society destroying the collectivity and instead promoting the interests of capital, private property, and individuality (Ball, 2013). This construction of rights is in juxtaposition to the *Sui Generis* nature of Indigenous rights and the Indigenous view of honoring Mother Earth.

The flexibility and evolving nature of *Sui Generis* Indigenous rights act as an interdisciplinary and intercultural legal bridge. As an example, in interpreting Indigenous rights to fish for food, the court in the Supreme Court of Canada's Van der Peet case recognized the contemporary Indigenous legal understanding of the right to fish for food, as well as social and ceremonial purposes connected to cultural and physical survival exercised in a contemporary manner. The decision to recognize these rights protects an allocation of water to preserve their practice. The very act of defining Indigenous rights requires a 'form of inter-societal law that evolved from long-standing practices linking the various communities' (Borrows, 1997). Borrows (1997) state:

> *Therefore, a true sui generis conception of Indigenous rights will respect the existence within Canada of two vastly different legal cultures, European and Indigenous, and will incorporate both legal perspectives. A sui generis approach will place 'equal weight' on each perspective and thus achieve a true reconciliation between the cultures.*

4.3 INDIGENOUS WATER LAW IN CANADA AND THE UNITED STATES

After outlining the inherent Indigenous water rights (4.3.1), this section describes Indigenous water rights on reserve (4.3.2), Indigenous water rights by treaty (4.3.3), and Indigenous groundwater rights (4.3.4).

4.3.1 Indigenous water rights

Aboriginal water rights have been established in many court cases. In Canada, the Van der Peet case found the right to use the land and adjacent waters for traditional sustenance purposes to be a fundamental right. Even though there was no counterpart in the English law, Canadian courts have recognized Indigenous rights. In law, existing Indigenous laws and customs and their corresponding interests in the land and waters required for their practice must be regarded as continuing, in the absence of extinguishment. Phare (2009) argues that Indigenous water rights exist if First Nations used water for domestic purposes, used the river for food supply, recreational activities, waterways for travel, trade or meetings, conducted ceremonies in the water, and had practices dictated by the significance of water for their culture. Indigenous fishing and harvesting rights protect rights reasonably incidental to the activities, including preserving water quality and quantity, habitat protection, and watershed management for protection of hunting and trapping grounds, harvesting and gathering grounds, and transportation on waterways. Also protected are uses of water reasonably incidental to fulfilling purposes of treaty or the economic stability of Indigenous people including water for manufacturing, irrigation, or the production and sale of electricity (Phare, 2009). Indigenous rights to water stem from the use and occupation of land since time immemorial (Laidlaw & Passelac-Ross, 2010). As with other Indigenous rights, these rights are not frozen in time, but are interpreted flexibly in order to permit their evolution over time and their meaningful exercise (Mitchell, 2001).

In South America, there are similar struggles for the recognition of Indigenous water rights. Water legislation was introduced for general purposes (in the 1970s in Ecuador, Peru, Chile, Bolivia) and resulted in ancient irrigation systems co-existing with more recent community or state water management initiatives. Protests and struggles for recognition have only been partially successful in preserving Indigenous pre-existing rights (de Vos *et al.*, 2006).

4.3.2 Indigenous water rights on reserve

In the United States, Indigenous people have an implied reservation of the water associated with their lands. These rights pre-date all other water claims; they take precedence being first in time, and therefore first in right (Winters v. United States). Treaties only confirm these rights and are not a source of these rights. The Winters case has been applied in Canada by the Privy Council in the case of Burrard Power as well as in policy documents that recognized the rights of Indians to 'Take for domestic, agricultural purposes all such water as may be necessary, both now and in the future' (Williams, 1920).

One illustrative example is the Secwepemc people on the Kamloops and Neskonlith reserves who developed irrigated agriculture successfully while competing with nearby settlers for control of water and supporting their traditional livelihoods (Matsui, 2005). Although the Reserve Commission and previous water officials specifically recognized the 'rights of the Indians as the oldest owners and occupiers of the soil to all the water which they require or may require for irrigation and other purposes,' provincial and federal government jurisdictional conflicts prevented the timely recording of these rights on the province's water registry. As the province's water registry worked on the basis of a first in time, first in right system, the late registration disadvantaged the Secwepemc people's water claim. The British Columbia Court of Appeal refused to recognize this claim in 1921; to this day, the Secwepemc people continue to assert their claim to the ownership of the creek and its water as an inherent Indigenous water claim (see 4.3.1) (Matsui, 2005).

4.3.3 Indigenous water rights and treaties

Canadian treaties did not specifically extinguish Indigenous water rights. Inherent Indigenous rights find their source in the Creator's laws and responsibilities and cannot be narrowed by other humans or governments (including their laws) and cannot be shed by Indigenous people; recognition by colonial legal systems and governments does not negate or restrict inherent Indigenous rights (Phare, 2009). When the Federal government transferred jurisdiction over water and natural resources to the Prairie provinces in the 1930s, no acknowledgment of or engagement with Indigenous people was made. As a result, some scholars argue that the transfer did not include Indigenous water rights (Phare, 2009) and that Indigenous people might be able to restrict public access to waters (Bartlett, 1986, in respect of Treaty #3). This may include restrictions on non-Indigenous fishing and hunting of waterfowl, public access to headwaters, limits on or removal of manufacturing and industrial uses, mining, hydroelectricity and building of dams. In British Columbia, Indigenous water rights were specifically recognized in the Indian Water Claims Act of 1921, which recognized that water went with land when reserves were created.

Indigenous rights were further recognized in law by the Supreme Court of Canada in 1972, and received constitutional protection through the Constitution Act of 1982. Nevertheless, some modern treaties acknowledge provincial water law as binding on Indigenous First Nation signatories. For example, the Maa-Nulth First Nation Final Agreement provides that the storage, diversion, extraction or use of water and groundwater will be in accordance with provincial and federal law (INAC, 2021). Furthermore, Indigenous water rights and fishery claims in Canada are not always successful. The issue of whether water is included in an Indian reserve is determined based on the reserve's specific provisions and facts surrounding its creation (Saanchton Marina). For example, a detailed review of historical facts and the parties' intention was undertaken in *R.* v. *Lewis* where it was determined the Squamish River was not part of an adjacent reserve, nor did the Bands receive an exclusive fishery. Many other court decisions concern only Indigenous land or fishing rights, and do not necessarily consider water and inherent Indigenous rights holistically (Matsui, 2005).

4.3.4 Indigenous groundwater rights

Indigenous people have succeeded in claiming rights to groundwater. In the United States, the Winters doctrine (the presumption that when Congress served land for an Indian reservation, the water sufficient to fulfill the purpose of the reservation was also reserved) has been extended both in relation to the quantity of water required on reserve and to groundwater. Even so, often the question of water quantity has not been definitively determined. However, in Arizona vs. California, a standard known as 'practicably irrigable acreage' was adopted measuring a reservation's water requirement in terms of the maximum reasonable land area within the reservation. The Winters doctrine has also been applied to groundwater in the U.S. Court of Appeals for the Ninth Circuit decision of Caliente v. Coachella Valley Water District *et al.* The court ruled that the fact the Tribe had not historically used groundwater was irrelevant and did not affect the seniority of the Tribe's groundwater rights which preempted water rights granted by the state.

In Halalt First Nation v. British Columbia (Environment), the British Columbia Supreme Court held that the Halalt had not been adequately consulted when several wells were approved. The Halalt had 'an arguable case that the groundwater in the Aquifer was conveyed to the federal Crown in order to fulfil the objects for which the reserve lands were set aside' and that the Province could not purport to expropriate the groundwater. Although the Court of Appeal set the decision aside based on a finding that there had been adequate consultation, this groundwater finding arguably remains binding law (Gullason, 2018). Gullason (2018) argues that British Columbia's recent Groundwater legislation continues the status quo of ignoring Indigenous water rights and continues to allocate groundwater licenses which will infringe both inherent Indigenous water rights and constitutionally protected Indigenous and treaty rights.

A similar situation exists in Alberta where Indigenous people claim water on reserves, for which interests had not been registered in the provincial water laws starting in 1894. Again, by failure to provincially document the rights, there is a risk of not considering them in allocation decisions. Indigenous reserves and their lands have always been under federal Canadian jurisdiction and argue the province has no jurisdiction over their lands and water (Statt, 2003).

4.4 MOTHER EARTH, RELATIONS, AND BUEN VIVIR

The primordial relation of Indigenous people is with Mother Earth. The Assembly of First Nations expression of Indigenous laws and practices considers the Earth and our relations to it and describes it as 'Honoring Earth' (AFN, 2020):

> *From the realms of the human world, the sky dwellers, the water beings, forest creatures and all other forms of life, the beautiful Mother Earth gives birth to, nurtures and sustains all life. Mother Earth provides us with our food and clean water sources. She bestows us with materials for our homes, clothes and tools. She provides all life with raw materials for our industry, ingenuity and progress. She is the basis of who we are as 'real human beings' that include our languages, our cultures, our knowledge and wisdom to know how to conduct ourselves in a good way. If we listen from the place of connection to the Spirit That Lives in All Things, Mother Earth teaches what we need to know to take care of her and all her children. All are provided by our mother, the Earth.*

Sustainable strategies held by Indigenous communities are part of this worldview that identifies problems and determines pro-social solutions to collective action problems (for an example of this in respect to Indigenous communities, hunter gathering lifestyles, sharing (reciprocated and non-reciprocated) and food security, see Ziker *et al.*, 2016). It is not that Indigenous people have all the solutions for environmental problems of the 21st century, it is that their Indigenous knowledge allows them to envision the common resource of our world, the 'Mother Earth' and provide teachings and learning on how to share the Mother Earth sustainably. Indigenous knowledge offers what, in the words of Elinor Ostrom, are behavioral approaches to collective action problems that entail understanding 'the effects of structural variables on the likelihood of organizing for successful modes of collective action' (Ostrom, 2010). Indigenous legal traditions are not trite or easy, but complex, based on legal traditions that include intentional and deliberative collective processes to change law over time, transform implicit law into explicit law, and create legal precedent and a formal memory archive (Napoleon, 2009).

In Maori philosophy, *hau* is one of the basic concepts: 'the wind of life that activates human and non-human networks alike animated by reciprocal exchanges' (Salmond, 2017). Reciprocal exchanges are not material exchanges of commodities but are empirically construed entirely on observation. The intermingling starts with the Maori greeting by touching noses, includes intermingling of breath, could include familial trees, but more often starts with recitation of names of main mountains or rivers in their home territory, binding people together (ibid.). Reciprocal gifting obligates the recipient to participate in the reciprocal infrastructure that the gift itself sets up, which does not just include human relations but structures the world and is underpinned by a debt that can never be repaid; it is that of giving life. In the Maori tradition, belonging to particular lands is not about who one is, but about what one does (Salmond, 2017).

Indigenous people do not have all the answers. As John Borrows notes, trying to understand all our environmental troubles through Indigenous knowledge can potentially compound our confusion as what was successful in one time and place may not be translated appropriately to other settings. 'Self-interest and cultural blindness to the potential dangers of one's own group's practices can be found everywhere; and a healthy degree of scepticism should also accompany any groups claim to a better path of environmental preservation' (Borrows, 1997). But Indigenous contributions are not just evidence of better practices; to be fully appreciated, institutional change is needed. In 1997, John Borrows pointed out that for transformational change, there must be change in people and ideas, the ground upon which decisions are made, and the integral application of Indigenous legal knowledge in decision making.

In Latin America, traditional Indigenous worldviews have merged with critiques of capitalism and the modernist world order to create the idea of *Buen Vivir*. This idea represents a civilizatory transformative proposal to build a new world that addresses colonial and capitalist structures (Gudynas, 2011). *Buen Vivir* represents a collective construction that prioritizes ecological and community co-existence required to meet the demands of society and not those of capital (Acosta, 2018). While the idea of *Buen Vivir* has been incorporated into many Latin American legal systems, it has had ambiguous and contradictory results. Many such initiatives were undermined by expanded exportation of natural resources, hydroelectric power development, oil pipelines and an exclusion of Indigenous people (Sieder & Barrera Vivero, 2017). However, Latin American scholars continue to breathe new life into the idea, imagining economies of degrowth where traditional Andean knowledge based on reciprocal help and collective work can flourish (Acosta, 2018). From this tradition, and Indigenous traditions, the rights of nature have evolved.

The implications are firstly that recognizing these philosophies and traditions is essential for pluralistic water management; by embracing these values and traditions, decisions of allocation and sharing in the context of conflicting claims can be mediated and resolved by all parties focusing holistically on Mother Earth, relations, and *Buen Vivir*. The exact nature of future water management decisions cannot be predicted, but collective discussion and action, taking into account the local context of resources and practices, can allow for appropriate and socially just sharing.

4.5 RIGHTS OF NATURE

Important developments of the rights of nature in law have, in some instances, successfully addressed the overexploitation of water resources and moved law, practice and water governance closer to a fairer and more socially just sharing of water resources between Indigenous communities and competing water claims including irrigated agriculture.

4.5.1 Constitutional protection for Mother Earth

In 2008, Ecuador became the first country to recognize the legal rights of nature in its constitution. Article 71 of the Constitution of Ecuador recognizes the right of nature to exist. Bolivia followed suit in 2010, recognizing the Rights of Mother Earth and her constituent life systems, including human communities (Rühs & Jones, 2016; Law 071 of the Plurinational Bolivian State). Rights include those of life, diversity of life (for the variety of beings that comprise Mother Earth), water, clean air, restoration and to live free from contamination (with regard to toxic and radioactive waste). Based on these precedents, the Paris Accord established the International Rights of Nature Tribunal (UN, 2016).

The Ecuador constitution was successfully applied in the Wheeler case (2011), granting an injunction to protect a river and finding the provincial government liable for damages resulting from a project widening a highway and depositing debris in the nearby Rio Vilcabamba. The debris had narrowed the river, increased its speed and caused erosion and flooding. Article 71 was specifically cited and the generational injuries to Nature were outlined, including their magnitude and repercussions to current and future generations. The court decision quotes Alberto Acosta, President of the National Constituent Assembly and his statement that human beings must not bring about the extinction of other species or destroy the functioning of natural ecosystems.

The inclusion of Mother Earth and the environment into law has also expanded law's environmental methodology (Ebbesson, 2003). Prior to colonization, in New Zealand the Māori people communally owned and managed all of the land and water according to traditional laws and customs governing resource sharing and use. Because of British riparian laws precluding ownership of water, Indigenous claims focused on land underneath water. However, Māori water relationships have been accounted for as part of the planning processes in Resource Management Acts with environmental and cultural value protection that must be taken into account in decision making. In New Zealand, the *Environment Act of* 1986 establishes an Environment Commissioner, tasked with maintaining and improving the quality of the environment through review of the government. The commissioner is tasked with the maintenance and restoration of ecosystems of importance, especially those supporting habitats of rare, threatened, or endangered species of flora or fauna. This is an important movement from utilitarian and imperial legal modes towards an appreciation of the rights of everything that constitutes the earth, decolonization and pluralism (Charpleix, 2018).

4.5.2 River rights

Recognition of rivers as legal entities with rights is increasingly being recognized through a mixture of court decisions and statutory enactment. The High Court in the State of Uttarakhand, India determined in the Salim case (2017) that the Ganges and Yamuna rivers were juristic persons that therefore required extraordinary measures in order to protect their health and wellbeing as communities from mountains to the sea. Some of the reasoning of the court included the religiously significant aspect of the rivers and the important spiritual and physical functions these rivers represented for Hindus. As a juristic person, the rivers hold rights and obligations and having representative standing, the rivers act through an intermediary, an interstate agency of appointed representatives.

New Zealand became the first country to grant a specific river legal rights in 2017, passing the Te Awa Tupua (Whanganui River Claims Settlement) Act. Māori water conceptualizations have been affected by creation of legal status including the example of the Te Awa Tupua or the Whanganui River that flows from the foothills of Mt Tongariro through the remote wild King Country and Whanganui region out to the ocean at the city of Whanganui. The Environmental Tribunal found the river – defined as a water resource or a single and indivisible entity comprised of water, banks and bed – to be a single, indivisible and living whole incorporating all its physical and metaphysical elements. This finding rendered the parties in a stalemate in water management and ultimately the dispute settled in 2017 with the river being recognized as a legal person, Te Awa Tupua (Whanganui River Claims Settlement) Act 2017 (Te Awa Tupua Act).

Toledo, Ohio is one of several U.S. communities to pass laws recognizing the rights of nature. Its Lake Erie Bill of Rights protects the lake's shores. Concern has been raised from farmers and river communities about the law adversely affecting their livelihoods and possible liability for their actions (Westermann, 2019). Bangladesh has also granted its rivers status as living entities to protect them from further pollution. A government-appointed National River Conservation Commission has been established which can charge people harming the river in the same fashion as if they harmed our mother (Turag River Case 2016).

In the Atrato River case in Colombia (2017), a grassroots movement, *acción de tutela*, sought an injunction to halt illegal mining and logging activities dumping harmful chemicals into the river. The court allowed the action based on its interpretation of biocultural rights in the Colombian constitution that represent a unity between nature and humans that respects the role of Indigenous relationships with non-human natural entities that foster biodiversity. International legal instruments incorporated into Colombian law, including the Convention on Biological Diversity, the UN Declaration of the Rights of Indigenous Peoples (UNDRIP) were also cited. The court specifically determined that nature is a subject of rights.

Glaciers, or headwaters of rivers, have also received protection in the Miglani case (2017). The High Court of Uttarakhand, India recognized the right of glaciers not to be polluted and the right to exist and persist as their own vital ecologic system, their own legal entity as a juristic, moral person. The High Court issued an order recognizing the right to clean water. Kauffman and Martin (2016) conducted a comprehensive review of Rights of Nature Cases and interviewed lawyers and judges involved. Their conclusions are that these cases are increasingly successful partly due to the issue's politicization and civil society's efforts to advance these cases. As a result, laws such as Ecuador's, that might be perceived as 'weak' by some, do matter. These cases enhance the protection of rivers and water but are not without problems. Interjurisdictional issues still exist with interprovincial and international boundaries that give rise to enforcement issues; protection also raises issues of liability and ultimately responsibility. Court cases are expensive, take considerable time, and sometimes their results are orders that are largely ineffectual at being enforced. Often the most difficult issue is changing the worldview of the non-Indigenous population (Westermann, 2019). Court cases structure guardianship differently, some drawing from New Zealand's model; court designated guardians may or may not rise to their appointment. Operationalizing river protection may be subject to further legislation and court orders (Kauffman & Martin, 2019).

4.6 UNITED NATIONS DECLARATION OF THE RIGHTS OF INDIGENOUS PEOPLES

Addressing water challenges of the 21st century will require innovative solutions and approaches to water management. Including and giving full voice to Indigenous water laws and practices is a necessity both in law and water management practice. This chapter has reviewed Indigenous water law in Canada and the United States as well as considering Mother Earth, worldviews and values of Indigenous relations, and Buen Vivir that can inform and mediate water management solutions and competing claims, overexploitation, and water conflict. New legal developments including the rights of nature, protecting these rights in constitutions, and recognizing river rights in laws in Central and South America, the United States and New Zealand also moves law, practice and water governance closer to a fairer and more socially just sharing of water resources between Indigenous communities and competing water claims including irrigated agriculture.

The United Nations Declaration of the Rights of Indigenous Peoples (UNDRIP) advances international Indigenous law and rights. UNDRIP recognizes Indigenous sovereignty and, at the same time, augments the legal status of Mother Earth and Water, the lifeblood of the Earth (AFN, 2021). Duties of 'consultation' are raised to requirements of 'consent', introducing the Right to Free, Prior and Informed Consent (FPIC) for Indigenous Peoples. Article 32.1 provides that FPIC is to be obtained prior to the approval of any project affecting Indigenous lands or territories in connection with the development, utilization or exploitation of mineral, water or other resources (Boutillier, 2017). UNDRIP moves beyond the conception of the State granting and distributing rights to people in a Rawlsian distributivist conception of justice (with the state as arbitrator of conflict and protector of individual rights) and embraces recognition justice (Rawls, 1971). Recognition is key in engaging with the 'other' when two groups have fundamentally different ontological positions, aims and goals. Recognition in accordance with Indigenous law does not aim to overcome each other's position, but to achieve recognition of and respect for difference, leading to more meaningful engagement and justice (Maciel, 2014), applying the *Sui Generis* principles of Indigenous sovereignty. In the words of the Canadian Assembly of First Nations (AFN, 2021):

> *The Creator placed us on this earth, each on our own sacred lands, to care for the earth, environment and humankind. We stand united to follow and implement our knowledge, laws and self-determination to preserve and protect life's most sacred gift – water.*

REFERENCES

Acosta A. (2018). O Bem Viver: uma Oportunidade Para Imaginar Outros Mundos [Good Living: an Opportunity to Imagine Other Worlds]. Editora Elefante, São Paolo, Brazil.

AFN. (2020). Honouring Earth. http://www.afn.ca/honoring-earth/ (accessed 24 December 2020)

AFN. (2021). Assembly of First Nations National Water Declaration. https://www.afn.ca/uploads/files/water/national_water_declaration.pdf (accessed 28 December 2020)

Atrato River Case. Center for Social Justice Studies *et al.* v. Presidency of the Republic *et al.*, Judgment. T-622/16, Constitutional Court of Colombia. (2016). ('Atrato River'), available at http://www.corteconstitucional.gov.co/relatoria/2016/t-622-16.htm (accessed 18 October 2021)

Ball D. P. (2013). Aboriginal rights are not human rights: in defence of indigenous struggles peter kulchyski (arbeiter ring, 2013). *Canadian Dimension*, **47**(6), 45. https://link.gale.com/apps/doc/A354262786/ITOF?u=ureginalib&sid=ITOF&xid=46edd235 (accessed 18 October 2021)

Bartlett R. (1986). Aboriginal Water Rights in Canada: A Study of Aboriginal Title to Water and Indian Water Rights. Canadian Institute of Resources Law, Calgary, Canada, pp. 101–110.

Battiste M. and Barman J. (1995). First Nations Education in Canada: The Circle Unfolds. UBC Press, Vancouver, Canada.

Borrows J. (1997). Living between water and rocks: first nations, environmental planning and democracy. *The University of Toronto Law Journal*, **47**(4), 417–468. https://doi.org/10.2307/825948; https://www.jstor.org/stable/825948

Borrows J. (2010). Canada's Indigenous Constitution. University of Toronto Press, Toronto, Canada.

Borrows J., Chartrand L., Fitzgerald O. E. and Schwartz R. (2019). Braiding Legal Orders: Implementing the United Nations Declaration on the Rights of Indigenous Peoples. McGill-Queen's Press, Montreal, Canada.

Boutillier S. (2017). Free, prior, and informed consent and reconciliation in Canada: proposals to implement articles 19 and 32 of the UN declaration on the rights of indigenous peoples. *Western Journal of Legal Studies*, **7**(1), 1–21.

Burrard Power Co v The King. (1911). AC 87 [Burrard Power].

Charpleix L. (2018). The whanganui river as Te Awa tupua: place-based law in a legally pluralistic society. *The Geographical Journal*, **184**(1), 19–30. https://doi.org/10.1111/geoj.12238

Day R. (2001). Who is this we that gives the gift? Native American political theory and the western tradition. *Critical Horizons*, **2**(2), 173–201, https://doi.org/10.1163/156851601760001300

De Vos H., Boelens R. and Bustamante R. (2006). Formal law and local water control in the andean region: a fiercely contested field. *Water Resources Development*, **22**(1), 37–48. https://doi.org/10.1080/07900620500405049

Delgamuukw v British Columbia. (1997). 3 SCR 1010, 1997 CanLII 302 (SCC).

Ebbesson J. (2003). 'Lex pernis apivorus': an experiment of environmental law methodology. *Journal of Environmental Law*, **15**(2), 153–174. https://doi.org/10.1093/jel/15.2.153

Gudynas E. (2011). 'Buen vivir: Germinando alternativas al desarrollo'. *América Latina en Movimiento*, **462**, 1–20.

Gullason K. (2018). The water sustainabillity act, groundwater regulation, and indigenous rights to water: missed opportunities and future challenges. *Appeal*, **23**, 29–40.

Halalt First Nations v. British Columbia (Environment). (2011). BCSC 945 2012 BCCA 472.

Henderson J. Y. (2002) Postcolonial Indigenous Legal Consciousness. Indig. Law J. Available online: https://scholar.archive.org/work/3h3bmgkwsja3hgwebweypk2pa4 (accessed on 18 February 2022)

Howard-Hassmann R. E. (2014). A Defence of the International Human Rights Regime. CIGI The Internationalization of Indigenous Rights. UNDRIP in the Canadian Context. Special Report. Centre for International Governance Innovation, Waterloo, Canada. https://www.cigionline.org/sites/default/files/indigenous_rights_special_report_web.pdf (accessed 18 October 2021)

INAC (Indigenous and Northern Affairs Canada). (2021). 'Maa-Nulth First Nations Final Agreement', online: https://www.aadnc-aandc.gc.ca/DAM/DAM-INTER-BC/STAGING/texte-text/mna_fa_mnafa_1335899212893_eng.pdf archived at https://perma.cc/5764-A7PN (accessed 3 August 2022)

Kauffman C. M. and Martin P. L. (2016). Can rights of nature make development more sustainable? Why some Ecuadorian lawsuits succeed and others fail. *World Development*, **92**, 130–142. https://doi.org/10.1016/j.worlddev.2016.11.017

Kauffman C. M. and Martin P. L. (2019). How courts are developing river rights jurisprudence: comparing guardianship in New Zealand, Colombia, and India. *Vermont Journal of Environmental Law*, **20**, 260–290.

Laidlaw D. and Passelac-Ross M. (2010). Water rights and water stewardship: what about aboriginal peoples? *Resources*, **107**, 1–8. http://ablawg.ca/wp-content/uploads/2010/07/blog_dl_mpr_waterrights_july2010.pdf archived at https://perma.cc/CFS3-FCM7 (accessed 18 October 2021)

Littlechild D. B. (2014). Transformation and re-formation: first nations and water in Canada. Master's thesis, University of Victoria, Victoria, B.C., Canada. http://hdl.handle.net/1828/5826 (accessed 18 October 2021)

Maciel R. (2014). Conflicting Ontologies and Balancing Perspectives CIGI The Internationalization of Indigenous Rights. UNDRIP in the Canadian Context. Special Report. Centre for International Governance Innovation, Waterloo, Canada.

Matsui K. (2005). 'white man has no right to take any of it' secwepemc water-rights struggles in British Columbia. *Wicazo Sa Review*, **20**(2), 74–101, https://doi.org/10.1353/wic.2005.0023

McNeil K. (2013). Aboriginal rights in Canada: the historical and constitutional context. *International Journal of Legal Information*, **41**(1), 16–38. https://doi.org/10.1017/S0731126500011574

Merry S. E. (1988). Legal pluralism. *Law and Society Review*, **22**(5), 869–896, https://doi.org/10.2307/3053638

Miglani (Lalit) v State of Uttarakhand & Others, WPPIL 140/2015, High Court of Uttarakhand. (2017). ('Glaciers'), Available at https://indiankanoon.org/doc/92201770/ (accessed 18 2021)

Mitchell v Minister of National Revenue. (2001). SCC 33, at para 22.

Napoleon V. R. (2009). Ayook: Gitksan Legal Order, Law, and Legal Theory. PhD thesis, University of Victoria, Victoria, B.C., Canada.

Ostrom E. (2010). Analyzing collective action. *Agricultural Economics*, **41**, 155–166. https://doi.org/10.1111/j.1574-0862.2010.00497

Paul C. (2020). Stories From the Magic Canoe of Wa'xaid. Rocky Mountain Books, Victoria, B.C., Canada.

Phare M. (2009). Denying the Source. The Crisis of First Nations Water Rights. Rocky Mountain Books, Surrey, B.C., Canada.

Rühs N. and Jones A. (2016). The implementation of earth jurisprudence through substantive constitutional rights of nature. *Sustainability* (Basel, Switzerland), **8**(2), 174. https://doi.org/10.3390/su8020174

R. v. Sioui. (1990). 70 D.L.R. (4th) 427 (S.C.C.).

Rawls J. (1971). A Theory of Justice. Belknap Press of Harvard University Press, Cambridge, MA, USA.

Saanichton Marina Ltd. V. Claxton. (1988). 43 D.L.R. (4th) 481. (B.C.S.C).

Salim, (Mohd) v State of Uttarakhand & Others, WPPIL 126/2014, High Court of Uttarakhand. (2017). ('Ganges and Yamuna'), available at http://lobis.nic.in/ddir/uhc/. RS/orders/22-03-2017/RS20032017WPPIL1262014.pdf (accessed 18 October 2021)

Salmond A. (2017). Tears of Rangi: Experiments Across Worlds. Auckland University Press, Auckland, New Zealand.

Sieder R. and Barrera Vivero A. (2017). Legalizing indigenous self-determination: autonomy and Buen Vivir in Latin America. *The Journal of Latin American and Caribbean Anthropology*, **22**(1), 9–26. https://doi.org/10.1111/jlca.12233

Statt G. R. (2003). Tapping into water rights: an exploration of native entitlement in the treaty 8 area of Northern Alberta. *Canadian Journal of Law and Society*, **18**(1), 103–129. https://doi.org/10.1017/S0829320100007493

Te Awa Tupua Act. Te Awa Tupua (Whanganui River Claims Settlement) Act 2017.

Turag River Case. Nation River Conservation Commission. Writ Petition 3839 of 2016 (Supreme Court of Bangladesh, High Court Division). The judgment was delivered February 2019 and released July 2019. The

decision is available in Bengali at https://www.scribd.com/document/533961661/Writ-Petition-No-13989-of-2016-only-17-directions-River-Turag-Case (accessed 3 August 2022)
UN (2016). https://www.rightsofnaturetribunal.org/ (accessed 3 August 2022)
UNDRIP (United Nations Declaration of the Rights of Indigenous Peoples). Adopted by the General Assembly of the United Nations on September 13, 2007.
University of Melbourne. (2021). Australian Indigenous Law. Library Guides. Available at: https://unimelb.libguides.com/indigenous_law/australia (accessed 3 August 2022)
Westermann A. (2019). Should Rivers Have Same Legal Rights as Humans? A Growing Number of voices say Yes. NPR. Available at: https://www.npr.org/2019/08/03/740604142/should-rivers-have-same-legal-rights-as-humans-a-growing-number-of-voices-say-ye (accessed 18 October 2021)
Wheeler c. Director de la Procuraduria General Del Estado de Loja, Juicio No. 11121-2011-0010. (2011). ('Wheeler'), available at https://elaw.org/system/!les/ec. wheeler.loja_.pdf (accessed 3 August 2022)
White (Charleson) Hee Naih Cha Chist, E. (2021). Making Space for Indigenous Law. Available at: http://jfklaw.ca/making-space-for-indigenous-law/ (accessed ??)
Williams to Scott, Black (Western) Series. (1920). Ottawa, Library and Archives Canada (RG 10, vol 3660, file 9755-4), cited in Bartlett, R.H. *Aboriginal Water Rights in Canada: A Study of Aboriginal Title to Water and Indian Water Rights* (Calgary: Canadian Institute of Resources Law, 1986).
Wilson N. J. and Inkster J. (2018). Respecting water: indigenous water governance, ontologies, and the politics of kinship on the ground. *Environment and Planning E: Nature and Space*, **1**(4), 516–538. https://doi.org/10.1177/2514848618789378
Ziker J. P., Rasmussen J. and Nolin D. A. (2016). Indigenous Siberians solve collective action problems through sharing and traditional knowledge. *Sustainability Science*, **11**(1), 45–55. https://doi.org/10.1007/s11625-015-0293-9

doi: 10.2166/9781789062786_0049

Chapter 5

Allocations and environmental flows

Eric D. Stein[1], Michael E. McClain[2,3], Ashmita Sengupta[4], Theodore E. Grantham[5], Julie K. H. Zimmerman[6] and Sarah M. Yarnell[7]

[1]*Southern California Coastal Water Research Project, California, USA*
[2]*IHE Delft Institute for Water Education, Delft, Netherlands*
[3]*Delft University of Technology, Delft, Netherlands*
[4]*CSIRO, Black Mountain, ACT, Canberra, Australia*
[5]*University of California, Berkeley, California, USA*
[6]*The Nature Conservancy, Sacramento, California, USA*
[7]*University of California, Davis, California, USA*

ABSTRACT

Over the past 30 years, much has been learned from strategies used around the world to establish and implement environmental flow programs. Approaches vary from highly prescriptive regulatory requirements to largely voluntary programs. These examples have shown that allocating water to the environment does not necessarily constrain human uses and can have benefits for both agriculture and ecosystems. Some efforts attempt to reduce conflict between agriculture and the environment by limiting water allocations spatially, while others attempt to reconcile competing water demands through comprehensive, regional allocation schemes that vary with climate conditions over time. Here we summarize strategies for water allocation planning and implementation that can be used to balance environmental and agricultural water needs. Effective strategies incorporate: a holistic environmental water allocation approach that focuses on protecting overall ecological structure and functions; environmental flow protections at broad spatial and temporal scales; consideration of surface–ground water interactions and the relationships between flow, sediment, temperature, and water quality. From an implementation perspective, approaches that establish a volumetric water budget for the environment based on interannual variation in water availability, integrate across programs in a transparent manner, are broadly inclusive, and incorporate traditional values and perspectives have the highest likelihood of success. Environmental flow strategies that consider technical solutions, establish clear objectives and anticipate how environmental water will be allocated under different water year types, and are sensitive to social issues and concerns will increase certainty in how much water is allocated for agriculture and the environment. Beyond reconciling conflicts between competing demands, emerging technical and institutional approaches to environmental flows can improve resiliency of water management programs to climate change by preventing the over-exploitation of water supplies, enhancing flexibility, and providing a framework for adaptation.

Keywords: Ecological effects, functional flows, governance, Indigenous uses, surface–groundwater connections, water budgets

5.1 INTRODUCTION

Conflicts between agriculture and the environment often stem from the perception that water allocation decisions are a zero-sum game. These tensions are exacerbated by 'one-dimensional' approaches to environmental flows that establish minimum thresholds to meet the needs of individual species and fail to consider the inherent ecological complexity and physical dynamism of river systems (Grantham *et al.*, 2020; Mount *et al.*, 2019). Consequently, in many rivers with environmental flow protections, populations of native fishes and other sensitive aquatic species continue to decline (Howard *et al.*, 2015; Mezger *et al.*, 2019; Moyle *et al.*, 2011). Such approaches have generally failed to achieve the desired goal of restoring and preserving ecosystem health while intensifying conflicts with agricultural water allocations. This tension has negatively affected agricultural production, increased social conflict, and inadequately served the needs of communities.

The need for environmental water allocations is based on the recognition that the ecological health and social value of rivers can only be maintained if the quantity and timing of water flows is sufficient to support essential ecosystem functions (Arthington *et al.*, 2018). New approaches also suggest that environmental flows managed for ecosystem functions can provide greater flexibility in how water is allocated and yield greater benefits for both the environment and agriculture (Grantham *et al.*, 2020; Mount *et al.*, 2019; Yarnell *et al.*, 2015). Protecting the flow necessary to sustain ecological functions across seasons and variable conditions across years will in many cases leave less overall water available for other uses, including agriculture. However, a focus on the functionality of flow left instream ensures that available water is used more effectively. Moreover, the recovery of overall ecosystem health should decrease the likelihood of further species decline and reduce future conflicts between agricultural and environmental water allocations (Howard *et al.*, 2015). Formalizing water allocations for the environment can also reduce uncertainty about how much water will be available for agriculture, allowing for long-term, adaptive planning that reduces conflicts.

These concepts have been used to guide water allocation schemes in different parts of the world, including the EU Water Framework Directives, Murray–Darling Basin Plan, and Nile River E-Flows/ Lower Mara River Plan. These, and other examples, provide lessons that can be applied to better balance water for agriculture and the environment for those seeking to develop or refine water allocation approaches that are more predictable over time, sustainable, and resilient to climate change.

In this chapter, we discuss emerging technical and institutional advances in planning and defining environmental flows, and we describe mechanisms for implementing environmental flow programs. We highlight examples from ongoing efforts to balance human and ecosystem needs. Lessons learned can guide the evolution of water allocation decision-making towards frameworks that enhance water supply reliability for agriculture, restore ecosystem health, and offer flexibility in adapting to uncertain future conditions. Ultimately, successful environmental flow programs will play a critical role in ensuring the allocation of river flows is sustainably managed across variable water year types and changing climate.

5.2 PLANNING AND DEFINING ENVIRONMENTAL FLOWS: EMERGING SCIENCE AND RECOMMENDED APPROACHES

Satisfying environmental water needs can be challenging, especially in over-allocated systems where natural water supplies have been fully claimed by legal entitlements. However, it is possible to provide water in ways that improve ecological conditions while reducing potential conflicts with agriculture. This section highlights emerging approaches to planning and defining the allocation of environmental flows and describes how such efforts could improve ecological conditions while increasing the certainty of when and where water will be available for irrigation. These advances include increased focus on flows that support a broad suite of ecological functions, recognizing the connections between surface and groundwater management, and planning in an inclusive transparent manner.

5.2.1 Increased focus on ecological functions

Water management strategies that aim to sustain overall ecosystem health through maintenance of ecological function (or processes) are emerging worldwide (Horne *et al.*, 2017; Reid *et al.*, 2019). These approaches aim to support a broad array of habitats and species and enhance the ability of ecosystems to adapt to changing conditions. Past environmental flow practices have focused on single iconic or endangered species protection, key high value sites, and/or concepts such as a minimum baseflow. These simplistic standards allocate fixed volumes of water and are typically expressed as a monthly or seasonal minimum flow threshold, above which water can be diverted or stored. However, there is evidence that such approaches tend to result in at least one ecological community losing out in favor of the species or habitat of management priority (Tonkin *et al.*, 2021). This species-centric approach also inherently focuses debate on trade-offs between the value of species versus that of agricultural products, rather than the broad benefits that healthy ecosystems provide to society.

A minimum or static flow approach also tends to mute natural flow variability due to storms, changing seasons, or differences between wet and dry years. Consequently, ecosystem health tends to decline. Decades of study have shown that riparian and aquatic communities rely on variability in flow and sediment discharge over time and across watersheds (Bendix and Hupp, 2000; Naiman *et al.*, 2008; Poff *et al.*, 1997). The fluctuation in flows across time and space interacts with the surrounding landscape to drive ecosystem processes, such as movement of organic matter and nutrients, scour and erosion of sediment, and hydrological connectivity enabling vegetation growth or fish migration (Palmer & Ruhi, 2019; Yarnell *et al.*, 2015). When such ecological functions are disrupted through the stabilization of flow regimes in time and space, long-term resiliency and biodiversity of the river system is reduced (Auble *et al.*, 1994; Merritt *et al.*, 2010; Tonkin *et al.*, 2021).

A focus on the functionality of flow in water management is expected to improve ecological outcomes and provide more flexibility than prescriptive, minimum flow requirements. Using a 'functional flows approach' (Grantham *et al.*, 2020; Yarnell *et al.*, 2020), environmental flow allocations can be targeted towards those components of the flow regime that most directly relate to ecological functions (e.g., periodic high flows that inundate floodplains), while allowing withdrawal for human uses during other times (e.g., winter baseflow periods). For example, in California, several 'functional flow components' have been identified that support essential ecosystem functions and collectively sustain a broad suite of species and ecosystem services (Figure 5.1; Yarnell *et al.*, 2020). In this way, functional flows and species-based management are not mutually exclusive. The transition from single species management to multi-species management through the lens of ecological functions provides a more holistic approach to water allocation.

A functions-oriented approach also targets environmental water allocations at locations and times of year where they will produce the greatest benefit for the ecosystem, leaving water available for agricultural and other human uses in other places and periods. Over longer timescales, the approach also provides flexibility to adjust environmental water allocations in different water year types, maximizing allocations in wet years to enhance ecosystem conditions and limiting allocations in drought years to those necessary to avoid catastrophic ecosystem impacts (Figure 5.2).

For example, the Drought Management Planning approach used in Spain under the Water Framework Directive takes both annual water availability and critical ecosystem locations into account, whereby 'drought flows' are anticipated through a risk-based approach and designed to support refuge habitats and connectivity patterns that are critical in dry years (Paneque, 2015). Similarly, in the Murray–Darling Basin in Australia, water resource availability scenarios are used to prioritize environmental water actions in the upcoming year. These scenarios ranging from 'very dry' to 'very wet' are generated using the previous year's climate conditions and surface water available in public dams, as well as modeling forecasts. In California (USA), the functional flows approach has been applied to develop ecological flow criteria that vary by type of rainfall year, but still include all major components of the annual hydrograph (Figure 5.3).

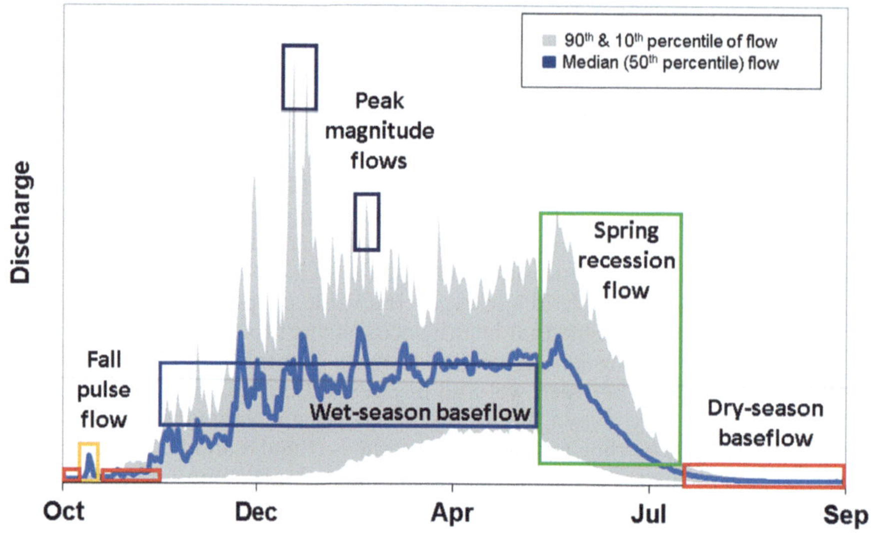

Figure 5.1 Functional flow components depicted on a representative hydrograph. Blue line represents median (50th percentile) daily discharge. Gray shading represents 90th to 10th percentiles of daily discharge over the period of record (Yarnell *et al.*, 2020).

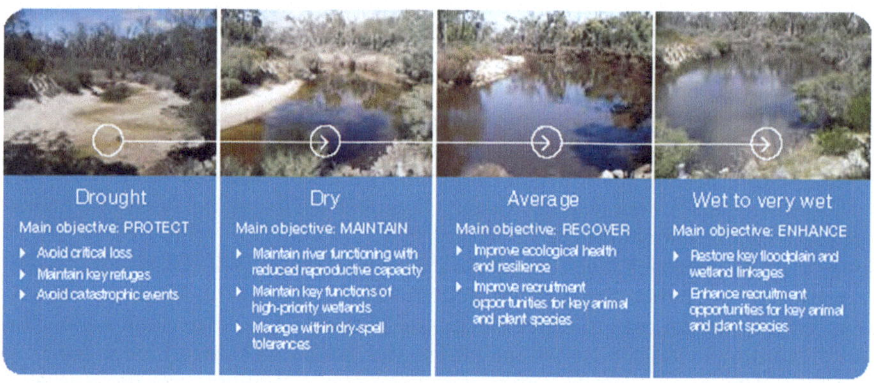

Figure 5.2 Examples of water allocation objectives under different climatic conditions. (Source: Victorian Environmental Water Holder, 2015. Seasonal Watering Plan 2015–2016.)

Tonkin *et al.* (2021) also demonstrated that decadal scale 'designer flows' that consider temporal variations in life history requirements were able to accommodate multiple species assemblages while providing water for other uses. Such proactive, long-term approaches will become more important in the future as global temperatures are expected to rise and increase the intensity and spatial extent of drought in much of the Western Hemisphere (Woodhouse *et al.*, 2010). Ultimately, this will likely force difficult decisions regarding water for the environment versus agriculture during extended dry periods; however, flexible approaches that aim to maximize ecosystem functionality, especially during wetter years, will help to build the resiliency of ecosystems to future droughts.

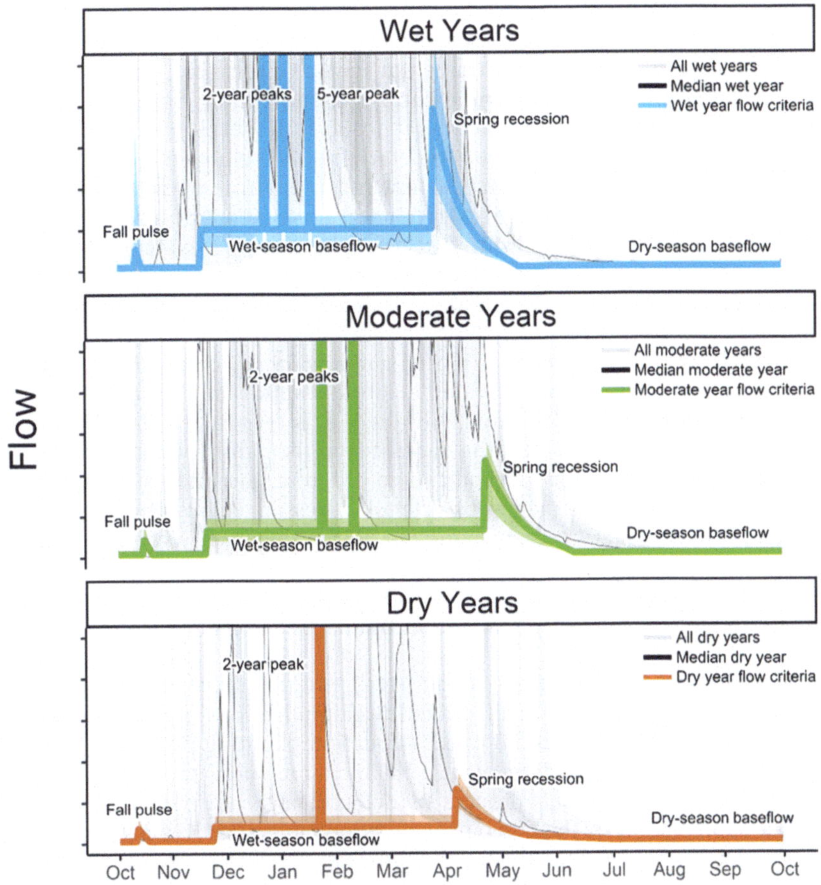

Figure 5.3 Application of functional flows approach from California (USA) to produce ecological flow criteria for different water year types. Environmental water allocations are indicated by shaded bands.

5.2.2 Recognizing connections between surface and groundwater management

Although rarely considered in environmental flow programs, groundwater can play an important role in sustaining streamflow, particularly in arid and semi-arid regions of the world (Brown *et al.*, 2010). Groundwater pumping can cause streamflow depletion, particularly in the dry season when groundwater may provide a substantial portion of baseflow and is crucial for maintaining stream temperatures and connectivity for aquatic species (Barlow & Leake, 2012; Zipper *et al.*, 2019). There is evidence that extensive groundwater pumping is causing declines in groundwater contributions to stream baseflow in many areas globally and that losing streams – those that lose flow to the surrounding aquifer – are becoming more common, particularly in arid climates (Jasechko *et al.*, 2020; Mukherjee *et al.*, 2018; Stromberg *et al.*, 2007). Groundwater pumping can also lower groundwater levels below the rooting depth for streamside or wetland vegetation, resulting in loss of riparian habitat (Barlow & Leake, 2012; Rohde *et al.*, 2021), and impacts to groundwater-dependent ecosystems (GDEs) (Howard & Merrifield, 2010).

Environmental flow programs can best achieve ecological goals when they consider all sources of water and all actions that can affect flows. Yet, existing watershed models used in development of environmental flows (e.g., SWMM, HSPF, HEC)[1] seldom account for relationships between surface and groundwater. Moreover, groundwater management programs are typically focused on urban and agricultural uses and do not consider environmental flows. GDEs have received some attention in groundwater management policy in California, the European Union, South Africa, and Australia, but methods to fully address connections have yet to be fully developed (Rohde *et al.*, 2017). In South Africa, a novel approach was developed to determine the volume of groundwater available for extraction annually without lowering stream baseflow below environmental flow targets (Ebrahim & Villholth, 2016). France and New Zealand also have programs that define a combined volumetric and flow-based approach to water extraction limits, including from surface water and aquifers to account for interactions between groundwater and surface water.

Integrated management of surface water and groundwater would expand the suite of management approaches that could be considered to balance environmental and agricultural needs. Managed aquifer recharge (MAR) is a strategy being explored globally for regions with groundwater overdraft (Dillon *et al.*, 2019) and involves storing water in groundwater aquifers during high flow events for use in the dry season or during dry years. Although it is recognized that MAR can be used to supplement streamflow during dry periods, the current focus is mainly on maximizing water supplies for agricultural use (Alam *et al.*, 2020). Despite some progress, the integration of groundwater and surface water management remains a major obstacle to addressing environmental flow needs globally (Gleeson & Richter, 2018; Ross, 2018). Improving our ability to quantify surface water–groundwater connections, account for the effects of groundwater recharge and pumping, and conjunctively manage surface water and groundwater are critical for developing water allocation strategies that effectively protect ecosystems and satisfy agriculture and other human demands.

5.2.3 Planning in an inclusive, consistent, structured, and transparent manner

Decisions about water allocation have social and cultural implications in addition to ecological ones. These values need to be integrated into the water allocation planning process to ensure the success of environmental flow programs (Richter *et al.*, 2003). A common vision and clearly articulated goals must be established early in the process of determining environmental flows based on an understanding and appreciation of the ecological functions and social values (including traditional and Indigenous values) provided by healthy rivers (Runge *et al.*, 2015). Environmental flow programs require prioritization, trade-offs, and compromises that can only be established through inclusive and transparent stakeholder participation (Conallin *et al.*, 2017). The support of local communities, who are most directly affected by alterations in flows, is essential to long-term success of environmental flow programs (Anderson *et al.*, 2019).

Initiatives incorporating knowledge, expertise, and perspectives of local communities and Indigenous people on environmental flows and ecosystem restoration occur in several places globally. In Australia, governments have developed processes to incorporate knowledge, expertise, and perspectives of Aboriginal nations into water resources planning. This approach recognizes the importance of traditional practices and can reduce conflicts through collaborative governance that incorporates cultural knowledge and priorities into the decision-making process (Jackson & Moggridge, 2019). Similarly, in New Zealand, treaty settlement agreements have been signed by the New Zealand Crown and Māori iwi groups that establish a co-governance mechanism which reassert Māori authority over and knowledge about wai (water) and awa (rivers). This system provides formal recognition of Indigenous knowledge systems to establish environmental flow recommendations (Parsons *et al.*, 2021). In the US, as part of their analysis of water releases from Glen Canyon Dam, the US Geological Survey used a multi-criterion structured decision-making approach to consider factors such as preservation of tribal, kiva group and clan history, respect for life, sacred integrity

[1]SWMM: USEPA Stormwater Management Model, HSPF: USEPA Hydrologic Simulation Loading Program in Fortran, HEC: ACOE Hydrologic Engineering Center Modeling System.

of the river, health of the river as a sentient being, sacred stewardship and education, sustenance, economic opportunity and tribal water rights and supply. Anderson *et al.* (2019) provide a series of case studies from Honduras, India, Canada, New Zealand and Australia that illustrate multidisciplinary, collaborative efforts where recognizing and meeting diverse flow needs of human populations was central to establishing environmental flow recommendations. These cases offer a lens for a better understanding of power relations among stakeholders and the importance of trust in supporting and developing dynamic relationships between humans, river flow regimes, and aquatic ecosystems through relationships that are sustainable, just, and inclusive. Including diverse stakeholder perspectives and values is still an emerging area for decisions about water management and environmental flows and most decisions still primarily involve natural resource agencies and stakeholders who own water rights or contracts. The above efforts recognize the importance of traditional practices and can reduce conflicts through collaborative governance that incorporates cultural knowledge and priorities into the decision-making process (Jackson & Moggridge, 2019).

5.3 IMPLEMENTATION OF ENVIRONMENTAL FLOWS: TOOLS AND APPROACHES

Implementation of environmental flow programs in agricultural settings varies globally based on local governance structure, socioeconomic contexts, and cultural values. Approaches range from those that attempt to avoid conflict by prioritizing agricultural uses in some areas and the environment in others (e.g., Mexico Water Reserve System) to those that attempt to reconcile conflict by making the environment an equal stakeholder to agricultural and urban water users (e.g., Murray–Darling Basin, Australia). Similarly, approaches range from highly prescriptive to entirely voluntary and require various amounts of capital and social investment. This section discusses innovative approaches and tools that can be used to implement 'next generation' environmental flow management strategies.

5.3.1 Ecosystem water budgets

Ecosystem water budgets establish fixed allocation volumes for the environment and provide a mechanism for flexibly managing environmental flows at a watershed scale (Grantham *et al.*, 2020). Water budgets allow flexibility in environmental water allocations by making it possible to shift water allocations to different periods and/or locations to optimize ecosystem benefits, such as through a functional flows approach (see above). Budgets should be developed through a stakeholder-driven process and account for ecosystem management priorities, agricultural needs, existing water sharing arrangements, and community priorities. Once established, budgets can be managed by an ecosystem trustee or river basin management authority. An example water budget for an agricultural system dominated by water withdrawals (vs. dam releases) is shown in Figure 5.4. In this case, the environmental water budget is managed by limiting water withdrawal magnitudes and rates during specific periods to restore functional flows. Ideally, budgets would be created over multi-year periods to allow allocation across different climatic years to better balance environmental and agricultural needs.

An ecosystem water budget has been established in the Murray–Darling Basin (Australia) and is defined by a Sustainable Diversion Limit (SDL) that sets thresholds on the volume of water to be extracted every year and which is then distributed among different users based on priority. While the Murray–Darling Basin Authority (MDBA) oversees the process, states are responsible for ensuring their water withdrawals are within limit. Since the environment is also a stakeholder, it must abide by the same allocation rules as the other users, with reductions during the dry years. Use of the ecosystem budget through 'environmental watering' actions is carefully navigated to maximize beneficial outcomes and is well-documented for transparency and legitimacy. Allocations are also seasonally dynamic and evaluated every two weeks based on the rainfall in the catchment. The overall water budget is driven by entitlements and allocations which are first completed at a state level, then the state decides how much will be provided to each catchment and imposes extraction limits with further granularity to protect environmental water.

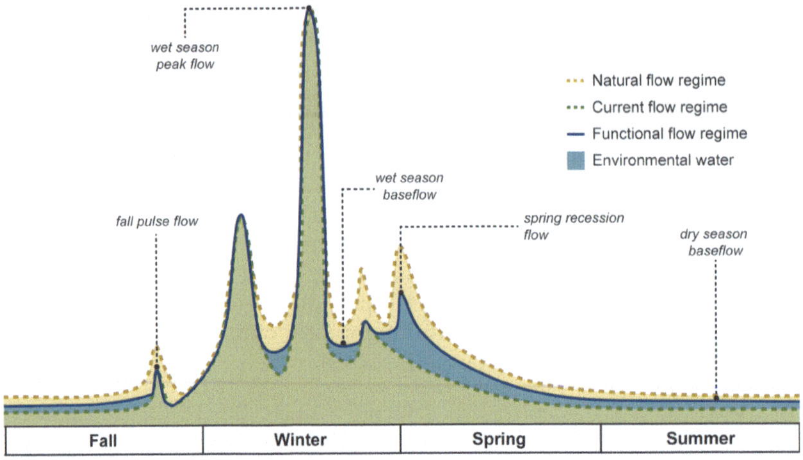

Figure 5.4 Example water budget approach for an undammed river dominated by agricultural diversions. Teal area represents the volume of water managed for the environment. (From Grantham *et al.*, 2020.)

As Horne *et al.* (2017) illustrate in the Olinda Catchment, the Victorian state issues two types of license, with the first permitting extraction all year and the second specifically for dam filling during the high flow season. The all-year license has seasonal caps with cease-in-extraction if the streamflow drops below a threshold. Similar systems have been established in New Zealand and France where environmental flow needs are aggregated across sub-basins and maximum allowable extraction volumes are established. These systems are somewhat unique in that they integrate groundwater and surface water resource needs when determining allocation limits and provide a mechanism for managing volumes over a series of years to better account for interannual climate variability. A shortcoming of these methods is that limits are established based on percentages of natural (or current) flows and are not directly linked to ecological functions. However, the water budget approach is a promising mechanism for supporting ecosystem needs and satisfying human demands in a changing climate.

5.3.2 Innovative governance structure of water allocation

Water resources are often managed by numerous government programs with differing objectives and are seldom coordinated in meaningful ways. This fragmented approach leads to inefficiencies as each program aims to maximize its ability to meet its mandates. Conflicts could be reduced by incorporating environmental flows into a broad range of regulatory and non-regulatory programs, including those focused on environmental restoration, river basin and groundwater management, 'working lands management', water quality, water quality credit trading, and community engagement.

Coupling programmatic strategies with the use of a consistent governance framework can distribute allocation burdens across numerous programs and build a broad consensus on implementation strategies. It can also help leverage funding and spread program management costs and responsibilities. The California Water Quality Monitoring Council[2] is an example of a legislatively mandated intergovernmental council where numerous Federal, State, and local agencies work collaboratively to balance water resource management programs. Such cooperative approaches have the advantage of building support and consensus while distributing costs and responsibilities. However, they also

[2]https://mywaterquality.ca.gov/monitoring_council/index2.html

require improved mechanisms for ongoing coordination, data sharing, and co-development of tools. This requires commitment and dedicated funding to institutionalize these partnerships.

The Australian water market primarily developed in the MDB is another illustration of innovative governance. Prior to the establishment of the market, transactions were limited to intradistrict trade and interstate trading was non-existent, with states guarding access to their water (Horne & Grafton, 2019). Implementation of the water market has not been without conflict about the role of brokers and exchange platforms, and the actual environmental impacts of the water buybacks (Ryan *et al.*, 2021). However, the number of sales and transactions, both in the water entitlement and allocation market, have been increasing with participation from irrigators, environmental non-governmental organizations, and investors. Data show that the market has provided a flexible mechanism of managing water, and through temporary/allocation trade and permanent/entitlement trades introduced changes in agricultural practices (Horne & Grafton, 2019; Wheeler *et al.*, 2020). During the millennium drought, allocation trading allowed the irrigators with permanent plantings to purchase water from irrigators in other regions reducing the overall negative impact and promoting resiliency (Kirby *et al.*, 2014).

Cross-program coordination is typically more difficult, yet more important, when water is transported across state or national boundaries to meet agricultural demands far from the original water source. In these cases, intergovernmental councils can improve opportunities for collaboration in water allocation decisions. Mechanisms for transnational governance exist through legal instruments such as the *Danube River Protection Convention*[3] and the *Agreement on the Cooperation for the Sustainable Development of the Mekong River Basin*.[4] Both these agreements include commitments for sustainable use and environmental protection which could be enhanced through establishment of environmental flow targets and establishment of an entity to advocate, monitor, and manage environmental water across national boundaries.

Perhaps the most well-established multinational governance framework is the European Union Water Framework Directive (WFD), which harmonizes and streamlines water policy and legislation across the Union's 27 member countries with the objective of increasing and maintaining the water quality and ecological condition for all surface- and ground-water bodies. It was launched in 2000 and called on all countries to assess the status of their water bodies and develop and implement River Basin Management Plans including a program of measures to achieve 'good status', which is defined as a condition of water quality and ecology that varies only slightly from undisturbed conditions. The goal was to achieve good status in water bodies by the year 2015, but a 2018 assessment conducted by the European Environment Agency (EEA, 2018) found only 40% of surface waters and 74% of groundwater areas met the good status goal. European Commission assessments (EC, 2012) have found that insufficient attention had been given to pressures of hydrological alteration across the Union and called for increased attention to this pressure and explicit incorporation of environmental flows into all river basin planning. Existing guidance on setting environmental flows under the WFD (EC, 2015) calls on all countries to assess the ecological impacts of hydrological alteration, establish appropriate monitoring programs, determine environmental flow requirements based on impacts on biological indicators that are sensitive to alterations in the hydrological regime, and implement needed measures to protect or restore environmental flow levels. The recent EC Biodiversity Strategy has called for greater attention to this issue (EC, 2021).

5.3.3 Holistic management by reducing silos between programs

Environmental flow programs are often either siloed from other water management programs or distributed across multiple programs with different jurisdictions and authorities. This leads to

[3]http://www.icpdr.org/main/icpdr/danube-river-protection-convention

[4]https://www.mrcmekong.org/assets/Publications/policies/agreement-Apr95.pdf

inefficiencies, inconsistencies, and conflicts that can increase uncertainty among agricultural water users and erode support for environmental flow programs. Flow management should be integrated with management of temperature, water quality, sediment, and invasive species, as the ability to support healthy ecological communities depends on interactions between flows, temperature, sediment, and biologically relevant water quality parameters (e.g., specific conductance, turbidity). Accounting for these interactions allows for more holistic flow management and for the inclusion of other management measures (e.g., increased stream shading, reconnecting floodplains) that may provide for increased allocations while still supporting stream ecosystems (Nilsson & Renöfält, 2008). For example, coupling flow management with stream restoration may allow for requisite functions to be supported with less water by providing overbank storage areas, deeper pools and so on. This in turn could allow for more water to be allocated to agricultural uses. Programs aimed at improving water quality often employ management measures such as retention of runoff, which may alter streamflow in undesirable ways. Improved coordination between water quality and environmental flow programs could improve management efficiency and reduce conflicts. Similarly, considering flow and temperature in tandem can improve climate change resiliency. For example, Letcher *et al.* (2016) showed that as streamflow increased, there was a diminishing impact of air temperature on stream temperature, and Van Vliet *et al.* (2012) showed that decreasing streamflow resulted in lower minimum temperatures in the winter and higher maximum temperatures in the summer due to the smaller thermal capacity. Failure to account for such interactions may increase sensitivity of streams to climate change effects under reduced flows. Finally, enhanced coordination between programs provides opportunities for sharing data, models, and assessment tools that can improve the overall ability to implement all aquatic resource management programs.

5.4 CONCLUSIONS

Providing water for the environment and agriculture requires compromises to support both uses with less overall water available than if only a single use was being supported. This balance can be achieved by adopting strategies that focus on maintaining ecosystem functions over long time periods and across watersheds, as opposed to strategies aimed at supporting specific species or single habitats. A function-based approach that accounts for the ecological needs of different stream types in different seasons and the natural variability of flows between wet and dry years ensures that the maximum benefit is obtained from environmental flow allocations. Coupling surface water and groundwater management will improve the effectiveness of environmental flow programs and also benefit agriculture. A water budget approach, in which the volume of water available for the environment is formally quantified and allowed to vary depending on water year conditions, would establish the environment as a legitimate user of water and provide flexibility for optimizing the benefits of environmental water allocations. Together, these approaches can ultimately provide more certainty about the amount, timing, and persistence of available water for agricultural uses. Finally, a function-based approach enables water allocation programs to adapt over time in response to both short-term droughts and long-term changes in precipitation and runoff patterns associated with long-term climate change. The science supporting such integrated, comprehensive, and adaptive approaches is relatively young and is rapidly advancing. Continued progress will depend on sustained support of this research and a commitment to sharing advances and lessons through an open science approach.

Successful environmental flow programs require full, inclusive, and transparent stakeholder involvement during both planning and implementation. Inclusive science builds trust and understanding among participants and increases the likelihood of identifying broadly acceptable solutions. Consideration of traditional and contemporary values and practices can produce innovative and sustainable programs that continue to enjoy public support. Such programs should explore pioneering approaches such as multi-year water budgeting and governance structures that account for multiple objectives while reducing compartmentalization between programs aimed at managing water quantity, quality, habitat restoration, sediment management, invasive species control, and groundwater. Only through integrated, holistic, and cooperative management can environmental flows be provided in a manner that minimizes conflicts and provides for sustainable and adaptable ecosystem management and agricultural production.

REFERENCES

Alam S., Gebremichael M., Li R., Dozier J. and Lettenmaier D. P. (2020). Can managed aquifer recharge mitigate the groundwater overdraft in California's central valley? *Water Resources Research*, **56**(8), e2020WR027244. https://doi.org/10.1029/2020WR027244

Anderson E. P., Jackson S., Tharme R. E., Douglas M., Flotemersch J. E., Zwarteveen M., Lokgariwar C., Montoya M., Wali A., Tipa G. T., Jardine T. D., Olden J. D., Cheng L., Conallin J., Cosen B., Dickens C., Garrick D., Groenfeldt D., Kabogo J., Roux D. J., Ruhi A. and Arthington A. H. (2019). Understanding rivers and their social relations: A critical step to advance environmental water management. *Wires Water*, **6**(6), e1381. https://doi.org/10.1002/wat2.1381

Arthington A., Bhaduri A., Bunn S., Jackson S., Tharme R., Tickner D., Young B., Acreman M., Baker N., Capon S., Horne A., Kendy E., McClain M., Poff L., Richter B. and Ward S. (2018). The Brisbane declaration and global action agenda on environmental flows. *Frontiers in Environmental Science*, **6**, 45. https://doi.org/10.3389/fenvs.2018.00045

Auble G. T., Friedman J. M. and Scott M. L. (1994). Relating riparian vegetation to present and future streamflow. *Ecological Applications*, **4**, 544–554. https://doi.org/10.2307/1941956

Barlow P. M. and Leake S. A. (2012). Streamflow depletion by wells–understanding and managing the effects of groundwater pumping on streamflow. *U.S. Geological Survey Circular*, **1376**, 84.

Bendix J. and Hupp C. R. (2000). Hydrological and geomorphological impacts on riparian plant communities. *Hydrological Processes*, **14**, 2977–2990. https://doi.org/10.1002/1099-1085(200011/12)14:16/17<2977::AID-HYP130>3.0.CO;2-4

Brown J. B., Bach L. B., Aldous A. R., Wyers A. and DeGagne J. (2010). Groundwater dependent ecosystems in Oregon: an assessment of their distribution and associated threats. *Frontiers in Ecology and the Environment*, **9**(2), 97–102. https://doi.org/10.1890/090108

Conallin J. C., Dickens C., Hearne D. and Allan C. (2017). Stakeholder Engagement in Environmental Water Management in Water for the Environment: From Policy and Science to Implementation and Management. A. C. Horne, (eds.), Elsevier Publishers, pp. 129–150.

Dillon P., Stuyfzand P., Grischek T., Lluria M., Pyne R. D. G., Jain R. C., Bear J., Schwarz J., Wang W., Fernandez E., Stefan C., Pettenati M., van der Gun J., Sprenger C., Massmann G., Scanlon B. R., Xanke J., Jokela P., Zheng Y., Rossetto R., Shamrukh M., Pavelic P., Murray E., Ross A., Bonilla Valverde J. P., Palma Nava A., Ansems N., Posavec K., Ha K., Martin R. and Sapiano M. (2019). Sixty years of global progress in managed aquifer recharge. *Hydrogeology Journal*, **27**, 1–30. https://doi.org/10.1007/s10040-018-1841-z

Ebrahim G. Y. and Villholth K. G. (2016). Estimating shallow groundwater availability in small catchments using streamflow recession and instream flow requirements of rivers in South Africa. *Journal of Hydrology*, **541**, 754–765. https://doi.org/10.1016/j.jhydrol.2016.07.032

EC (European Commission). (2012). A Blueprint to Safeguard Europe's Water Resources /COM/2012/0673 final/ https://eur-lex.europa.eu/legal-content/EN/TXT/?uri=CELEX:52012DC0673 (accessed 15/01/2022).

EC (European Commission). (2015). Ecological flows in the implementation of the Water Framework Directive. Guidance Document No. 31. https://op.europa.eu/en/publication-detail/-/publication/b2369e0f-d154-11e5-a4b5-01aa75ed71a1 (accessed 15/01/2022).

EEA (European Environment Agency). (2018). European waters – assessment of status and pressures 2018. https://www.eea.europa.eu/themes/water/european-waters/water-quality-and-water-assessment/water-assessments (accessed 15/01/2022).

European Commission, Directorate-General for Environment, EU biodiversity strategy for 2030: bringing nature back into our lives, Publications Office of the European Union, 2021, https://data.europa.eu/doi/10.2779/677548

Gleeson T. and Richter B. (2018). How much groundwater can we pump and protect environmental flows through time? Presumptive standards for conjunctive management of aquifers and rivers. *River Research and Applications*, **34**(1), 83–92. https://doi.org/10.1002/rra.3185

Grantham T., Mount J., Stein E. D. and Yarnell S. M. (2020). Making the most of water for the environment: a functional flows approach for California's rivers, Southern California Coastal Water Research Project Technical Report #1142, Public Policy Institute of California, San Francisco, CA, USA.

Horne A. C., O'Donnell E. L. and Tharme R. E. (2017). Mechanisms to allocate environmental water. In: Water for the Environment, A. Horne, A. Webb, M. Stewardson, B. Richter and M. Acreman (eds.). Academic Press, pp. 361–398.

Horne J. and Grafton R. Q. (2019). The Australian water markets story: incremental transformation. *Successful Public Policy*, 165.

Howard J. and Merrifield M. (2010). Mapping groundwater dependent ecosystems in California. *PLoS One*, **5**, e11249.

Howard J. K., Klausmeyer K. R., Fesenmyer K. A., Furnish J., Gardali T., Grantham T, Katz J. V. E., Kupferberg S., McIntyre P., Moyle P. B., Ode P. R., Peek R., Quinones R. M., Rehn A. C., Santos N., Shoenig S., Serpa L., Shedd J. D., Slusark J., Viers J. H., Wright A. and Morrison S. A. (2015). Patterns of freshwater species richness, endemism, and vulnerability in California. *PLoS ONE* **10**(7), e0130710. https://doi.org/10.1371/journal.pone.0130710

Jackson S. and Moggridge B. (2019). Indigenous water management. *Australasian Journal of Environmental Management*. **26**(3), 193–196. https://doi.org/10.1080/14486563.2019.166164

Jasechko S., Seybold H., Perrone D., Fan Y. and Kirchner J. W. (2020). Widespread potential loss of streamflow into underlying aquifers across the USA. *Nature*. **591**, 391–397. https://doi.org/10.1038/s41586-021-03311-x

Kirby M., Bark R., Connor J., Qureshi M. E. and Keyworth S. (2014). Sustainable irrigation: How did irrigated agriculture in Australia's Murray–darling basin adapt in the millennium drought? *Agricultural Water Management* **145**, 154–162. https://doi.org/10.1016/j.agwat.2014.02.013

Letcher B. H., Hocking D. J., O'Neil K., Whiteley A. R., Nislow K. H. and O'Donnell M. J. (2016). A hierarchical model of daily stream temperature using air-water temperature synchronization, autocorrelation, and time lags. *PeerJ*, **4**, e1727. https://doi.org/10.7717/peerj.1727

Merritt D. M., Scott M. L., LeROY P., Auble G. T. and Lytle D. A. (2010). Theory, methods, and tools for determining environmental flows for riparian vegetation: riparian vegetation-flow response guilds. *Freshwater Biology*, **55**, 206–225. https://doi.org/10.1111/j.1365-2427.2009.02206.x

Mezger G., De Stefano L. and Del Tánago M. G. (2019). Assessing the establishment and implementation of environmental flows in Spain. *Environmental Management*, **64**(6), 721–735. https://doi.org/10.1007/s00267-019-01222-2

Mount J., Gray B., Bork K., Cloern J. E., Davis F. W., Grantham T., Grenier L., Harder J., Kuwayama Y., Moyle P., Schwartz M. W., Whipple A. and Yarnell S. (2019). A Path Forward for California's Freshwater Ecosystems. Public Policy Institute of California, San Francisco, CA, USA.

Moyle P., Katz J. and Quiñones R. (2011). Rapid decline of California's native inland fishes: A status assessment. *Biological Conservation*, **144**, 2414–2423. https://doi.org/10.1016/j.biocon.2011.06.002

Mukherjee A., Nath Bhanja S. and Wada Y. (2018). Groundwater depletion causing reduction of baseflow triggering Ganges river summer drying. *Scientific Reports*, **8**, 12049. https://doi.org/10.1038/s41598-018-30246-7

Murray Darling Basin Plan. https://www.mdba.gov.au/basin-plan/plan-murray-darling-basin (accessed 15/01/2022).

Naiman R. J., Latterell J. J., Pettit N. E. and Olden J. D. (2008). Flow variability and the biophysical vitality of river systems. *Comptes Rendus Geoscience*, **340**, 629–643. https://doi.org/10.1016/j.crte.2008.01.002

Nilsson C. and Malm Renöfält B. (2008). Linking flow regime and water quality in rivers: a challenge to adaptive catchment management. *Ecology and Society*, **13**(2): 18. http://www.ecologyandsociety.org/vol13/iss2/art18 (accessed 15/01/2022). https://doi.org/10.5751/ES-02588-130218

Palmer M. A. and Ruhi A. (2019). Linkages between flow regime, biota, and ecosystem processes: implications for river restoration. *Science* **365**, 1264. https://doi.org/10.1126/science.aaw2087

Paneque P. (2015). Drought management strategies in Spain. *Water*, **7**, 6689–6701. https://doi.org/10.3390/w7126655

Parsons M., Fisher K. and Crease R. P. (2021). Transforming river governance: The Co-governance arrangements in the Waikato and Waipa rivers. In: Decolonising Blue Spaces in the Anthropocene. Palgrave Studies in Natural Resource Management. Palgrave Macmillan, Cham, Switzerland, pp. 283–323. https://doi.org/10.1007/978-3-030-61071-5_7

Poff L. N., David Allan J., Bain M. B., Karr J. R., Prestegaard K. L., Richter B. D., Sparks R. E. and Stromberg J. C. (1997). The natural flow regime. *BioScience*, **47**, 769–784. https://doi.org/10.1073/pnas.0911197107

Poff L. N., David Allan J., Bain M. B., Karr J. R., Prestegaard K. L., Richter B. D., Sparks R. E. and Stromberg J. C. (1997). The natural flow regime. *BioScience*, **47**, 769–784.

Reid A. J., Carlson A. K., Creed I. F., Eliason E. J., Gell P. A., Johnson P. T. J., Kidd K. A., MacCormack T. J., Olden J. D., Ormerod S. J., Smol J. P., Taylor W. W., Tockner K., Vermaire J. C., Dudgeon D. and Cooke S. J. (2019). Emerging threats and persistent conservation challenges for freshwater biodiversity. *Biological Reviews*, **94**(3):849–873. https://doi.org/10.1111/brv.12480

Richter B. D., Mathews R., Harrison D. L. and Wigington R. (2003). Ecologically sustainable water management: managing river flows for ecological integrity. *Ecological Applications*, **13**, 206–224. https://doi.org/10.1890/1051-0761(2003)013[0206:ESWMMR]2.0.CO;2

Rohde M. M., Froend R. and Howard J. (2017). A global synthesis of managing groundwater dependent ecosystems under sustainable groundwater policy. *Groundwater*, **55**(3), 293–301. https://doi.org/10.1111/gwat.12511

Rohde M. M., Stella J. C., Roberts D. A. and Singer M. B. (2021). Groundwater dependence of riparian woodlands and the disrupting effect of anthropogenically altered streamflow. *Proceedings of the National Academy of Sciences*, **118**(25). https://doi.org/10.1073/pnas.2026453118

Ross A. (2018). Speeding the transition towards integrated groundwater and surface water management in Australia. *Journal of Hydrology*, **567**, e1–e10. https://doi.org/10.1016/j.jhydrol.2017.01.037

Runge M. C., LaGory K. E., Russell K., Balsom J. R., Butler R. A., Coggins L. G., Jr., Grantz K. A., Hayse J., Hlohowskyj I., Korman J., May J. E., O'Rourke D. J., Poch L. A., Prairie J. R., VanKuiken J. C., Van Lonkhuyzen R. A., Varyu D. R., Verhaaren B. T., Vesekla T. D., Williams N. T., Wuthrich K. K., Yackulic C. B., Billerbeck R. P. and Knowles G. W. (2015). Decision analysis to support development of the glen canyon dam long-term experimental and management plan, U.S. Geological Survey Scientific Investigations Report 64, 2015–5176. http://doi.org/10.3133/sir20155176

Ryan A., Colloff M. J. and Pittock J. (2021). Flow to nowhere: the disconnect between environmental watering and the conservation of threatened species in the murray–darling basin, Australia. *Marine and Freshwater Research*, **72**, 1408–1429. http://doi.org/10.1071/MF21057

Stromberg J. C., Beauchamp V. B., Dixon M. D., Lite S. J. and Paradzick C. (2007). Importance of low-flow and high-flow characteristics to restoration of riparian vegetation along rivers in arid south-western United States. *Freshwater Biology*, **52**, 651–679. https://doi.org/10.1111/j.1365-2427.2006.01713.x

Tonkin J. D., Olden J. D., Merritt D. M., Reynolds L. V., Rogosch J. S. and Lytle D. A. (2021). Designing flow regimes to support entire river ecosystems. *Frontiers of Ecology and the Environment*, **19**(6), 326–333. https://doi.org/10.1002/fee.2348

Van Vliet M. T. H., Yearsley J. R., Franssen W. H. P., Ludwig F., Haddeland I., Lettenmaier D. P. and Kabat P. (2012). Coupled daily streamflow and water temperature modelling in large river basins. *Hydrology and Earth System Sciences*, **16**, 4303–4321. https://doi.org/10.5194/hess-16-4303-2012

Wheeler S. A., Haensch Y., Xu J. and Seidl C. (2020). Water market literature review and empirical analysis, Report for the Australian Competition and Consumer Commission, p. 207.

Woodhouse C. A., Meko D. M., MacDonald G. M., Stahle D. W. and Cook E. R. (2010). A 1200-year perspective of 21st century drought in southwestern North America. *Proceedings of the National Academy of Sciences*, **107**(50), 21283–21288. https://doi.org/10.1073/pnas.0911197107

Yarnell S. M., Petts G. E., Schmidt J. C., Whipple A. A., Beller E. E., Dahm C. N. and Viers J. H. (2015). Functional flows in modified riverscapes: hydrographs, habitats, and opportunities. *Bioscience*, **65**(10), 963–972. https://doi.org/10.1093/biosci/biv102

Yarnell S. M., Stein E. D., Webb J. A., Grantham T., Lusardi R. A., Zimmerman J., Peek R. A., Lane B. A., Howard J. and Sandoval-Solis S. (2020). A functional flows approach to selecting ecologically relevant flow metrics for environmental flow applications. *River Research and Applications*, **36**, 318–324. https://doi.org/10.1002/rra.3575

Zipper S. C., Carah J. K., Dillis C., Gleeson T., Kerr B., Rohde M. M., Howard J. K. and Zimmerman J. K. H. (2019). Cannabis and residential groundwater pumping impacts on streamflow and ecosystems in Northern California. *Environmental Research Communications*, **1**, 125005. https://doi.org/10.1088/2515-7620/ab534d

doi: 10.2166/9781789062786_0063

Chapter 6

Economics and water allocation reform

C. Dionisio Pérez-Blanco[1,2]

[1]Associate Professor, Universidad de Salamanca, Salamanca, Spain
[2]Associate Researcher, Centro Euro-Mediterraneo sui Cambiamenti Climatici, Venice, Italy

ABSTRACT

This chapter discusses major economic challenges and outlines key economic considerations in the design and implementation of water allocation reforms where agriculture is a major water use. It first examines key economic issues in water allocation reform, focusing on the challenges brought by the trade-offs between efficiency, social (distributive) justice, environmental sustainability and institutional reform, and the inherent uncertainties in managing these trade-offs, which are amplified by the complex nature of social-ecological systems. Based on this analysis, the chapter outlines a series of principles towards a water allocation reform that achieves sustainable, equitable and robust economic growth, and discusses the role of economic instruments in such reform. Finally, it identifies persistent research gaps towards delivering actionable science for informed water (re)allocations.

Keywords: Economic instruments, institutional reform, water scarcity

6.1 INTRODUCTION

Water crises are among the greatest global societal risks in terms of potential impacts (WEF, 2020). If current water use patterns continue, 'water demand will exceed supply by 40% by 2030, decreasing the growth rate of the GDP by up to 6%' in water-stressed regions (2030 Water Resources Group, 2019). Avoiding this 'misery in slow motion' calls for systemic and paradigm shifts in water resources management (World Bank, 2017). In mature water economies with inelastic supply (i.e., supply cannot be readily expanded and eventually becomes fixed), such transformation will need water allocation reforms that redistribute available water resources among uses (including redistribution to/from the environment). In order to address present and future challenges, most countries will need to modify their water allocation regimes (Young, 2014).

Some authors argue that water, being an economic good, should be freely reallocated through exchanges in water markets. The economic value of water arises spontaneously from the actions of willing buyers and willing sellers through prices, which ensures the good's reallocation towards uses that are more highly valued in that market and the maximization of economic efficiency (Mendelsohn, 2016). This is the rationale behind the development of water markets in several parts of the world

(Wheeler, 2021). Yet, water is not like most goods or services that are usually found in a market (Savenije, 2002). Water is essential for life and society. It is simultaneously employed for private and public uses, including essential drinking water and sanitation services. Water is bulky, variable across space and time, and part of a complex system of water bodies (e.g., river–aquifer dynamics). This means that alternative water uses (environmental or economic) are interrelated and affect each other.

This chapter discusses major economic challenges and outlines key economic considerations in the design and implementation of water allocation reforms where agriculture is a major water user. In contrast to the classical principles of a market economy which assume that existent allocation problems can be addressed by transforming collective concerns into private decisions and public authorities only have to properly define and protect/enforce property rights (Mendelsohn, 2016), some scholars argue that the unique characteristics of water (e.g., conjunctive use; return flows and reuse; public, private and common pool goods) call instead for collective action where governance and institutions (and the reform thereof) play a decisive role (Gómez *et al.*, 2017). Rather than providing panaceas (Meinzen-Dick, 2007), the economics of water (re)allocation involves institutional changes and high transaction costs. Re(allocation) decisions must be implemented with a thorough understanding of human–water systems dynamics, uncertainties, and limits in order to avoid future surprises and crises (Marchau *et al.*, 2019; Sivapalan *et al.*, 2012).

Hence, this chapter first examines economic issues in water allocation reform, focusing on the challenges brought by the trade-offs between efficiency, social (distributive) justice, environmental sustainability, institutional reform, and uncertainties in complex social-ecological systems (Section 6.2). The chapter then sets out a series of principles for water allocation reform (Section 6.3) and discusses the role of economic instruments in reforming agricultural water allocations (Section 6.4). Finally, the chapter discusses the role of scientific research in informing efficient and effective water reallocations that are robust–that is, low-regret/no-regret (re)allocations that are capable of tolerating and adapting to perturbations (Section 6.5).

6.2 ECONOMIC ISSUES IN WATER ALLOCATION REFORM

6.2.1 Reforming allocation regimes results in large transaction costs

Numerous water (re)allocation reforms with the potential to lead to improvements in economic efficiency have been identified in the literature. Yet, very few have been implemented (Gómez *et al.*, 2017). Several factors explain this resistance and the barriers to water allocation reform.

- *First*, users can exert pressures on public institutions to build new infrastructures to expand the amount of water available for use to increase private gains, even if the total costs (largely paid by the public) exceed the benefits–which is typically the case in mature water economies. There is abundant evidence that building new infrastructures towards expanding supply is significantly costlier than water (re)allocations, but the former is politically expedient (while reallocations can be politically very costly–see below) (Garrick, 2015).
- *Second*, water (re)allocations will often create winners and losers. Naturally, those who lose will have incentives to oppose these (re)allocations. This can lead to regulatory capture, where those benefiting from the current allocation regime increase their capacity to co-opt decision-makers to serve their private interests, usually through lobbying and corporatism (Lopipero *et al.*, 2007; Wiarda 1996). Regulatory capture can be exerted through regulations (*de iure*) or in a *de facto* manner through insufficient or non-existent enforcement of adopted regulations. Notably, it is estimated that between 30% and 50% of water supply worldwide is appropriated by irrigators without a formal license; yet, despite increasing detection rates, prosecution rates remain as low as 2.2% in many developed economies, which undermines compliance (Loch *et al.*, 2020).
- *Third*, users and decision-makers may not accept economic efficiency as a relevant factor to assess water (re)allocations (or at least not as the only factor). In their view, water allocation

reform should be (at least in part) based on other criteria, such as equity. These alternative views explain why certain (re)allocation mechanisms based on economics, notably markets, are rejected by users, which leads to inefficient allocations (Hérivaux *et al.*, 2020).

- *Fourth*, institutions have their own dynamics, and changing their trajectories is costly. As a result, while conventional economic analysis argues that incremental improvements in performance (efficiency, robustness) are sufficient to drive the adoption of superior (re)allocations, the costs of the design and implementation of new institutional organizations can be high and can require improvements in economic efficiency to make the reform viable (Unruh, 2002).

Together, these barriers add additional costs to policy reform in the form of transaction costs, which add up to neoclassical abatement costs (Marshall, 2013). Transaction costs consist of the costs of arranging a (re)allocation ex-ante and then monitoring and enforcing it ex-post, as opposed to neoclassical abatement costs which are the costs of executing the (re)allocation (Matthews, 1986).

Transaction costs can be further subdivided into private and institutional transaction costs. Private transaction costs include search and information costs, bargaining costs (e.g., time to negotiate a reallocation), as well as policing and enforcement costs incurred by human agents in economic trades. Institutional transaction costs include institutional investments (rules and regulation capacity) and organizational investments (people and knowledge capacity) required to achieve water policy objectives (McCann, 2013).

Recent research has monetized the transaction costs of water market reallocations in the United States and Australia, showing that private transaction costs can represent up to 30% of total policy costs (i.e., transaction costs plus abatement costs), institutional transaction costs up to 35% of total policy costs, and aggregate public and private transaction costs up to 53% of total policy costs (Garrick *et al.*, 2013; Loch *et al.*, 2018; McCann, 2013).

These non-trivial magnitudes mean that transaction costs will affect the optimal choice and design of (re)allocation policies. Economic analyses (such as cost-effectiveness) must consider abatement and transaction costs when comparing alternative water (re)allocations. Ignoring any of these considerations would underestimate the cost of (re)allocations, leading to only a partial understanding of the system as well as misleading policy recommendations. However, the empirical base on transaction costs of water policy reform beyond water markets in the United States and Australia is virtually non-existent. As a result, transaction costs are typically excluded from empirical assessments of water or other environmental policies (Garrick, 2015).

6.2.2 Water allocation reforms require compromises between economic efficiency, environmental performance, and social justice

Population growth, higher standards of living, and changing societal perceptions of environmental problems increasingly constrain policymakers in efforts to reallocate increasing amounts of water resources to urban and environmental uses. Water resources to meet these needs are typically sourced from agriculture, the largest water use (about 70% of total global water withdrawals) (FAO, 2021). As water becomes scarcer and its value increases, agriculture and irrigation are progressively transformed from traditionally extensive (high water input to output ratio) to intensive (high output to water input ratio) systems through the adoption of new more technically efficient irrigation technologies (e.g., drip irrigation, greenhouses) (Pérez-Blanco *et al.*, 2021). This process is asymmetric and can create non-trivial equity issues, where the remaining traditional farmers are targeted to achieve further water savings at the least cost which are then made available for the environment or higher value-added economic uses (Gómez *et al.*, 2017). In addition, intensive agricultural systems often demand less labor than traditional extensive systems, which can lead to rural depopulation and migration to urban areas that, if left unaddressed, can further aggravate other environmental and social problems, such as inequity, marginalization, and violence (Todaro, 1969).

Farmers will likely oppose this transition to water (re)allocation and in many places they already do. However, the question is not whether water will be reallocated from agriculture towards urban and environmental uses (as it inevitably will be), but rather how the specific transition process will occur (organized v. disorganized transition) and consequently, what the subsequent social impacts of the specific transition process will be (Perry, 2019). Under disorganized (re)allocation of water from agriculture, farmers and urban users will be negatively affected by reduced water availability, including the more productive users. Uncertain supply will reduce on-farm investments and the production of higher value-added crops (e.g., greenhouses). Loss of flexibility through reduced buffer stocks will often test systems to breaking points and lead to irreversible losses (e.g., disinvestments through loss of permanent crops) (Adamson & Loch, 2021). Unconventional water resources such as treated wastewater and desalinated water will be introduced in agriculture (at a higher cost). Initially, these unconventional resources will be used mostly as a buffer stock during drought years–in normal years, cheaper conventional resources will be preferred. As conventional water resources are exhausted, non-conventional resources will be progressively added to the regular irrigation supply base. This will lead to inequitable impacts (another option is that these sources are subsidized, which may partly correct inequity but lead to inefficient allocation) (March *et al.*, 2014). In the urban sector, under disorganized reallocation of water, water will be unreliable at the margin during part of the transition period, creating costs. Non-conventional resources and transfers from distant areas (if possible) will be progressively added to stabilize supply, while (overexploited) agricultural resources will be intermittently transferred to urban uses. This will create supply costs and rising prices, and affordability issues. This disorganized transition is inefficient from an economic standpoint, ineffective from an environmental standpoint, and inequitable from a social standpoint even if it may be more practical from a policy standpoint.

Organized reallocation can be planned over time to help ease the transition for less productive farmers to minimize overall losses, conserve buffer stocks to minimize capital losses and facilitate investments, and involve progressive introduction of non-conventional resources, which are only added where they are financially sustainable (i.e., high value-added uses) (Strosser *et al.*, 2013). Urban organized reallocation buffers consumers from major price shocks and sudden water conservation emergencies and enables continuation of existing mechanisms that provide a basic supply of water to the poor (e.g., low-cost water fees and low-cost basic water supply blocks). A major issue is how to create compensation mechanisms to address equity issues (including reduced income from the users relinquishing their water rights, reduced employment, reduced GDP and services–hospitals, schools, etc.) arising from organized reallocation (e.g., buy-back, markets). This alternative will involve significant opposition and institutional transaction costs and conflict but offers superior environmental and economic performance.

6.2.3 Reallocations may result in externalities, which are poorly accounted for in conventional water policy

A water reallocation from an agent with a low technical efficiency (e.g., flood irrigation at 50% efficiency) to an agent with high technical efficiency (e.g., drip irrigation at 95% efficiency) will increase the amount of water consumed, and significantly reduce return flows back into the environment, which are often reused downstream (see Figure 6.1). These potential impacts on environmental and other third-party uses are typically not internalized by markets and lead to externalities, namely costs to third parties not directly involved in the market exchange (Pérez-Blanco *et al.*, 2020a, 2020b). Internalizing externalities (such as an increase in technical efficiency upstream that reduces return flows, water available, and income to third parties downstream–including environmental income through amenities) in complex socio-ecological systems is far from simple and requires complex valuation and modeling exercises.

Over the past decades, environmental economics has developed and improved techniques to estimate the environmental values of water (e.g., the sheer enjoyment of a free-flowing river) (Dasgupta, 2021).

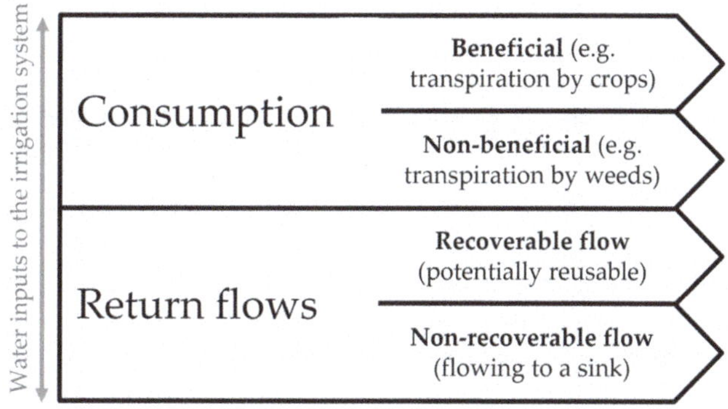

Figure 6.1 Water accounting balance.

By estimating environmental values and modeling trade-offs, a more comprehensive measurement of abatement costs (i.e., the cost of implementing a reallocation) and benefits of water (re)allocations is feasible. The environmental value of water is significant and sometimes can greatly outweigh other commercial benefits (UN, 2021). Yet, despite its advantages in terms of transparency and visibility of environmental values, environmental valuation faces many technical challenges and it is rarely used to inform economic analyses (e.g., cost–benefit analyses) supporting water allocation reforms (Arrow *et al.*, 1993; UN, 2021).

All water entering the irrigation system goes to either: (1) productive consumption, water that is purposefully converted to water vapor, primarily crop transpiration; (2) unproductive consumption, water that is not purposefully converted to vapor, such as through transpiration by weeds or evaporation from wet surfaces; (3) reusable return flows, water reaching a usable aquifer or stream with downstream demand; and (4) non-reusable return flows, water flowing without benefit to a sink such as the sea, and therefore not usable (adapted from Pérez-Blanco *et al.*, 2020b).

6.2.4 Reallocations must account for large uncertainties intrinsic to complex social-ecological systems

Even with significant information on the costs and benefits of water (re)allocations, decision-makers should exert caution due to the uncertainty involved in most estimations. Most cost and benefit estimates use hypothetical scenarios and a complete set of probabilities to produce point predictions (i.e., a single benefit/cost estimate) using a model. However, some scenarios that are deemed implausible and are accordingly allotted very low probabilities may happen. Also, some futures that were completely unknown may be realized down the road. Finally, models can be wrong, creating (significant) biases in predictions.

In this context, point predictions may attribute a relatively minor importance to, or directly ignore, plausible tipping points leading to catastrophic outcomes with major environmental and economic implications. This is the case with several irrigation modernization policies worldwide, some of which initially aimed to abate the impacts of water overallocation through 'water savings'. There, we encounter the paradoxical outcome of a project with an expected positive welfare gain (reduced water inputs, higher income, water saving) turning into minor economic gains or even losses (higher incomes from adopters are obtained by cannibalizing return flows previously used by downstream users, see Figure 6.1) and significant environmental damage (downstream users attempt to mitigate losses by appropriating resources previously allotted to the environment) (Pérez-Blanco *et al.*, 2020b).

Risk and uncertainty should thus feature as a basis for any target setting and water allocation choice. At a minimum, alternative plausible futures/scenarios must be considered, including future inflows, increased demand, transformational industry, and so on. In addition, the precautionary principle should drive decisions (Bishop, 1978). The precautionary principle prescribes the protection and conservation of renewable natural resources so that they cannot go below a minimum threshold beyond which uncertainty is deemed too high (e.g., minimum environmental flows in rivers). The precautionary principle doctrine pervades Integrated Water Resource Management, as exemplified by the European Union (EU) Water Framework Directive (WFD), whose objective is to achieve the 'good ecological status' of water bodies, for which the policies with the lowest economic cost should be chosen (i.e., robustness first, followed by economic performance) (OJ, 2000).

When considering uncertainties, the priority in the economics of water (re)allocation shifts from efficiency first (cost–benefit analysis) to the achievement of robustness (i.e., no-regret); efficiency is then pursued conditioned to this priority target (UN, 2012). This involves an economic trade-off since prioritizing robustness will typically come at the expense of sacrificing uncertain (re)allocations with a potentially higher payoff. Thus, decisions may lead to second-best solutions in economic terms (Hino & Hall, 2017).

6.3 A ROBUST BASIS FOR ECONOMICALLY-SOUND WATER ALLOCATION REFORM

Based on the previous analysis, it becomes clear that any water allocation reform that promotes sustainable, equitable and robust economic growth and welfare must be based on a number of preconditions. We highlight the following principles, which can respond to some of the challenges highlighted above.

First, the water allocation reform should be underpinned by a unified, *transdisciplinary* accounting framework that integrates knowledge and methods from different disciplines (Willardson *et al.*, 1994) (see Figure 6.1). A readily available, but often overlooked, accounting framework is the fractions approach (Willardson *et al.*, 1994), which is represented in Figure 6.1. The fractions approach avoids value-laden terminology such as modernization, losses, or inefficiencies; all of which can lead to divisiveness and a loss of focus in water allocation and institutional choices. The accounting framework should be supported by innovations that can enhance its accuracy such as data provided by satellites, smart meters, or citizen science (e.g., involving citizens in water-use monitoring). Accounting and monitoring should be complemented with a single public register of all allocations within the basin.

Second, total allocable resources must be decided conditional on available supply, and individual allocations defined so that their sum does not exceed these total harvestable resources. This 'unbundling' process of total harvestable resources and water allocations has two key advantages (Rouillard & Rinaudo, 2020; Young, 2014). First, it complies with the theory of economic policy, or Tinbergen Principle, which states that in order to achieve a number of policy objectives, an equal number of interventions is needed (Tinbergen, 1952). In an unbundled scheme, environmental sustainability (target 1) is achieved by defining the total volume of harvestable resources (intervention 1); while water is distributed to economic uses (target 2) through an allocation mechanism (intervention 2).

Next, through unbundling, risk and uncertainty are fully transferred to users. This sets an efficient framework where the institutional level does not necessarily interfere with the operational decisions of private users (Ciriacy-Wantrup & Bishop, 1975). This occurs when authorities focus on defining harvestable volumes and then leave it to users to define individual (re)allocations. Instead of investing public funds to protect users against growing scarcity (e.g., by building additional storage), and deciding in the process what farm-level interventions are desirable (e.g., irrigation modernization), the decision on how to adapt at the farm-level is left to farmers. Note that robust institutions that are capable of tolerating and adapting to perturbations, including extreme ones, will nevertheless still be required to support the allocation decision-making (e.g., water users associations).

Third, hydrological integrity must be ensured. The process of defining harvestable resources requires objective indicators in order to avoid the vested interests of economic users undermining hydrological integrity by setting too-high harvestable caps (Pérez-Blanco & Gómez, 2014). It is crucial that allocations account for return flows (to surface and sub-surface water bodies), interconnections between water bodies and any remaining informal/unlawful use. It is also important to revise allocations where autonomous adaptation responses by users affect the consumption rate—most notably, in those cases when farmers adopt modern irrigation systems that increase technical efficiency and the fraction of water consumed (Pérez-Blanco *et al.*, 2020b).

Young (2014) proposes two mechanisms to address the challenge of growing technical efficiencies: (1) defining harvestable resources as net harvestable resources (i.e., based on consumption rather than withdrawals), which seems the most straightforward option but presents non-trivial technical difficulties; or (2) reductions in the amount of water received via individual allocations as the consumed fraction increases. This change is necessary to avoid rent-seeking behavior, where technology adopters extract rent from the appropriation of water resources of downstream users.

Fourth, users must be provided with institutional certainty to ensure they undertake the necessary investments to successfully adapt to water scarcity (Young, 2014). This means having pre-determined allocation mechanisms, defined for instance as shares of the harvestable resources, fixed volumes defined with a security of supply and modulated on actual available resources in any one year, or as pre-defined priority rights based, for example, on seniority (Rouillard & Rinaudo, 2020; Santato *et al.*, 2016; Young, 2014). Once the indicators used to define the total amount of harvestable resources and the allocation mechanisms are defined, they can only be changed based on clearly established and fair rules that are understood by all users (e.g., through predetermined formulas, through regulated market (re)allocations).

Fifth, gathering longitudinal data on transaction costs over time and across places can help identify and evaluate reforms that were successful in changing the trajectory of institutions towards sustainable and inclusive economic growth; this data can also illustrate the annealing forces that gave change momentum and inform the development of successful water allocation reforms elsewhere (Garrick, 2015). Information on transaction costs can be valuable when assessing the suitability of proposed water allocation regimes. That is, if the decision-maker is unwilling to commit to the required transaction cost investment required to support the (re)allocation, then that proposal could easily be removed from a choice set or shelved until such time as it was feasible to invest as necessary.

The principles above conform to the building blocks for a robust water allocation regime that can tolerate and adapt to uncertainty, while supporting water bodies' ecological integrity, and enabling economic growth. In the next section we explore how, building upon this set of principles, economic instruments can be used to underpin and enhance economic, social, and environmental performance.

6.4 THE ROLE OF ECONOMIC INSTRUMENTS IN REFORMING AGRICULTURAL WATER ALLOCATIONS

6.4.1 Defining economic instruments

Water challenges, such as scarcity, pollution, and increased water insecurity, are driven by inadequate economic incentives that promote responses such as using more water than is available or getting private benefits while transferring costs to third parties (externalities). These incentive-driven drawbacks cannot be sorted out by defining regulations alone. Users within the water allocation regimes must comply with the regulations, as has been repeatedly shown in cases of widespread non-compliance with water allocation rules (often referred to as 'water theft') (Loch *et al.*, 2020). It is also necessary to define economic instruments, namely, 'incentives designed and implemented with the purpose of adapting individual decisions to collectively agreed goals' (Strosser *et al.*, 2013). Thus, while economic incentives are at the source of the problem, they are also called to be an essential part of the solution.

Several taxonomies of economic instruments are available in the literature (eds. Gómez *et al.*, 2017; Lago *et al.*, 2015; Rey *et al.*, 2019). The most relevant categories in relation to water allocation reforms include markets of water use rights (Wheeler, 2021), sanctions for non-compliance with water allocations (Figureau *et al.*, 2015; Loch *et al.*, 2020; Rouillard & Rinaudo, 2020), subsidies (e.g., to adopt water-efficient technologies or to compensate the economic impact of reduced allocations) (Pérez-Blanco *et al.*, 2020b) and charges to incentivize water savings (eds. Dinar *et al.* 2015). More recent studies also explore the role of insurance (e.g., drought insurance that substitutes natural capital from aquifers with financial capital) (Gómez-Limón, 2020; Pérez-Blanco & Gómez, 2013), as well as pecuniary (i.e., Payments for Ecosystem Services–PES) (Asbjornsen *et al.*, 2015) and non-pecuniary voluntary agreements (Gómez *et al.*, 2014).

6.4.2 Designing appropriate economic instruments to support water allocation reforms

Early attempts to mainstream economic instruments into water resources management were primarily concerned with enhancing efficiency; for example, by allowing water trading to expand the economic surplus of willing buyers and sellers or by setting water charges (commonly referred to as prices) at the 'right' level (Briscoe, 1996; Dinar & Subramanian, 1997). However, water management problems are complex and do not result from the lack of efficiency, but rather from the lack of coordination that results from an incompatibility between multiple individual actions that pursue private benefits on the one hand, and the long-term sustainability of water resources and the economic activities that depend on them on the other (Gómez *et al.*, 2017). Thus, while water markets clearly maximize the returns for those involved in the transaction, they are not necessarily beneficial to third parties potentially affected by the transaction (including the environment) or to society as a whole (Connor & Kaczan, 2013). Similarly, while setting the 'right' water charge level can contribute to cost-recovery, much more can be achieved by setting the right *type* of charges. For example, volumetric levies can induce water conservation and contribute to restore the balance in overallocated basins (Caswell *et al.*, 1990).

The vision of economic instruments changes radically when we put the focus on the problem and not on the instrument. In this case addressing water management challenges is equivalent to making the multitude of decisions people make about water compatible with collective water governance objectives such as curbing water depletion and pollution trends or building inclusive water security for the future. This approach provides the basis for assessing the performance of existing economic instruments such as charging and trading schemes, as well as for reshaping economic instruments to serve the objectives of Integrated Water Resources Management (IWRM).

Economic instruments differ from regulatory instruments in their capacity to achieve IWRM objectives in an efficient manner, that is, at a lower economic cost and/or higher benefit for society. This capacity manifests when two preconditions are met: appropriate rationales for participation and incentive compatibility (Laffont & Tirole, 1991). The former precondition indicates that individuals will only engage in a specific action when they expect a positive return from said action (or a negative return from non-compliance). The latter precondition indicates that these positive returns should be made available to individuals only when their actions contribute to IWRM objectives (including economy-wide efficiency, but also inclusivity and sustainability). Accordingly, use of economic instruments is advised when 'there are welfare-improving opportunities that can be transformed into private benefits for water-users', and 'collective gains then follow' (Gómez *et al.*, 2017).

The use of economic instruments that only achieve private benefits is not warranted under this framework. For instance, the use of water markets that can potentially create externalities with a negative impact on the environment and other economic uses is not advised, *unless complemented with a sound regulatory and water allocation regime that ensures hydrological integrity and addresses other third-party impacts.* Conversely, where only collective benefits are expected, economic instruments will not be capable of putting in place the necessary incentives to drive individual actions towards the desired objectives and other instruments such as regulations will be necessary. In this vein, it should be noted that economic instruments are not a substitute for norms and regulations; rather, they should be designed to complement them, as a part of a comprehensive water allocation regime that serves the objectives of IWRM (eds. Lago *et al.*, 2015).

Table 6.1 Definition, advantages, and disadvantages of economic instruments for water allocation reform. Adapted from Rey *et al.* (2019).

Instrument	Definition in the context of water allocation reform	Advantages	Disadvantages	Example(s) of economic instruments that align individual choices with collectively agreed IWRM goals
Charges	Levies on water use related to conveyance and storage services and the opportunity cost of the resource (resource and environmental costs). They can be earmarked (tariff) or not (tax).	• Effective, compliance with polluter-pays • principle • Revenue-raising	• Resistance (lobbying) and related • transaction costs • Willingness and ability to pay may lead to equity issues	Charging mechanism that recovers, beyond financial costs, the environmental and resource costs of water. In 2017, the Piedmont Region in Italy introduced a pioneering reform that stated that in order to access the EU's Common Agricultural Policy (CAP) funding, irrigators should pay the environmental and resource costs of water use. However, the implementation of such conditionality is still pending.
Trading	Institutional setting that allows transferring of (marketable) water allocations across agents, places, and time, in exchange for a pecuniary compensation	• Efficient reallocation of water • Enables buy-back/ PES	• High transaction costs • Institutional/legal complexity • Limited acceptance across stakeholder groups • Asymmetric impacts • (e.g., selling areas with negative impact on food industry) and potential externalities • Incentive towards increasing consumptive use • Illegal markets	Water markets that observe hydrological integrity (rarely achieved; in some legislations hydrological integrity must be *de iure* observed, but *de facto* it is often bypassed– e.g., Spain)
Sanctions	Pecuniary penalty imposed by the judiciary when a specific behavior is observed (e.g., overdraft)	• Effective, compliance with polluter-pays • principle • Revenue-raising	• Monitoring can be expensive • Limited enforcement	Sanctions for non-authorized water uses in Spain can amount to EUR 500 000, but are often not enforced

(*Continued*)

Table 6.1 Definition, advantages, and disadvantages of economic instruments for water allocation reform. Adapted from Rey *et al.* (2019) (*Continued*).

Instrument	Definition in the context of water allocation reform	Advantages	Disadvantages	Example(s) of economic instruments that align individual choices with collectively agreed IWRM goals
Insurance	Insurance is the most commonly used instrument for financial protection against risk contingent losses, in which 'the insured party or policyholder transfers the cost of potential loss to the insurer in exchange for monetary compensation known as a premium' (DRMKC, 2017). The insurer thus absorbs, pools, and diversifies the individual risks acquired from policyholders, making them assessable and manageable.	• Deters groundwater withdrawals that are difficult to monitor • Privately funded (largely)	• Willingness and ability to pay may necessitate subsidies to become feasible (particularly systemic risks such as drought), budgetary issues (public reinsurance), high institutional uncertainty (allocation mechanism during scarcity). As a result, drought insurance in irrigated agriculture is very rarely made available to farmers.	The EU CAP funds an income stabilization tool through mutual funds (non-profit, cooperation and self-help organizations that gather groups of farmers who conform their own contingency fund and assume responsibility for their own risk management) that addresses both production and market risks, including drought risk; yet it has limited uptake, since insurance policies are typically issued by commercial insurers.
Subsidies	Financial aid or support. Can be explicit (price supports, subsidized loans and direct payments) or implicit (reduced regulation and tax/charges relief)	• Limited transaction costs, can enhance equity	• Infringement of polluter-pays principle • Budgetary constraints • Coupled subsidies may aggravate scarcity (e.g., subsidization of irrigation modernization) • Low effectiveness and cost-effectiveness	Subsidies to farmers that agree to practices that mitigate scarcity (e.g., forested infiltration areas in Northern Italy that recharge groundwater aquifers by channeling surface waters during non-irrigation months)

(*Continued*)

Table 6.1 Definition, advantages, and disadvantages of economic instruments for water allocation reform. Adapted from Rey *et al.* (2019) (*Continued*).

Instrument	Definition in the context of water allocation reform	Advantages	Disadvantages	Example(s) of economic instruments that align individual choices with collectively agreed IWRM goals
Payments for Ecosystem Services	Conditional payments offered to water users in exchange for the voluntary provision of some sort of ecological service (e.g., water reacquisitions to enhance environmental flows)	• Limited transaction costs • Flexible • Cost-effective	• Infringement of polluter-pays principle • Budgetary constraints • Crowding-out of intrinsic • motivations to protect ecosystems • Equity issues	Water buy-back in the Murray–Darling Basin (Australia)
Voluntary agreements	Non-pecuniary and voluntary incentives, based on opportunities for individual profit or loss mitigation, to enhance negotiated arrangements among agents to achieve public policy objectives	• Flexible • Acceptable • Inexpensive	• Low performance if incentives not properly defined • Limited to win-win situations • Technological, institutional and/or legal barriers, can delay action during droughts • Exclusion of some users	Voluntary agreement to release pulse flows in the Lower Ebro River (Spain) between the hydropower company (which benefits from reduced costs of removing macrophytes from the dam outlet, and from enhanced social corporate responsibility) and the river basin authority (which achieved an improved ecological status of the river)

6.4.3 Economic instruments and water allocation reforms: some examples

There is a wide array of economic instruments at the disposal of policymakers concerned with water allocation reform. Some have been extensively tested in multiple real-life settings, most notably subsidies (e.g., subsidies for irrigation modernization) or penalties for non-compliance. Others like PES are becoming increasingly relevant, although some legal and other barriers towards their implementation remain. For instance, water-related PES have rehabilitated an area one and a half times the size of India, with a total investment of \$25 billion (10^9); however, in the EU context for example, public authorities often issue lawsuits against PES since they do not comply with the polluter-pays principle (Bhaduri *et al.*, 2021). A third group of economic instruments includes markets and charges which, despite the promising performance suggested by a growing research body (Rey *et al.*, 2019), have a limited geographical scope (full-fledged water markets have predominately been implemented in Australia, Chile and the western United States), or their implementation is incomplete (water charges typically recover financial costs but not opportunity costs, including resource costs – foregone income from alternative uses of the resource – and environmental costs). Finally, some economic instruments are very uncommon (e.g., voluntary agreements and drought insurance for irrigated agriculture). Table 6.1 presents how each type of economic instrument can contribute to water allocation reforms.

6.5 THE WAY FORWARD: ACTIONABLE SCIENCE FOR INFORMED (RE)ALLOCATIONS

Science has supported the production of increasingly sophisticated data and models to deal with water (re)allocation challenges, including high-granularity earth observations (FAO, 2020; IMPEL, 2017), digital twins (very high precision digital models of socio-ecological systems to monitor and predict impacts, such as the European initiative Destination Earth), improved integration between human and water systems (Sivapalan *et al.*, 2014) and enhanced understanding and management of uncertainty (Cloke *et al.*, 2013; Marchau *et al.*, 2019), among others. Such mechanistic frameworks have undoubtedly improved our understanding of water (re)allocation challenges and performance and delivered valuable insights into their design. However, any analysis of water (re)allocations must go beyond the boundaries of quantitative outcomes provided by mechanistic models and rely also on heuristic frameworks and the expertise of stakeholders, so as to make the best possible use of available information and adequately balance the multiple targets pursued by IWRM (efficiency, sustainability, equity) and the trade-offs between them. Stakeholders' expertise can be highly instrumental for the solution of complex problems, particularly under uncertainty, for example, from *ad-hoc* interpretations of their experience that are applied to speculate upon the consequences of alternative water allocation regimes. Notably, stakeholder expertise can contribute to robust decision making, guided by frameworks such as 'deliberation with analysis' (Groves *et al.*, 2015) or 'bounded rationality' (Quiggin, 2007), which leverage on stakeholder knowledge and experience to complement conventional mechanistic approaches (i.e., modeling) to decision making. The resultant combination of models and expertise is critical to understand the functioning of socio-ecological systems, explore plausible futures, and anticipate potential surprises and vulnerabilities to proposed (re)allocations, as well as identify and exploit spill-over effects and synergies across instruments and sectors.

REFERENCES

2030 Water Resources Group. (2019). The 2030 Water Resources Group Annual Report, World Bank, Washington D.C., USA.

Adamson D. and Loch A. (2021). Incorporating uncertainty in the economic evaluation of capital investments for water use efficiency improvements. *Land Economics*, **0**(1), 1–15.

Arrow K., Solow R., Portney P., Leamer E., Radner R. and Schuman H. (1993). Report of the NOAA Panel on Contingent Valuation, National Oceanic and Atmospheric Administration, Washington D.C., USA.

Asbjornsen H., Mayer A. S., Jones K. W., Selfa T., Saenz L., Kolka R. K. and Halvorsen K. E. (2015). Assessing impacts of payments for watershed services on sustainability in coupled human and natural systems. *BioScience*, **65**(6), 579–591. https://doi.org/10.1093/biosci/biv051

Bhaduri A., Pérez-Blanco C. D., Rey D., Iftekhar M. S., Kaushik A., Escriva-Bou A., Calatrava J., Adamson D., Palomo-Hierro S., Jones K. W., Asbjornsen H., Altamirano M. A., Lopez-Gunn E., Polyakov M., Motlagh M. and Bekchanov M. (2021). Economics of water security. In: Handbook of Water Resources Management: Discourses, Concepts and Examples, J. Bogardi, T. Tingsanchali, K. D. W. Nandalal, J. Gupta, L. Salamé, R. R. P. Nooijen, A. G. Kolechkina, N. Kumar and A. Bhaduri (eds.), Springer International Publishing, Berlin, Germany.

Bishop R. (1978). Endangered species and uncertainty: the economics of a safe minimum standard. *American Journal of Agricultural Economics*, **60**, 10–18 https://doi.org/10.2307/1240156

Briscoe J. (1996). Water as an Economic Good: The Idea and What It Means in Practice. The World Bank, Washington D.C.

Caswell M., Lichtenberg E. and Zilberman D. (1990). The effects of pricing policies on water conservation and drainage. *American Journal of Agricultural Economics*, **72**(4), 883–890. https://doi.org/10.2307/1242620

Ciriacy-Wantrup S. and Bishop R. (1975). Common property as a concept in natural resources policy. *Natural Resources Journal*, **15**(4), 713–727.

Cloke H. L., Pappenberger F., van Andel S. J., Schaake J., Thielen J. and Ramos M. H. (2013). Hydrological ensemble prediction systems. *Hydrological Processes*, **27**(1), 1–4. https://doi.org/10.1002/hyp.9679

Connor J. D. and Kaczan D. (2013). Principles for economically efficient and environmentally sustainable water markets: the Australian experience. In: Drought in Arid and Semi-Arid Regions, K. Schwabe, J. Albiac, J. D. Connor, R. M. Hassan and L. M. González (eds.), Springer, Netherlands, pp. 357–374.

Dasgupta P. (2021). The Economics of Biodiversity: The Dasgupta Review. HM Treasury, Government of the UK, London, UK.

Dinar A. and Subramanian A. (1997). Water Pricing Experiences: An International Perspective. The World Bank, Washington D.C.

Dinar A., Pochat V. and Albiac J. (eds.) (2015). Water Pricing Experiences and Innovations. Springer International Publishing, Zurich, Switzerland.

FAO. (2020). WaPOR, WaPOR Database.

FAO. (2021). FaoStat, Food and Agriculture Organization of the United Nations.

Figureau A.-G., Montginoul M. and Rinaudo J.-D. (2015). Policy instruments for decentralized management of agricultural groundwater abstraction: a participatory evaluation. *Ecological Economics*, **119**, 147–157. https://doi.org/10.1016/j.ecolecon.2015.08.011

Garrick D. E. (2015). Water Allocation in Rivers Under Pressure: Water Trading, Transaction Costs and Transboundary Governance in the Western US and Australia. Edward Elgar Pub, Cheltenham, UK: Northampton, MA.

Garrick D., McCann L. and Pannell D. J. (2013). Transaction costs and environmental policy: taking stock, looking forward. *Ecological Economics*, **88**, 182–184. https://doi.org/10.1016/j.ecolecon.2012.12.022

Gómez C. M., Pérez-Blanco C. D. and Batalla R. J. (2014). Tradeoffs in river restoration: flushing flows vs. Hydropower generation in the Lower Ebro River, Spain. *Journal of Hydrology*, **518** Part A, 130–139.

Gómez C. M., Pérez-Blanco C. D., Adamson D. and Loch A. (2017). Managing water scarcity at a river basin scale with economic instruments. *Water Economics and Policy*, **04**(01), 1750004. https://doi.org/10.1142/S2382624X17500047

Gómez-Limón J. A. (2020). Hydrological drought insurance for irrigated agriculture in southern Spain. *Agricultural Water Management*, **240**, 106271. https://doi.org/10.1016/j.agwat.2020.106271

Groves D. G., Evan B., J. L. R. R., F. J., Jennifer N. and Brandon G. (2015). Developing key indicators for adaptive water planning. *Journal of Water Resources Planning and Management*, **141**(7), 05014008. https://doi.org/10.1061/(ASCE)WR.1943-5452.0000471

Hérivaux C., Rinaudo J.-D. and Montginoul M. (2020). Exploring the potential of groundwater markets in agriculture: results of a participatory evaluation in five French case studies. *Water Economics and Policy*, **06**(01), 1950009. https://doi.org/10.1142/S2382624X19500097

Hino M. and Hall J. W. (2017). Real options analysis of adaptation to changing flood risk: structural and nonstructural measures. *ASCE-ASME Journal of Risk and Uncertainty in Engineering Systems, Part A: Civil Engineering*, **3**(3), 04017005. https://doi.org/10.1061/AJRUA6.0000905

IMPEL. (2017). Water Over-abstraction and Illegal Water Abstraction Detection and Assessment. European Union Network for the Implementation and Enforcement of Environmental Law (IMPEL), Brussels, Belgium.

Laffont J.-J. and Tirole J. (1991). The politics of government decision-making: a theory of regulatory capture. *The Quarterly Journal of Economics*, **106**(4), 1089–1127. https://doi.org/10.2307/2937958

Lago M., Mysiak J., Gómez C. M., Delacámara G. and Maziotis A. (eds.) (2015). Use of Economic Instruments in Water Policy. Springer International Publishing, Cham, Switzerland, vol. **14**.

Loch A., Pérez-Blanco C. D., Carmody E., Felbab-Brown V., Adamson D. and Seidl C. (2020). Grand theft water and the calculus of compliance. *Nature Sustainability*, **3**(12), 1012–1018. https://doi.org/10.1038/s41893-020-0589-3

Loch A., Wheeler S. A. and Settre C. (2018). Private transaction costs of water trade in the Murray–darling basin. *Ecological Economics*, **146**, 560–573. https://doi.org/10.1016/j.ecolecon.2017.12.004

Lopipero P., Apollonio D. E. and Bero L. A. (2007). Interest groups, lobbying, and deception: the tobacco industry and airline smoking. *Political Science Quarterly*, **122**(4), 635–656. https://doi.org/10.1002/j.1538-165X.2007.tb00612.x

March H., Saurí D. and Rico-Amorós A. M. (2014). The end of scarcity? Water desalination as the new cornucopia for Mediterranean Spain. *Journal of Hydrology*, **519**(Part C), 2642–2651. https://doi.org/10.1016/j.jhydrol.2014.04.023

Marchau V. A. W. J., Walker W. E., Bloemen P. and Popper S. W. (2019). Decision Making under Deep Uncertainty: From Theory to Practice, 2019th edn, Springer, Cham, Switzerland, vol. **1**.

Marshall G. R. (2013). Transaction costs, collective action and adaptation in managing complex social–ecological systems. *Ecological Economics*, **88**, 185–194. https://doi.org/10.1016/j.ecolecon.2012.12.030

Matthews R. C. O. (1986). The economics of institutions and the sources of growth. *The Economic Journal*, **96**(384), 903–918. https://doi.org/10.2307/2233164

McCann L. (2013). Transaction costs and environmental policy design. *Ecological Economics*, **88**(C), 253–262. https://doi.org/10.1016/j.ecolecon.2012.12.012

Meinzen-Dick R. (2007). Beyond panaceas in water institutions. *Proceedings of the National Academy of Sciences*, **104**(39), 15200–15205. https://doi.org/10.1073/pnas.0702296104

Mendelsohn R. O. (2016). Adaptation, climate change, agriculture, and water. *Choices*, **3**, 1–3.

OJ. (2000). Water Framework Directive 2000/60/EC, Council Directive.

Pérez-Blanco C. D. and Gómez C. M. (2013). Designing optimum insurance schemes to reduce water overexploitation during drought events: a case study of La Campiña, Guadalquivir river basin. Spain. *Journal of Environmental Economics and Policy*, **2**(1), 1–15. https://doi.org/10.1080/21606544.2012.745232

Pérez-Blanco C. D. and Gómez C. M. (2014). Drought management plans and water availability in agriculture: a risk assessment model for a Southern European basin. *Weather and Climate Extremes*, **4**, 11–18. https://doi.org/10.1016/j.wace.2014.02.003

Pérez-Blanco C. D., Essenfelder A. H. and Gutiérrez-Martín C. (2020a). A tale of two rivers: integrated hydro-economic modeling for the evaluation of trading opportunities and return flow externalities in inter-basin agricultural water markets. *Journal of Hydrology*, **584**, 124676. https://doi.org/10.1016/j.jhydrol.2020.124676

Pérez-Blanco C. D., Hrast-Essenfelder A. and Perry C. (2020b). Irrigation technology and water conservation: a review of the theory and evidence. *Review of Environmental Economics and Policy*, **14**(2), 216–239. https://doi.org/10.1093/reep/reaa004

Pérez-Blanco C. D., Loch A., Ward F., Perry C. and Adamson D. (2021). Agricultural water saving through technologies: a zombie idea. **16**(11), 114032.

Perry C. (2019). Learning from water footprints. *Policy Quarterly*, **15**(3), 70–74. https://doi.org/10.26686/pq.v15i3.5690

Quiggin J. (2007). Complexity, Climate Change and the Precautionary Principle. University of Queensland, School of Economics, Brisbane, Australia.

Rey D., Pérez-Blanco C. D., Escriva-Bou A., Girard C. and Veldkamp T. I. E. (2019). Role of economic instruments in water allocation reform: lessons from Europe. *International Journal of Water Resources Development*, **35**(2), 206–239. https://doi.org/10.1080/07900627.2017.1422702

Rouillard J. and Rinaudo J.-D. (2020). From state to user-based water allocations: an empirical analysis of institutions developed by agricultural user associations in France. *Agricultural Water Management*, **239**, 106269. https://doi.org/10.1016/j.agwat.2020.106269

Santato S., Mysiak J. and Pérez-Blanco C. D. (2016). The water abstraction license regime in Italy: a case for reform? *Water*, **8**(3), 1–15. https://doi.org/10.3390/w8030103

Savenije H. H. G. (2002). Why water is not an ordinary economic good, or why the girl is special. *Physics and Chemistry of the Earth, Parts A/B/C*, **27**(11), 741–744. https://doi.org/10.1016/S1474-7065(02)00060-8

Sivapalan M., Konar M., Srinivasan V., Chhatre A., Wutich A., Scott C. A. and Wescoat J. L. (2014). Socio-hydrology: use-inspired water sustainability science for the Anthropocene. *Earth's Future*, **2**, 225–230. https://doi.org/10.1002/2013EF000164

Sivapalan M., Savenije H. H. G. and Blöschl G. (2012). Socio-hydrology: a new science of people and water. *Hydrological Processes*, **26**(8), 1270–1276. https://doi.org/10.1002/hyp.8426

Strosser P., Delacámara G., Gómez C. M., Lago M. and Maziotis A. (2013). Changing current practice in the application of EPIs to achieve the objectives of the WFD.

Tinbergen J. (1952). On the Theory of Economic Policy. North-Holland Pub. Co., Amsterdam, The Netherlands.

Todaro M. P. (1969). A model of labor migration and urban unemployment in less developed countries. *The American Economic Review*, **59**(1), 138–148.

UN. (2012). UN-Water Status Report on the Application of Integrated Approaches to Water Resources Management for Rio+20, United Nations.

UN. (2021). The United Nations World Water Development Report 2021, United Nations.

Unruh G. C. (2002). Escaping carbon lock-in. *Energy Policy*, **30**(4), 317–325. https://doi.org/10.1016/S0301-4215(01)00098-2

WEF. (2020). Global Risks 2020. World Economic Forum, Geneva, Switzerland.

Wheeler S. A. (2021). Water Markets: A Global Assessment. EE Elgar, Cheltenham, UK.

Wiarda H. J. (1996). Corporatism and Comparative Politics: The Other Great 'Ism'. Routledge, Armonk, NY, USA.

Willardson L., Allen R. and Frederiksen H. (1994). Eliminating Irrigation Efficiencies, Conference Proceedings, 1–15, USCID, Denver, Colorado.

World Bank. (2017). Uncharted Waters: The New Economics of Water Scarcity and Variability. World Bank Publications, Washington D.C., USA.

Young M. D. (2014). Designing water abstraction regimes for an ever-changing and ever-varying future. *Agricultural Water Management*, **145**, 32–38. https://doi.org/10.1016/j.agwat.2013.12.002

doi: 10.2166/9781789062786_0079

Chapter 7

England and Wales: countering 'unsustainable abstraction' with the catchment based approach

David Benson[1], Hadrian Cook[2], M. Yasir Ak[1] and Burcin Demirbilek[3]

[1]Department of Politics and International Relations, University of Exeter, Penryn, Cornwall, UK
[2]Harnham Water Meadows Trust, Harnham, Salisbury, UK
[3]Department of Political Science and Public Administration, Faculty of Economics and Administrative Sciences, Çankırı Karatekin University, Çankırı, Turkey

ABSTRACT

England and Wales have a long-established abstraction licensing regime for determining water allocations amongst economic sectors, particularly agriculture. This regime is implemented by the Environment Agency (EA) and Natural Resources Wales (NRW), primarily through Catchment Abstraction Management Strategies (CAMS) and attendant Abstraction Licensing Strategies (ALS), to support policy requirements for environmental sustainability. Over time, water licensing has been increasingly linked to water availability in catchments while licence trading now provides greater flexibility in allocations. Ongoing reforms will further seek to integrate resource sustainability and the catchment based approach (CaBA) to management into this evolving regime. Yet a critical question concerns whether such policy commitments to countering 'unsustainable abstraction' have been achieved, particularly by the agricultural sector. Here, we define sustainability in terms of the environmental, social and economic outcomes of governance.

Keywords: Abstraction licensing, England and Wales, environmental sustainability, water allocation

7.1 INTRODUCTION

England and Wales face significant threats to water resources (Cook, 2017) (Figure 7.1). Ground and surface waters are subject to growing abstraction pressures from industrialisation, population growth and agricultural production. Up to 20% of surface waters experience over-abstraction (HM Government, 2018). Around 19.7% of total abstraction is from groundwater (Defra, 2019a), particularly in drier eastern regions where spray irrigation is practised. Climate change is predicted to increase such pressures, with a 15% fall in river flows anticipated by 2050 (Houses of Parliament, 2017). Countering 'unsustainable abstraction' is therefore a central guiding principle of national water abstraction policy alongside a progressive shift to catchment based approaches (CaBA) in water management (Defra and Welsh Government, 2013). However, the degree to which this institutional framework is securing sustainable water supply to sectors such as agriculture is an important concern.

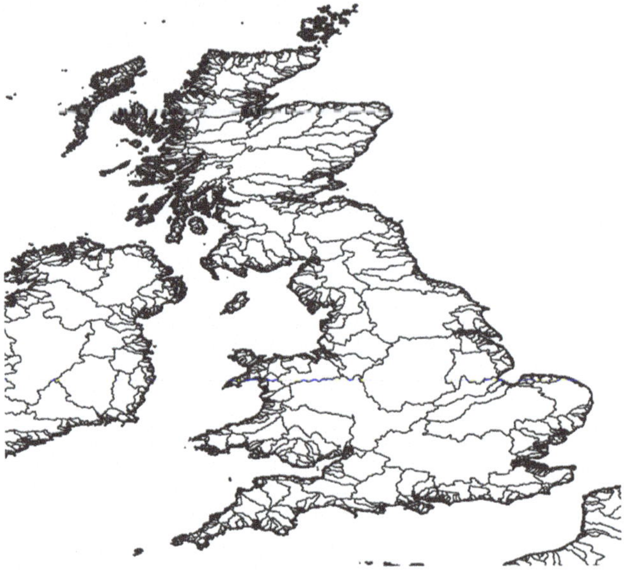

Figure 7.1 Hydrological boundaries of river basins in England and Wales.

By first setting out the main institutional framework for water allocation in England and Wales, this chapter will then provide an overview of related law and policy, management targets, volumetric caps, formal allocation and re-allocation rules, monitoring and compliance mechanisms and related policy instruments. Analysis of environmental effectiveness, economic efficiency, social equity, and resilience suggests that water abstraction by the agricultural sector is sustainable, yet significant risks remain. Evidence is drawn from national-scale trends and catchment-scale data.

7.2 THE OVERARCHING INSTITUTIONAL FRAMEWORK

7.2.1 The nature of water rights

The abstraction regime evolved from common law rights initially developed in the late medieval period (Cook, 2017). Under English common law, riparian rights to abstract surface water relate to ownership of land over which water flows. Holders of riparian rights are entitled to 'ordinary' proprietary use of waters flowing on their land for domestic or agricultural needs (McGillivray, 2013). However, as there has 'never been absolute ownership' of surface water flows or riparian rights to percolating water, securing judicial remedy for disruption to such flows has consequently proved difficult for riparian owners under the common law system[1] (Cook, 2017). Problematically, there is an absence of a 'common law right to receive any particular flow of surface water and, in the case of groundwaters… no right to receive even a reasonable flow' (McGillivray, 2013).

Over-abstraction emanating from this common law system became increasingly apparent in the 19th century, leading to the creation of water conservation institutions such as the Thames Conservancy Board in 1857 (Cook, 1999). Demands for national controls then emerged in the early

[1]In *Chasemore* v. *Richards* (1859), the riparian owner sought legal redress through the courts to stop Croydon Board of Health abstracting groundwater for public supply, as it reduced river flows to his mill. The action and subsequent appeal to the House of Lords failed since it was ruled that there was no common law right to percolating water.

20th century. A serious drought in 1921 prompted the government Ministry of Health's Advisory Committee on Water to recommend controls on groundwater abstraction, subsequently adopted in the 1945 Water Act (Cook, 1999). Limited central government licensing for groundwater abstraction in vulnerable zones was introduced, thereby initiating a gradual expansion of administrative regulation through water law. Regulation has however not removed common law rights entirely. What has emerged incrementally is a 'stewardship' system that limits 'private rights in a water law context in order to advance important public interests' such as environmental protection through licensing abstraction (McGillivray, 2013).

7.2.2 Current legal and policy context

The current water abstraction licensing system effectively dates from the Water Resources Act 1963 (Downing, 1993). Key institutional innovations introduced were firstly a Water Resources Board to steer national water resource development, plus 29 regional river authorities to undertake the water management functions of pre-existing local river boards, including abstraction licensing and control. Licensing functions were then transferred to 10 multi-function Regional Water Authorities after the Water Act 1973. Responsibility for abstraction management subsequently passed to the Environment Agency (EA) in 1995, after the Water Resources Act 1991. The EA's equivalent, Natural Resources Wales (NRW) was created in 2013. The Water Act 2003 then introduced three abstraction licence categories (see below), a simplified licensing application process and an abstraction limit of 20 cubic metres per day below which licenses were not required. Further reforms were adopted following the government white paper *Water for Life* in 2011 (HM Government, 2011) and were formalised in the Water Act 2014.

These legal innovations are implemented through Defra (Department for Environment, Food and Rural Affairs) and EA strategy, plans and programmes. A strategic steer is set by broader government environment policy; currently Defra's 25 Year Environment Plan (HM Government, 2018)[2]. The key implementing measure is Catchment Abstraction Management Strategies (CAMS). Introduced in 2001, CAMS assess water availability in each catchment and identify where demand affects the water balance. Each CAMS relates to a specific hydrological unit: for example, the Stour (Environment Agency, 2003). Abstraction Licensing Strategies (ALS) are produced for each catchment based on CAMS assessments. They in turn determine abstraction licensing within the catchment boundaries. Both strategies also integrate with River Basin Management Planning objectives, originally introduced under the EU Water Framework Directive (WFD).

Reflecting the Directive's environmental protection emphasis, a critical driving principle of national policy has been avoidance of 'unsustainable abstraction', deemed to occur where (Defra and Welsh Government, 2013):

> … *abstraction can alter the natural flow regime either directly changing surface water flows or indirectly by lowering groundwater levels and consequently affecting flows to springs, wetlands, lakes and rivers.*

Where abstraction was considered unsustainable, the EA implemented the Restoring Sustainable Abstraction (RSA) programme (2008 to 2018). The programme allowed the Agency to investigate, amend and revoke abstraction licenses where they were considered environmentally damaging. These powers were significantly enhanced by the removal of pre-existing compensation measures for revoking and amending water company licences in the Water Act 2014. A programme was also adopted in 2017 whereby the EA can revoke unused or underused licences, plus review time-limited licence renewals.

---

[2]Chapter 2 of the 25 Year Plan identifies maintaining 'sustainable supplies of water for future generations' as a central aim (HM Government, 2018).

Water companies[3] additionally have statutory requirements relating to water abstraction through production of Water Resources Management Plans (WRMPs) and drought plans. The Water Act 2003 made completion of WRMPs every five years a statutory requirement. To ensure sustainable abstraction, company plans must show how future abstraction demand will be met alongside environmental protection through supporting River Basin Management Plan objectives. The annual Water Industry National Environment Programme (WINEP) also sets out industry requirements for meeting statutory environmental obligations, thereby steering company investment plans (United Utilities, 2020). The Programme is supported by WISER (Water Industry Strategic Environmental Requirements), constituting water company actions requested by government agencies for meeting environmental obligations. Water companies are legally required to produce plans detailing drought response actions.

This system is currently under reform to enhance sustainability and the catchment based approach. After consultation with the UK and Welsh Governments, Defra published its Water Abstraction Plan for sectoral reform in 2017 (Defra, 2017a, 2019b). A key Plan goal is 'to end damaging abstraction of water from rivers and groundwater whenever it is cost beneficial to do so', to increase the number of 'sustainably abstracted water bodies from 82% to 90% for surface water and from 72% to 77% for groundwater by 2021' (Defra, 2017a). Actions to achieve this goal include: using WINEP to ensure sustainable abstraction by water companies; reviewing 50% of time-limited licences by 2021 to address their environmental impacts; adjusting permanent licences to prevent environmental damage; and completing the Restoring Sustainable Abstraction Programme by revoking unused licences, removing under-used licences and regulating previously exempt abstraction (ibid.). To achieve these actions, the Plan refers to updating abstraction licensing strategies to better protect the environment in conjunction with a 'stronger catchment focus' that involves 'engagement in catchments facing particular environmental pressures from abstraction' (ibid.). Subsequently 10 priority catchments were designated by the Environment Agency 'for developing and testing innovative solutions to achieve greater access to water and address unsustainable abstraction' through catchment based approaches involving multiple stakeholders (Environment Agency, 2019a). The first four priority catchments were introduced in 2018, with another six announced in 2019[4].

Additional reforms have since been undertaken. Firstly, the abstraction licensing service is being modernised by digitising licence information. A 'Manage your water abstraction and impoundment licence' service allows users to submit abstraction records online, access their licence details and share licence management with the EA (GOV.UK, 2018). By early 2019, 3000 licence holders had registered for this service (Defra, 2019b). Secondly, a future reform will shift the current abstraction licensing system into the EA's wider Environmental Permitting Regime[5], to ensure consistency with industrial permitting. Finally, water trading is being expanded through revisions of Abstraction Licensing Strategies (see below).

7.2.3 Controlling access to water: a catchment based approach

Access to water for abstraction is determined by CAMS and their associated Abstraction Licensing Strategies. Catchment Abstraction Management Strategies involve a technical assessment of water resources available for abstraction using data collected within defined CAMS boundaries. This assessment determines Abstraction Licensing Strategies within each catchment – 82 in England

[3]The water industry was privatised under the Water Act 1989, creating 33 companies to provide drinking water supply and sewerage.

[4]The 10 priority catchments are the Till and Tweed, Idle and Torne, South Forty Foot (Witham catchment), Cam and Ely Ouse, East Suffolk, Arun and Western Streams, Alt and Crossens, Wye, Brue and Otter.

[5]The Environmental Permitting Regulations (EPR) combined pre-existing Waste Management Licensing (WML) and Pollution Prevention and Control (PPC) regulations, becoming operational in 2008. Permits are required for activities including waste management and discharging pollutants to water bodies.

and 14 in Wales. These documents specify how the EA and NRW will manage licensing within each catchment. Each strategy contains information on water resources available for abstraction, the conditions for applying for new licences and conditions attached to existing licences (Environment Agency, 2016). In issuing licences, the Environment Agency states that it will examine 'the local impacts of the proposed abstraction or impoundment and ensure that we protect the rights of existing water users in addition to protecting the environment' (ibid.), reflecting the 'stewardship' notion by balancing private riparian rights for water abstraction with public interests in environmental protection.

7.2.4 Permitting requirements

Part IV, 23 (1) of the Water Resources Act 1963 originally required that any person abstracting water from a supply source had to apply for a licence from the requisite river authority, with charges for licenses specified in the legislation. Exceptions included abstractions below one thousand gallons (4.5 m^3), plus water for domestic and agricultural use, apart from spray irrigation because of its higher water demands. Reflecting common law, only landowners or those they permitted to access the land were entitled to apply.

Under the Water Act 2003, the pre-existing single abstraction licence system was changed to encompass three categories (Environment Agency, 2014a). A *full licence* is required to abstract over 20 m^3 per day, a *transfer licence* is required to transfer over 20 m^3 per day between sources. A *temporary licence* is needed for abstracting over 20 m^3 per day for less than 28 days. An additional licence may be needed for impounding water. Separate consent is required to abstract groundwater from a borehole. New licences must also be granted as time limited (Ofwat, 2015). An application charge is payable to the Environment Agency or NRW, in addition to the annual licence charge (Environment Agency, 2020a). If granted, licence fees are calculated according to volume of water abstracted annually multiplied by factors including the source type, season, the degree of loss to the water source and whether the source is tidal (ibid.). Trading of licences is also permitted, as discussed below. Further amendments to the licensing system are likely as it becomes incorporated into the Environmental Permitting System (Defra, 2019b).

7.2.5 Collaborative programmes and decision-making

Catchment-scale collaboration of actors is encouraged under current government water policy. Although the Water Resources Act 1963 first created catchment based river authorities, until recently water management agencies were characterised by technocratic decision-making with limited input from user groups (Benson *et al.*, 2014). As such, CaBA was established by Defra in 2013 to promote collaborative partnership working. To date, over 100 catchment based partnerships have come under CaBA coordination (CaBA, 2021). Links are also being created with the Catchment Sensitive Farming partnership between Defra, Natural England and the Environment Agency which integrates on-farm advice and training around sustainable abstraction to enable farmers to protect water resources (Defra, 2019b).

A collaborative approach was also promoted in the 2017 Water Abstraction Plan through its 10 Water Resources Priority Catchments (WRPC). For example, the CamEO Partnership encompasses the Rivers Cam, Lark, Little Ouse and Thet, Wissey and South Level within the Cam & Ely Ouse WRPC in East Anglia (CamEO, 2021). The Partnership is co-hosted by Anglian Water and the Rivers Trust, partnering with the National Farmers Union, local councils and wildlife protection non-governmental organisations (NGOs). CamEO has undertaken several initiatives to support sustainable abstraction in collaboration with industry actors.

Another collaborative partnership model is Water Abstractor Groups (WAGs), comprised of farmers in eastern England (Holman & Trawick, 2011). These groups represent agricultural and horticultural members in protecting their water rights through working collaboratively with government agencies. For example, the Broadland Agricultural Water Abstractors Group (BAWAG), is an association of 170 members, formed in 1997, based within the Broadland and North Norfolk CAMS areas (BAWAG, 2021). Key aims include lobbying on behalf of their members regarding water licenses and promoting best practice in sustainable agricultural water management.

Finally, water companies are engaged in parallel regional-scale partnership processes. Water Resources East (WRE), a not-for-profit company established by Anglian Water in 2014, engages water companies, energy companies, internal drainage boards, public bodies, NGOs, universities, civil society and farmers groups in collaboratively managing regional water resources (WRE, 2021). Other regional bodies are Water Resources North (WReN), Water Resources West and Water Resources South East.

7.3 DEFINING THE AVAILABLE RESOURCE POOL

7.3.1 Setting and meeting the volumetric cap in catchments

The volumetric cap is set on a catchment basis by the Abstraction Licensing Strategy. For each sub-catchment, surface and groundwater interactions may be investigated using numerical modelling techniques enabling limits to be placed for individual sources. The EA or NRW estimates the amount of water available for abstraction throughout the year, calculated at four different flow parameters: 'Q95 (the flow of a river which is exceeded on average for 95% of the time i.e., low flow), Q70, Q50, and Q30 (higher flow)' (Environment Agency, 2020b). These are used to assess water availability but defining full Environmental/Ecological Flow Regimes remains difficult (Cook, 2017). While it is desirable to modulate flow with human and natural requirements, in many situations and particularly in chalk streams, variables such as in-stream ecology and abstraction support cause complexity (Klaar *et al.*, 2014).

Such data are mapped across catchments for each parameter, with colours indicating potential water abstraction availability: green for water available, yellow for restricted water available, and red for water not available. The same system is used for groundwater availability. Licences are set according to the water available in a sub-catchment, with new licences considered by the EA in 'green' areas depending on the impacts on downstream flows and protected areas. In 'yellow' areas, where Environmental Flow Indicators (EFIs – see below) show environmental impacts, no new licences are issued and existing licences may be subject to volumetric reduction. In 'red' zones, licences are not issued as flows cannot ensure 'good ecological status'. However, licenses can still be traded within this overall cap, under certain conditions (see below).

7.4 DEFINING ALLOCATION AND RE-ALLOCATION RULES

Water allocations are subject to EA or NRW licensing, but various exemptions have applied. The 1963 Act had permitted holders of abstraction licences to pass them on after their death through succession of land ownership, that is in perpetuity. The historical exemption for 'grandfathering' abstraction rights was then changed, through 2017 amendments to the Water Resources Act 1991[6]. Existing licences may also be split or transferred if land is sold or leased, subject to EA agreement, meaning the licensing is not now tied to absolute land ownership and trading is also possible within the same catchment or groundwater unit (Environment Agency, 2014b). Water suppliers also received government compensation for revocation and variation of licences, but this was ended by the 2014 Water Act. Other historical exemptions removed by legal amendments in 2017 included the rights of irrigators and The Crown to take unlimited amounts of water, and authorities to transfer water between inland sources (Defra, 2017b). 'Low risk' abstraction activities are still exempted from

[6]Four pieces of legislation amend the 1991 Act: The Water Abstraction (Transitional Provision) Regulations 2017; The Water Abstraction and Impounding (Exemptions) Regulations 2017; The Water Abstraction (Revocations etc.) (England) Order 2017; The Water Abstraction (Specific Enactments) Regulations 2017.

licensing, for example dredging (ibid.). Agriculture has no priority licensing exemption over other sectors, with applications assessed on their individual merit.

7.5 MONITORING AND COMPLIANCE

7.5.1 Hydrological monitoring

Government agencies conduct continuous resource assessment through monitoring networks. Surface water is monitored at Assessment Points (APs), covering river basins. In the Hampshire Avon catchment there are 18 APs, situated mainly along the Avon and Wylye rivers (Environment Agency, 2020b). Water situation reports are published monthly, while river flow and rainfall reports are issued weekly. Situation reports provide data on rainfall, soil moisture deficit, river flows and groundwater levels plus projections on future water availability.

Environmental impacts of abstraction are measured through modelling Environmental Flow Indicators (EFIs) (Defra, 2020). Both the EA and NRW use these indicators to estimate the water required to sustain ecological health of rivers. Here, the EA defines 'surface water bodies with flow greater than the EFI as supporting Good Ecological Status under the Water Framework Directive' (ibid.). Groundwater abstraction is monitored using quantitative tests for groundwater levels, surface water flow protection, water quality and spring discharge (ibid.). In meeting these tests, agencies determine whether groundwater complies with WFD 'good' status for groundwater bodies. Where water levels drop to unsustainable levels, a 'hands off flow' condition can be imposed on licensing. According to Defra (2016) a 'hands off flow (HoF) or level (HoL) condition allows the Regulator to reduce or stop abstraction when flows at a gauging station, or levels in a borehole, pass a specified threshold'.

7.5.2 Enforcement and ensuring compliance

Infringements to licences can result in enforcement action (Environment Agency, 2019b). Here, an 'outcome-focused enforcement' approach is pursued (ibid.). Enforcement intervention options start with (i) advice and guidance to suspected transgressors aimed at behaviour modification, followed by (ii) warnings, (iii) enforcement notices, (iv) civil sanctions and finally (v) commencement of criminal proceedings. Here, the EA can issue compliance notices, restoration notices, stop notices, enforcement cost recovery notices and non-compliance penalty notices. For minor infringements, a fixed monetary penalty can be issued. More serious offending can elicit a variable monetary penalty. Significant infringements can involve criminal proceedings, which could result in fines or even imprisonment. Although no data exist on numbers of criminal actions undertaken relating to water abstraction, the EA recently stated that illegal activity was increasing in the West Country through drilling of unlicensed boreholes (Environment Agency, 2019c). Compliance can be enforced through other areas of government policy, for example through cross-compliance conditions for government funding, described below.

7.6 THE BROADER POLICY INSTRUMENT MIX

7.6.1 Drought policy

Water abstraction policy links to national drought policy (Environment Agency, 2017). In terms of environmental, agricultural and domestic water supply, UK Government policy classifies droughts as civil emergencies involving a multi-actor response. Droughts are managed centrally by a National Drought Group in Defra, comprising the EA, government ministries and key stakeholders. The EA has also adopted drought management plans for its 14 operational areas, identifying actions for drought events that are reviewed annually. Under drought management conditions, the Environment Agency can change licence conditions for agricultural irrigators (Defra, 2015a). Under a Section 57 Restriction, spray irrigators within a specific catchment can have their water allocations reduced or even stopped.

7.6.2 Other regulatory instruments

Water abstraction in England and Wales is dominated by industrial concerns. The principal abstraction sector is the privatised water industry, thereby necessitating interrelated national policy for integrating abstraction objectives into water company policy through statutory obligations to produce Water Resources Management Plans and drought plans. However, agricultural abstraction is more significant in specific regions, particularly where spray irrigation is used by farmers, and is covered by parallel rules. For example, farmers receiving financial support from agri-environment agreements and the Basic Payment Scheme, Countryside Stewardship (CS) and other grants must also comply with cross-compliance conditions for water abstraction related to meeting licensing conditions (GOV. UK, 2020; NFU, 2021a). These rules are comprised of Statutory Management Requirements (SMRs) and Good Agricultural and Environmental Conditions (GAECs) (Defra, 2015b). A set of conditions, GAEC2, are established for water abstraction.

Other areas of policy important for agricultural water abstraction are regulations for nature protection. Under the EU Habitats and Birds Directives, the Environment Agency and Natural Resources Wales have been legally compelled to intervene where abstractions impact on protected areas; particularly Natura 2000 sites, including Special Areas of Conservation and Special Protection Areas (Defra, 2013; JNCC, 2020a). Applications for water abstraction licences are also assessed for any impacts on Sites of Special Scientific Interest (SSSIs), designated by Natural England under powers in the Wildlife and Countryside Act 1981. The Act forbids landowners to undertake water abstraction that could damage an SSSI without first obtaining consent.

7.6.3 Economic instruments

The licensing system also permits the trading of abstraction rights for surface waters, which has proved popular in agricultural areas. Parties can secure a licence by securing agreement with an existing licence holder for all or part of their rights, subject to EA permission. Water trading applications are considered according to guidance criteria for preventing water body deterioration. In East Anglia they relate to location, season, quantity, rate, purpose and catchment (Environment Agency, 2020c). Both the donor and recipient should 'abstract within the same surface watercourse and have comparable effects on other surface water features' (ibid.). A preference for trading water abstractions from downstream sources is maintained by the EA, since it potentially results in lower environmental impacts. Trading should also occur for abstraction in the same season: shifting licenses from winter to summer could result in greater environmental pressures. Where water is also determined unavailable for abstraction by the ALS, water trading is unlikely to be permitted. Trades should also occur for the same purpose, with the 'consumptiveness' of abstraction comparable (ibid.).

7.7 ASSESSING PERFORMANCE

7.7.1 Environmental effectiveness

Environmental effectiveness can be gauged by examining national and catchment level evidence on water abstraction levels. In assessing the ongoing reform process, Defra (2019b) reflected back on natural flow level targets in its Water Abstraction Plan 2017. Some progress was discernible: while 82% of surface waters met the required flow standards in 2017 (using 2016 data), by 2019 this figure had marginally increased to 84% (Defra, 2019b). Surface water bodies classified as 'potentially unsustainably abstracted' identified in 2017 was 10% but by 2019 it had declined to 7% (ibid.). The number of unsustainably abstracted water bodies, however, increased from 8 to 9 during this period (ibid.). In addition, the report predicts that reductions in groundwater abstractions in wetland areas should allow an increase in groundwater bodies classified as sustainable, from 72% recorded in 2019 to 77% in 2021 (ibid.). In attributing reasons for these positive environmental trends, Defra refers to implementation of the Restoring Sustainable Abstraction programme: by 2019, 282 licences had been either amended or revoked, making up to 40 billion (10^9) litres of water available for the environment (ibid.).

Data from individual priority catchments support the broader national environmental trends. For example, the Cam and Ely Ouse catchment was defined as a 'priority catchment' in the Defra Water Abstraction Plan 2017. While intensively farmed, with 60.5% of its land under arable cultivation, the catchment also encompasses 167 designated protected areas, including Special Areas of Conservation, Special Protected Areas, SSSIs and Ramsar sites (Environment Agency, 2020d). Actions to reduce water abstraction under the Restoring Sustainable Abstraction programme resulted in savings of 18 443 006 m^3/year, in addition to a further 333 957 m^3/year from revoking unused and underused licences (ibid.). Another factor for improvements in environmental flows may be the nascent partnership approach, for example CamEO.

Given this evidence, to what extent does the sector exhibit environmental unsustainability, that is restricted flows? National levels of actual abstraction of groundwater and non-tidal surface water in England had fallen between 2001 and 2011, from 11.6 billion m^3 to 8.2 billion m^3, but started to rise up to 2017 when it reached 10.4 billion m^3 (Defra, 2019a). No data are published for the period since, making reform impacts difficult to assess. However, the WFD river basin management planning assessment shows that surface water bodies meeting high or good ecological status has not improved since 2000: 36% or 3322 UK surface water bodies were classified as being high or good status in 2019 but in England this figure fell to only 16%; down from 25% in 2009 (JNCC, 2020b). The picture in individual catchments is also concerning. Priority catchments are still experiencing unsustainable flows despite licence modifications or revocations, according to recent revisions of Abstraction Licensing Strategies. For example, the Cam and Ely Ouse – already under high abstraction pressures from agriculture – shows no water availability at Q70 flows and above; even at Q50 water availability is restricted or unavailable (Environment Agency, 2020d).

7.7.2 Economic efficiency

The economic efficiency of water abstraction has arguably increased under the reform process for some sectors, particularly through digital transformation and water trading. By 2019, 3000 licence holders were registered on the EA's digital licence system. Registered licence holders can access their licences online and submit abstraction data; representing a significant increase in efficiency over the previous paper-based system (Defra, 2019b). For the EA, the digital service also involves efficiencies in managing licence applications and implementation. Licences containing hands off flow or hands off level conditions can be actioned online during low river flows, thereby freeing up time for Agency staff.

Water rights trading is also creating economic efficiencies in agricultural sector allocations. Barriers to water trading were eased by the Water Act 2003 but uptake was low, with only 51 trades before 2011 (Lumbroso *et al.*, 2014). Despite initial concerns over potential challenges (see Houses of Parliament, 2017), trialling of the new trading system amongst farmers, coordinated online by the EA, has been positively received. Initial assessment shows that 84% of users were satisfied with the system, with Defra (2019b) quoting positive feedback from farmers. Similarly, a joint National Farmers' Union (NFU) and Cranfield University report (Rey *et al.*, 2018) identifies strengths and weaknesses in water trading, as expressed by farmers. Respondents highlighted the opportunities offered by better water allocations in catchments, greater water efficiencies and the potential to generate money. But concerns included the potential for water companies and industry to buy up agricultural allocations and the erosion of historic rights to water (ibid.). The NFU also highlights examples of successful trading such as the Wheatley Watersource pilot project, jointly developed by Anglian Water and Essex & Suffolk Water (NFU, 2021b). Success of this online trading platform has led to the inclusion of abstractors in the Cam and Ely Ouse and adjacent catchments (ibid.). While some ongoing concerns over the system remain amongst farmers, the NFU has demonstrated its support by establishing an online Water Bank to fast-track short-term licence trades (NFU, 2021c). Donors of licences can register online with their local Environment Agency offices and be matched with recipients within their catchment.

7.7.3 Social equity

Another important aspect of sustainability is its social dimension. Social equity could be measured through conditions for setting licences and priorities given for different sectoral allocations, although these – as described previously – are assessed against individual applications on merit and could be considered procedurally fair between social actors. However, we could also consider the shift towards collaboration in water abstraction as another indicator of equity. As Wondolleck and Yaffee (2000) argue '[i]n successful collaborative efforts, considerable effort is made to craft decision-making processes that are fair and outcomes that are judged equitable'. When applied to catchment partnerships this principle of procedural justice or fairness is considered critical to ensure the sustainability of institutional arrangements and improvement actions (Smith *et al.*, 2015).

National-level data show that social equity is apparent in catchment partnerships. Although less than 10% of CaBA catchment coordinators stated in 2017 that the approach was effective in addressing water resources issues, primarily because of the difficulties in engaging abstractors, the government has sought to counter this problem through promoting collaboration around abstraction in the Water Resources Plan (Defra, 2019b). Here, the CaBA abstraction working group is co-chaired by the Rivers Trust and the Environment Agency, to provide a steer on the integration of abstraction objectives into catchment planning and programmes.

Social equity is also being increased through the EA's priority catchments. According to Defra (2019b), within the four initial catchments, based in agricultural areas, 'abstractors and stakeholders… are keen to work together' to better share water. It further states that the Environment Agency has 'made significant progress on engagement and is moving towards co-development of solutions' (ibid.). The approach in the South Forty Foot catchment has been to 'Ask the people who farm the land' through workshops designed to inform options (ibid.). Ideas generated included: 'artificial recharge schemes, and improved sharing of surface water resources by better communication amongst abstraction sectors' (ibid.). Collaborative initiatives adopted in the Cam and Ely Ouse include a trial Water Abstraction Alert system, involving the NFU and internal drainage boards in information sharing, and the Water Resource Advisory Farm Visits scheme that provides outreach advice for farmers from the NFU, Norfolk Rivers Trust and Natural England (Environment Agency, 2020d).

7.7.4 Climate resilience

An evident risk to water allocation is climate change, particularly for the agricultural sector. In England, only one per cent of abstracted water is used by agriculture and most comes via mains supply as well as direct abstraction from rivers and boreholes. In the agricultural sector, the largest user is dairy production particularly in the South West (Defra, 2017c), although in the east of England and the Midlands spray irrigation use is considerable (Defra, 2011). Drought incidences are predicted to increase significantly over coming decades (Water UK, 2016). Attendant restrictions on abstraction licensing may then occur, with the Environment Agency predicting an average reduction of 15% in river flows by 2050 (Houses of Parliament, 2017). Groundwater recharge in Southern England may also experience a 40% decline by 2080 under a 'worst case' climate change scenario (ibid.). Significant impacts are predicted for the agricultural sector (Defra, 2018).

National policy has consequently promoted resilience in determining water allocations. Defra has signalled to water companies that Water Resource Management Plans should reflect increasing challenges from climate change (Defra, 2019b). Regional scale collaborations such as Water Resources East should allow greater strategic responses. A National Framework for Water Resources, adopted in 2020, prioritises future regional water planning in conjunction with industry stakeholders (Environment Agency, 2020e). The UK National Adaptation Programme (Defra, 2018) also identifies risks to water supply for agriculture from climate change, also citing the licensing reforms as integral to enhancing sector resilience:

Agriculture is a key consumer of water, most notably during what tend to be drier months, when there is an increased public demand for water. We are [therefore] working with farmers and other abstractors to ensure that abstraction licences are sustainable… and that the agriculture sector has access to water and uses it efficiently.

Irrigators are at particular risk, leading to the UK government designating this sector in terms of 'essential water need' in planned future reforms of water abstraction (see Knox *et al.*, 2020). Future water scarcity may have a disproportionate effect on high-value irrigated vegetable production. Potato growing for example accounts for 25% of irrigated farming but is primarily located in already-water-stressed regions such as East Anglia (Watts *et al.*, 2015). A smaller allocation goes to meadow irrigation, which may be similarly impacted. This historical use of surface and (occasionally) spring water dates back typically some 400 years and represents both management of historical landscapes on valley floors and hillsides as well as supporting habitats and biodiversity (Cook & Williamson, 2007). Here, the 2003 legislation requires a transfer licence because water abstracted from a river is returned to the river system locally. This is the case for the Harnham Water Meadows in Salisbury, Hampshire where a licence has been granted (Wiltshire Times, 2020).

That said, implementation of sustainable abstraction in catchments is variable, particularly in intensively farmed areas. Examination of recently revised Abstraction Licensing Strategies shows that climate change is not generally a determining factor in allocations. One briefly pledges to review climate impacts in licensing (Environment Agency, 2020f). Another identifies likely climate impacts for the catchment but no related management measures (Environment Agency, 2020d). Meanwhile, some agricultural catchments continue to face water stress. East Anglian farmers faced challenging conditions in August 2020 due to low rainfall and higher than average temperatures, with low river flows, farm reservoirs depleted and additional abstraction pressure placed upon irrigators (NFU, 2020).

7.8 CONCLUSIONS

Water abstraction policy is undergoing significant reform in England and Wales to ensure greater sustainability. Reforms aim at better linking water availability to allocations in catchments through licensing, alongside the promotion of catchment based partnership working and instrumental innovations such as water trading and licensing system digitisation. The Water Act 2003 permits the revocation or variation of a licence in order to prevent 'serious damage' to the water environment and hence remains a mechanism for limiting abstraction at source. On the one hand, reforms are realising positive environmental impacts through reductions in water abstraction which restore river flows, enhance economic efficiency through online licence application management and water trading, and increase water available to agricultural users. Increased social equity is achieved through the inclusion of actor groups, such as farmers, in catchment decision-making. Benefits for future climate resilience are also discernible. On the other hand, such reforms have yet to reduce albeit historical levels of over-abstraction in many catchments, particularly those subject to intensive agricultural abstraction pressures. They are however a positive step, as such changes – particularly collaborative partnership working – can take time to produce long-term sustainability benefits (Smith *et al.*, 2015).

REFERENCES

BAWAG. (2021). Broadland Agricultural Water Abstractors Group. BAWAG, Norfolk. www.bawag.co.uk (accessed 27 March 2021)

Benson D., Fritsch O., Cook H. and Schmid M. (2014). Evaluating participation in WFD river basin management in England and Wales: processes, communities, outputs, outcomes. *Land Use Policy*, **38**, 213–222, https://doi.org/10.1016/j.landusepol.2013.11.004

CaBA. (2021). About the catchment based approach. https://catchmentbasedapproach.org (accessed 30 March 2021)
CamEO. (2021). Cam & Ely Catchment Partnership. www.cameopartnership.org (accessed 30 March 2021)
Cook H. F. (1999). Groundwater development in England. *Environment and History*, **5**(1), 75–96, https://doi.org/10.3197/096734099779568399
Cook H. F. (2017). The Protection and Conservation of Water Resources. Wiley Blackwell, Chichester, UK.
Cook H. and Williamson T. (eds) (2007). Water Meadows: History, Ecology and Conservation. Windgather Press, Macclesfield, UK.
Defra. (2011). Water Use on Farms: 2. Agricultural Water Abstraction for Irrigation. Defra, London, UK. https://adlib.everysite.co.uk/adlib/defra/content.aspx?id=000IL3890W.16NTC2BRGOY37W (accessed 01 April 2021)
Defra. (2013). Managing Abstraction and the Water Environment. Defra, London, UK.
Defra. (2015a). Manage Water on Land: Guidance for Land Managers. Defra, London, UK. https://www.gov.uk/guidance/manage-water-on-land-guidance-for-land-managers (accessed 27 March 2021)
Defra. (2015b). The Guide to Cross Compliance in England. Defra, London, UK. https://www.gov.uk/guidance/guide-to-cross-compliance-in-england-2021 (accessed 27 March 2021)
Defra. (2016). Changes to Water Abstraction Licensing Exemptions. Defra, London, UK. https://consult.defra.gov.uk/water/water-abstraction-licensing-exemptions/ (accessed 27 March 2021)
Defra. (2017a). Water Abstraction Plan. Defra, London, UK.
Defra. (2017b). Government Response to Consultation on Changes to Water Abstraction Licensing Exemptions in England and Wales: New Authorisations. Defra, London, UK.
Defra. (2017c). Water Usage on Farms: Results From the Farm Business Survey, England 2015/16. Defra, London, UK.
Defra. (2018). The National Adaptation Programme and the Third Strategy for Climate Adaptation Reporting. Defra, London, UK.
Defra. (2019a). Water Abstraction Statistics: England, 2000 to 2017. Defra, London, UK.
Defra. (2019b). Abstraction Reform Report: Progress Made in Reforming the Arrangements for Managing Water Abstraction in England. Defra, London, UK.
Defra. (2020). Water Abstraction Plan: Environment. Defra, London, UK.
Defra and Welsh Government. (2013). Managing Abstraction and the Water Environment. Defra, London, UK.
Downing R. A. (1993). Groundwater resources, their development and management in the UK: an historical perspective. *Quarterly Journal of Engineering Geology and Hydrogeology*, **26**, 335–358, https://doi.org/10.1144/GSL.QJEGH.1993.026.004.09
Environment Agency. (2003). The Stour Catchment Abstraction Management Strategy. Environment Agency, Worthing, UK.
Environment Agency. (2014a). Guidance: Apply for A Water Abstraction or Impoundment Licence. Environment Agency, Bristol, UK. https://www.gov.uk/guidance/water-management-apply-for-a-water-abstraction-or-impoundment-licence (accessed 12 April 2021)
Environment Agency. (2014b). Abstracting Water: A Guide to Getting Your Licence. Environment Agency, Bristol, UK.
Environment Agency. (2016). Managing Water Abstraction. Environment Agency, Bristol, UK.
Environment Agency. (2017). Drought Response: Our Framework for England. Environment Agency, Bristol, UK.
Environment Agency. (2019a). Water Resources Priority Catchments. Environment Agency, Bristol, UK.
Environment Agency. (2019b). Environment Agency Enforcement and Sanctions Policy. Environment Agency, Bristol, UK.
Environment Agency. (2019c). Abstracting Water? Think Before You Drill. Press Release 03.08.2019. Environment Agency, Bristol, UK. https://www.gov.uk/government/news/abstracting-water-think-before-you-drill (accessed 10 April 2021)
Environment Agency. (2020a). Environment Agency Scheme of Abstraction Charges 2020/21. Environment Agency, Bristol, UK.
Environment Agency. (2020b). Hampshire Avon Abstraction Licensing Strategy. Environment Agency, Bristol, UK.
Environment Agency. (2020c). East Anglia Area Approach to Abstraction Licence Trading. Environment Agency, Peterborough, UK.
Environment Agency. (2020d). Cam and Ely Ouse Abstraction Licensing Strategy (ALS). Environment Agency, Bristol, UK.

Environment Agency. (2020e). Meeting our Future Water Needs: A National Framework for Water Resources. Environment Agency, Bristol, UK.

Environment Agency. (2020f). South and West Somerset Abstraction Licensing Strategy. Environment Agency, Bristol, UK.

GOV.UK. (2018). Manage Your Water Abstraction and Impoundment Licence. Environment Agency, Bristol, UK. https://www.gov.uk/guidance/manage-your-water-abstraction-or-impoundment-licences-online (accessed 04 April 2021)

GOV.UK. (2020). Guide to Cross Compliance in England 2021. Rural Payments Agency, London, UK. https://www.gov.uk/guidance/guide-to-cross-compliance-in-england-2021 (accessed 04 April 2021)

HM Government. (2011). Water for Life. The Stationary Office, Norwich, UK.

HM Government. (2018). A Green Future: Our 25 Year Plan to Improve the Environment. Defra/HM Government, London, UK.

Holman I. P. and Trawick P. (2011). Developing adaptive capacity within groundwater abstraction management systems. *Journal of Environmental Management*, **92**(6), 1542–1549, https://doi.org/10.1016/j.jenvman.2011.01.008

Houses of Parliament. (2017). Reform of Freshwater Abstraction. POSTnote Number 546. The Parliamentary Office of Science and Technology, London, UK.

JNCC. (2020a). Special Areas of Conservation – Overview. JNCC, Peterborough, UK. https://jncc.gov.uk/our-work/special-areas-of-conservation-overview/ (accessed 05 April 2021)

JNCC. (2020b). UK Biodiversity Indicators: B7. Surface Water Indicators. JNCC, Peterborough, UK. https://data.gov.uk/dataset/95081440-7621-44aa-a3b2-84ebdc2e9ace/uk-biodiversity-indicator-b7-surface-water-status (accessed 05 April 2021)

Klaar M. J., Dunbar M. J., Warren M. and Soley R. (2014). Developing hydroecological models to inform environmental flow standards: a case study from England. *WIRES Water*, **1**(2), 207–217, https://doi.org/10.1002/wat2.1012

Knox J. W., Kay M. G., Holman I. P. and Hess T. M. (2020). Irrigation Water Strategy for UK Agriculture and Horticulture. Booklet Published by Cranfield University, UKRI and NFU. National Farmers Union, Stoneleigh, UK.

Lumbroso D. M., Twigger-Ross C., Raffensberger J., Harou J. J., Silcock M. and Thompson A. J. K. (2014). Stakeholders' responses to the use of innovative water trading systems in east anglia, England. *Water Resources Management*, **28**, 2677–2694, https://doi.org/10.1007/s11269-014-0633-z

McGillivray D. (2013). Water rights and environmental damage: an enquiry into stewardship in the context of abstraction licensing reform in England and Wales. *Environmental Law Review*, **15**(3), 205–224, https://doi.org/10.1350/enlr.2013.15.3.188

NFU. (2020). East Anglian Irrigators Feel the Heat as River Flows Drop. National Farmers Union, Stoneleigh, UK. https://www.nfuonline.com/archive?treeid=145457 (accessed 07 April 2021)

NFU. (2021a). Cross Compliance and Irrigation. National Farmers Union, Stoneleigh, UK. https://www.nfuonline.com/archive?treeid=119214 (accessed 07 April 2021)

NFU. (2021b). East Anglia Online Water Trading Platform Expanded. National Farmers Union, Stoneleigh, UK. https://www.nfuonline.com/updates-and-information/east-anglia-online-water-trading-platform-expanded/ (accessed 07 April 2021)

NFU. (2021c). NFU Water Bank. National Farmers Union, Stoneleigh, UK. https://www.nfuonline.com/archive?treeid=109668 (accessed 07 April 2021)

Ofwat. (2015). The Case for Change – Reforming Water Abstraction Management in England. Ofwat, Birmingham, UK. https://www.ofwat.gov.uk/publication/the-case-for-change-reforming-water-abstraction-management-in-england/ (accessed 29 March 2021)

Rey D., Hammett P., Salmoral G. and Montilla N. (2018). Assessing Opportunities for Secondary Markets for Water in Response to Proposed Abstraction Reforms: Key Findings. Cranfield University and National Farmers Union, Cranfield, UK.

Smith L., Porter K., Hiscock K., Porter M. J. and Benson D. (eds) (2015). Catchment and River Basin Management: Integrating Science and Governance. Earthscan, London, UK.

United Utilities. (2020). Evolving the Water Industry National Environment Programme to Deliver Greater Value. United Utilities, Warrington, UK.

Water UK. (2016). Water Resources Long-Term Planning Framework. Water Resources UK, London, UK. https://www.water.org.uk/publication/water-resources-long-term-planning/ (accessed 30 March 2021)

Watts G., Jenkins A., Hess T., Humble A., Olbert C., Kay M., Pope V., Stannard T., Storey M., Meacham T., Benton T. and Noble A. (2015). Agriculture's Impacts on Water Availability. Global Food Security, Wiltshire, UK.

Wiltshire Times. (2020). ENVIRONMENT AGENCY Water Resources Act 1991 (as Amended by the Water Act 2003) Notice of Application for A Transfer Licence to Abstract (Take) Water for A Previously Exempt Abstraction. Notice ID: MFN0584848. Wiltshire Times, Swindon, UK.

Wondolleck J. M. and Yaffee S. L. (2000). Making Collaboration Work: Lessons From Innovation in Natural Resource Management. Island Press, Washington DC, USA.

WRE. (2021). The Water Resources East Vision. WRE, Norwich. https://wre.org.uk (accessed 28 March 2021)

doi: 10.2166/9781789062786_0093

Chapter 8

Water allocation in Spain. Legal framework, instruments and emerging debates

Carles Sanchis-Ibor[1], Manuel Pulido-Velazquez[2], Juan Valero de Palma[3] and Marta García-Mollá[1]

[1]Universitat Politècnica de València, Centro Valenciano de Estudios del Riego (CVER), València, Spain
[2]Universitat Politècnica de València, Institute of Water Engineering and Environment (IIAMA), València, Spain
[3]Acequia Real del Júcar, Valencia, Spain

ABSTRACT

Spain has a tradition of water management and allocation based on a dual model of water rights, where surface water is public, but groundwater has historically been considered a private resource. The Water Law (1985) created a new concessional system for the assignment of water rights to users of both resources, controlled by the State, but preserving the historical rights. Water allocation, in practice, is determined by the resources and demands that are estimated in the water planning documents. Therefore, water allocation is granted by River Basin Authorities and depends on water availability, social and environmental priorities, and the system operating rules. It is an institutional model of water allocation, under State control, but partially open to participation and negotiation with users. The main weakness of the Spanish water allocation model derives from the application, during the 20th century, of an excessively generous policy of water rights allocation and the lack of control of water uses. This policy has led to the overallocation of water rights and groundwater overexploitation in some basins, generating structural deficits, dramatic environmental impacts and significant social and territorial tensions. This context, which hinders the implementation of environmental flows, has required the design of different programmes and plans to control water uses, and the introduction of new legal instruments to stimulate water rights temporal exchanges.

Keywords: Environmental flows, groundwater overexploitation, overallocation, water allocation, water markets, water rights

8.1 INTRODUCTION

Spain presents a marked spatial heterogeneity in the availability of water resources, which has historically led many authors to divide the country between the *humid* (Atlantic) and *dry* (Mediterranean) areas. At the beginning of the 20th century, the state launched an ambitious programme to transform the country based on the mobilisation of water resources for energy production and irrigation. Throughout the century, this national hydraulic mission increased the reservoir capacity from 0.1 to 53.2 km^3 and the irrigated area from 1.3 to 3.7 Mha (Figure 8.1). This expansive policy, developed in some regions through basin overbuilding and water resources overallocation processes, led most of the rivers in the

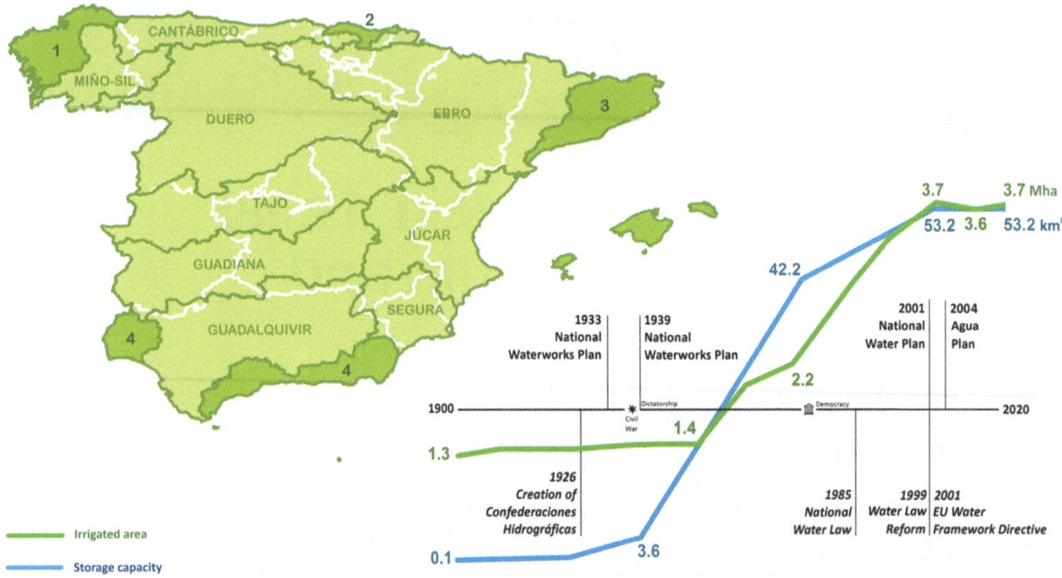

Figure 8.1 The current nine *Confederaciones Hidrográficas* are represented in clear green. White boundaries divide the regions (autonomous communities). In dark green, basins transferred to regional governments: 1. Galicia; 2. Basque Country; 3. Catalonia; 4. Andalusia. 20th century time-line for the main milestones of water planning and legal framework for water, also representing the expansion of irrigated area (Mha) and storage capacity (km^3).

southern half of the Iberian Peninsula to basin closure, and was the origin of various environmental problems and numerous social and territorial conflicts (Fornés *et al.*, 2021; García-Mollá *et al.*, 2019).

Regarding its overall water governance, Spain is a semi-federal state in which the central government shares power with 17 regional governments. The central government is responsible for water management through one of its ministries (currently the Ministry of Ecological Transition and Demographic Challenge – MET). MET's action on water management is executed by nine *Confederaciones Hidrográficas* (river basin authorities – RBAs), endowed with several participative organs. These RBAs, structured in different offices, assign water to users and control its quality and use, in compliance with the EU Water Framework Directive (WFD). However, regional governments are responsible for managing urban planning, agriculture and natural heritage; and in six regions, they govern the river basins that are entirely within their territorial scope (Figure 8.1). This complex administrative division requires continuous work to coordinate and has generated various territorial conflicts (López-Gunn, 2009).

Spain has a tradition of water management and allocation based on a dual model of water rights, formulated in the nineteenth century (Water Laws of 1866 and 1879). On the one hand, surface water is public and has traditionally been governed by a concessionary regime. On the other hand, groundwater has historically been considered a private resource and remained outside public control for a long time. This division, repeatedly referred to as 'hydro-schizophrenic' (Llamas, 1975), continues to this day, despite the fact that the Water Law of 1985 established a way to eradicate the private water regime.

Because water rights allocation and water resources allocation have different scales, procedures and problems in practice, this chapter has two parts. The first describes the nature of water rights and the concessional mechanisms. The second part examines water allocation in practice, considering some constraints such as the overallocation problem, groundwater overexploitation and environmental flows. Finally, the water rights transfer procedures are outlined and some brief concluding remarks are given.

8.2 THE LEGAL FRAMEWORK FOR WATER

The Water Law (1985), which repealed the antiquated 1879 law, is the cornerstone of the Spanish legal framework for water (Embid, 2010). Written to provide a response to the important social, political and territorial changes that took place in Spain after the advent of democracy, the Water Law created a concessional system for the assignment of water rights to users, controlled by the state. In practice, what these concessions establish is the annual maximum volume of water that each user can receive, but the final allocation of water is decided by the basin organisations at all times – through hydrological planning instruments or through negotiation with users. Water planning is the basis for all public and private actions in the public hydraulic domain. It is carried out through the River Basin Plans (RBPs) and the National Water Plan (NWP). The RBPs establish an official forecast of water allocation for each use, based on the estimation of demands and available resources, while the NWP resolves the imbalances between demand and supply that have not been solved by the RBP – through inter-basin water transfer or other means of water resources mobilisation.

In 1999, the Water Law was reformed without altering its legal essence or main articles. The reform sought to redress some deficiencies and grant the greatest level of protection to water as an environmental asset. It introduced some technical improvements and aspects not previously considered, such as desalination and some new rules for public works. Regarding water allocation, this reform introduced new rules and protocols for the temporary transfer of water rights (see Section 8.3.4).

8.2.1 The double nature of water rights

The Water Law places all continental water bodies in the public hydrological domain. Water use is subject to control, authorisation and regulation by the government. Users can only obtain water rights via administrative concession, which is bestowed by the RBAs through the Water Commissioner Office (*Comisaría de Aguas*), the administrative body responsible for the administration and control of water rights and the use of the hydraulic public domain.

However, the legislation that existed prior to the Water Law only considered surface water to belong to the public domain. Groundwater was subject to a regime of private appropriation. The current Water Law provided for the respect of rights acquired under the prior legislation with the following options:

(i) Holders of administrative concession in accordance with the previous legislation continue to enjoy their rights for a term of 75 years.
(ii) If users of public water without rights prove that they have enjoyed this water for a further period of 20 years, they obtain the right to use this water for 75 years from when the Law came into effect.
(iii) The holders of any private rights in accordance with the previous law may choose to prove their rights for inscription in the Water Registry as a temporary use of private water. In this case, the regime is respected for 50 years and entitles the owner to obtain an administrative concession at the end of this term. Those that did not exercise this option before 1 January 1989 remained under the previous framework – that of private water – in perpetuity.

Some authors (Moreu, 2002) estimated that only 10–20% of private users chose the public option. In the case of not accepting this option, users had to be included in the Private Water Catalogue, but the vast majority of private users did not fulfil this obligation. This left most groundwater out of public control, and forced the government to launch two consecutive programmes to register private uses. In 1994, the ARYCA programme made an inventory of private wells, but the results were insufficient. The programme closed without the inclusion of 80–90% of private users (Fornés *et al.*, 2005). In 2001, the government approved the ALBERCA programme to update both the registries of public and private uses. It was in force until 2012, produced uneven results among the different RBAs, and the ministry did not publish a report on the results of the programme. Since 2012, each RBA has been advancing at a different pace in the process of inscribing groundwater users in the catalogue and registry, and they also follow their own strategy to apply the Order ARM/1312/2009. This order obliged users to install and maintain water meters at their own expense, detailing their characteristics, and established a regime of sanctions in order to improve the control of groundwater bodies and discharges in the public hydraulic domain.

This process leaves two complicated questions unanswered: How many wells are yet to be registered? Probably no one knows exactly; Are there any illegal wells? The answer depends on the interviewee. The response of the RBAs' representatives is that there are none, as when one is identified, it is either sealed or registered. The complaint of many experts and environmental non-governmental organisations (NGOs) is that, due to insufficient governmental control, there are numerous illegal wells in certain Spanish regions, putting the sustainable management of the resource and the future of some valuable ecosystems at risk (De Stefano & Lopez-Gunn, 2012).

8.2.2 Allocation of water rights

In Spain, water allocation is based on the water concession regime and the cooperation of stakeholders with RBAs. For all public water concessions, according to the Water Law, the Water Commissioner Office sets the purpose, expiration period (no longer than 75 years), the maximum and continuous flow average allowed, as well as the municipality in which the water is taken, regardless of whether the water is from surface water or groundwater.

In the case of water concessions for irrigation, the authorisations for the use of public water define exactly the specific surface area that may be irrigated and the maximum volume of water to be used per hectare and year in order to calculate the instantaneous volume of the concession. The RBAs or MET can revise or annul water concessions for various reasons (Table 8.1).

Concession of water rights for irrigation can be granted for individual farmers who own the land or a group of landowners constituting an irrigation community (*Comunidad de regantes*). Water rights remain attached to land ownership. Hence, the acquirer of a previously irrigated land obtains the associated water right. Where land is divided, the acquirer of irrigable land without access to any water source can request a new access and withdrawal right as a public concession.

In the event that at a specific time the supply of water is not sufficient to meet demand, the Water Law establishes an order of preference among the different uses to allocate water if there is a deficit or total or partial incompatibility between two or more uses. All concessions are subject to compulsory expropriation in favour of another use that precedes it according to the order of preference established in the RBP. In the absence of this order of preference, the following order

Table 8.1 Reasons to revise or annul water concessions according to the Public Water Domain Regulation.

Reasons to Revise or Annul Water Concessions	Compensation
Non-fulfilment of the essential conditions that are indicated within the concession	No
Failure of the holder to carry out the stated activity for three consecutive years	No
Changes in circumstances that make it impossible to achieve the purpose of the concession	No
Forceful expropriation	Yes
Force majeure (at the request of the concession holder)	No
Requirement for compliance with the hydrological plans	Yes
The object of the concession can be fulfilled with a smaller supply	No
Technical improvement in the resource usage contributes to reduce its use	No
Committing of a second very serious infraction in areas and periods in which the government has declared the application of extraordinary measures (drought or other)	No

shall generally apply: (i) urban water supply; (ii) irrigation and agricultural uses; (iii) hydroelectric uses; (iv) industrial uses; (v) aquaculture; (vi) recreational uses; (vii) navigation and water transport; and (viii) other uses. The order of preference operates at the time of granting a concession and in the event of incompatibility between two uses.

Among agricultural uses, the general principle of the right of 'Earlier in time, stronger in Law' applies, and tribunals have repeatedly used this formula. Accordingly, the current Water Law indicates that 'every concession shall be understood to be made without prejudice to a third party'. This chronological priority responds to the greater need to provide legal certainty and guarantee the investments made. However, the Water Law also established in 1986 that in the event of incompatibility of uses, those of greater public or general utility will be given preference, along with those that introduce technical improvements that result in less water consumption or in the maintenance or improvement of its quality.

The concessional regime is not applied in some particular cases, such as small users, private waters and wastewaters. Small users include the use of rainwater that flows on a property, water that is held on it, water coming from springs located within it, or from groundwater when the annual volume does not exceed 7000 m^3 (this last clause has sometimes been perceived as an open door to the overexploitation of aquifers). In these cases, administrative concession is not required. The water can be used with the only requirement being notification to the RBA, which will record the use in the Water Registry. The only exception is when the water is located in an aquifer that is overexploited or at risk of becoming so, in which case administrative authorisation is required.

For private water, there are two possible situations. First, uses included in the Water Registry have to follow the same rules as those for public waters, including the possibility of cooperating in the participatory bodies of the RBA. The holder of this right may sell the right to third parties. This transfer of rights must involve the transfer of the land entitled to use the water, but it is not allowed in three situations: if aquifers are declared to be overexploited or at risk of becoming so; when resource availability demands it; and under circumstances of extraordinary droughts, situations of need, emergency or the simultaneous occurrence of anomalous or exceptional situations.

The second case refers to the unknown proportion of private uses currently included in the Catalogue of Private Waters. The use of private water is strictly limited to that which existed prior to the publication of the Law. Any increase in flows, any modification to the facilities apart from maintenance and preservation, or any change to the purpose of the usage (change of irrigated land, etc.) represent a 'change of flows, conditions or framework of use'. In these cases, the Water Law requires an appropriate administrative concession and its inscription in the Water Registry, thereby becoming public water. These private rights are limited by the Law in the case of overexploited aquifers, in times of droughts or other extraordinary circumstances, under the same conditions as public waters.

Finally, the Water Law also establishes a special legal framework for wastewater and its direct reuse. It determines that any urban or industrial discharge of wastewater into natural watercourses or irrigation channels requires authorisation from the RBA. For wastewater that flows directly to certain irrigated areas, the law distinguishes between two cases. If water is reused by the first user that produces it, the original concession must be modified if this did not anticipate its reuse. If water is reused by a person or persons other than the first user, both uses shall be considered separate, and the reuse must be the object of a concession. Since 2007, a specific legal framework has regulated both the application and quality of the resource being reutilised, and the RBAs have delivered administrative concessions to users. From 2023 onwards, the Regulation (EU) 2020/741 of the European Parliament on minimum requirements for water reuse will define these conditions of use in terms of quality.

The concessions for urban water supply and irrigation can be legally reviewed in the cases in which it is proven that the object of the concession can be fulfilled with a lesser endowment or a more efficient use of the resources that contributes to saving it. For these purposes, RBAs can carry out audits and controls of the concessions in order to verify the efficiency of the management and use of the water resources object of the concession. In practice, some RBAs have used this instrument more frequently than others. More temporary restrictions can also apply in the case of drought, but in these situations, decisions are made through a negotiation process between RBA technicians and users' representatives (Andreu *et al.*, 2013; Carmona *et al.*, 2017).

8.3 WATER ALLOCATION IN PRACTICE. RULES AND PROCESSES

In Spain, as mentioned above, the right to use water is defined by the administrative concessions, but water distribution is determined by the demands that are estimated in the water planning documents. Therefore, in practice, water allocation is granted by RBAs and depends on the balance between available resources and current and future demands. In the case of surface water, each RBA has one or various (one per river or exploitation system) Water Allocation Commissions (*Comisiones de Desembalse*) that propose the allocation of water for each user and time period. These Commissions are composed of RBA technicians, the different users of the basin, environmental organisations and representatives of other government departments involved in water management.

In circumstances of extraordinary droughts, of serious overexploitation of aquifers, or in similar exceptional situations, the Government, by means of a Decree agreed in the Council of Ministers, and after hearing statements from the RBA, may adopt restrictive measures for all users, even if they had been the subject of a concession. Both situations – droughts and overexploitation – are difficult to manage due to the problems caused by the prolonged application of an excessively generous policy of water rights allocation and the lack of interest in the control of water uses.

8.3.1 The overallocation problem

Since the 1980s, Spanish water policy has combined measures to control water use with actions to mobilise water resources to meet new demands or consolidate some developments carried out in precarious conditions. The persistence of these measures in promoting irrigation or their laxity in controlling its expansion has led many basins to a situation of overallocation of water resources and, consequently, to basin closure. 'Paper water' is more abundant than real water in some RBPs in the southern half of the Iberian Peninsula.

In the Júcar Basin, for example, this overallocation is explicitly recognised by the current planning instruments. In the Eastern Mancha aquifer, the current Water Plan (2016–2021) establishes 260 Mm^3/year as recommended sustainable exploitation, allocates 320 Mm^3/year, and recognises the rights of 460 Mm^3/year. Instead of declaring the overexploitation of the aquifer, an attempt was made to meet these demands with water from the Júcar River, causing a deficit of 195 Mm^3, as recognised by the 2014 (art. 33) and 2016 (art. 25) Water Plans. Further south, in the Segura Basin, the overallocation of water resources was a consequence of the enormous expectations created by the Tajo-Segura Transfer, against the advice of RBA technicians (Sanchis-Ibor *et al.*, 2011). The registered agricultural water rights total a volume of 1603 Mm^3, while the resources annually available in the basin are, according to the RBA, 854 Mm^3. The Murcia Regional Government called for an 'amnesty' to legalise all these uses and, finally, the 2015 RBP legalised all the unregistered agricultural water uses developed before 2013, provided that they had some infrastructure available to obtain water from coastal desalination plants.

In other more recent cases, water planning has contributed to the overallocation of resources by overestimating the saving capacities of pressurised irrigation technologies. Although the real water savings generated by these techniques have been widely questioned by recent studies (Grafton *et al.*, 2018), the Guadalquivir RBA may assign up to 45% of the resources saved by the modernisation of irrigation systems to future expansions within the basin.

Overall, the RBAs have not regularly followed the legal principles, priorities and preferences for water allocation. The experiences of the southern river basins in the last few decades clearly show that new demands are processed due to 'political' pressures, particularly from the regional governments, even though there are neither available resources nor supply guarantees that justify the investments necessary to make the new use effective. New concessions have been granted without available water, indirectly altering the prior concessions without observing the requirements established in the water legislation. They also jeopardise the consideration of the environmental conservation as a restriction for the water system, as established by the WFD.

8.3.2 Groundwater overexploitation

The 1985 Water Law gave the RBAs enforcement authority to control groundwater use and impose sanctions for illegal wells and excessive abstractions, although this authority was implemented only to a limited extent due to a lack of monitoring and control (Closas *et al.*, 2017). According to the Water Law, the RBAs can also declare that a body of groundwater is at risk of not achieving a good quantitative or chemical status. In this event, the following measures will be carried out: (a) Within six months, the RBA will constitute a community of users in the absence of one, or will entrust its functions on a temporary basis to an entity representing the concurrent interests. (b) After consulting with the users' community, the Governing Board will approve an action programme for the recovery of the good state of the body of water within a maximum period of one year from the date the declaration has taken place. Although these periods have rarely been met, the approval of this action programme and the creation of the community of users have taken place in all the overexploited aquifers, with uneven impacts.

The action programme aims to order the allocation regime to achieve a rational exploitation of resources in order to bring about the good status of the groundwater bodies and to protect and improve the associated ecosystems. The authorities can reduce concessions and private rights to meet the sustainable extractable volume, blocking all new groundwater abstraction concessions. The action programme may establish the replacement of pre-existing individual intakes with community intakes and transform the individual titles with their inherent rights into a collective one, although this has only taken place voluntarily after the creation of collective institutions for irrigation management. It may also provide the contribution of external resources to the groundwater body and can define protection areas where water usage and other activities are restricted to certain conditions controlled by the administrations.

These measures have been successful in some cases, but clearly insufficient in others. The region of La Mancha provides very varied results. On the one hand, in the Eastern Mancha, groundwater pumping for agriculture has been efficiently controlled through annual remote sensing campaigns (Calera *et al.*, 2017; Castaño *et al.*, 2009). The water users' association of this aquifer (*Junta Central de Regantes de la Mancha Oriental*) has played a key role, cooperating with the RBA in order to stabilise the water tables. This has been an example of a collective action successfully imposing common interest over individual short-sighted decisions on groundwater management (López-Gunn & Rica, 2012; Sanz *et al.*, 2015).

On the other hand, in the Western Mancha, overexploitation began in the mid-1970s when new irrigation systems were introduced under the private regime of groundwater extractions prior to the 1986 Water Law, causing the degradation of the National Park of the Tablas de Daimiel wetland (López-Gunn *et al.*, 2012). During the last 30 years, the RBA has launched various programmes with different formulae (quotas in 1991, income compensations in 1993, rights purchase in 2008, direct inspections in 2020), but the aquifer levels have not yet recovered, and most of the Tablas de Daimiel wetland currently remains dry. This case is not an isolated example. Other valuable wetlands, such as the Doñana National Park, have also recently been affected by groundwater overexploitation as a result of the incapacity or the disinterest of both the RBA and the regional government to strictly control irrigation water use (Díaz-Paniagua & Aragones, 2015). In short, in Spain, the so-called 'silent revolution' of groundwater is more like an 'opaque revolution', which has caused irreparable damage to many ecosystems and has altered the balance of some river basins.

8.3.3 Environmental flows

The concept of environmental flow (EF) appeared for the first time in Spain in the Water Law of 1985, obliging all RBPs to consider the allocation of water resources for the conservation of the natural environment. Initially, each RBA used different criteria in the 1998 RBPs, mostly in a lax or arbitrary fashion. The approval of the European WFD in 2001 forced improvements to be made in the definition of these EFs, first through a modification of the Water Law passed in 2005 (Law 11/2005), and later through the 2008 approval of the Water Planning Instruction (WPI), which defines in detail how RBPs should be prepared.

The WPI explicitly established that EFs are not a use but rather a restriction prior to water use, and defined EFs as the flow that contributes to achieving good status or good ecological potential in rivers or transitional waters to maintain, at minimum, the aquatic life that may naturally inhabit the river, as well as the riverside vegetation. To establish these EFs, the WPI required the establishment of minimum flows, maximum flows, generating flows (ordinary and natural flood that conditions the morphology and structure of the channel and the river habitats) and rates of change (in order to avoid the negative effects of a sudden variation in river flow) in all basins. In addition, the WPI proposed the application of habitat modelling methods in representative sections based on hydraulic simulation, coupled with the use of habitat preference curves for some target species, allowing curves to be obtained that relate the potential useful habitat to the flow. To this end, since then, numerous applied research studies have been carried out in different RBAs (Magdaleno, 2017).

The application of EF under the WPI began in 2013 and has generated a tense debate between different actors. The latest RBP monitoring report, published in 2017 by MET, admitted that although 76.9% of the rivers had established minimum flows, only 9% had set maximum flows, 8.7% generating flows and 11.4% change rates. For some authors, the reality is different from the legislation. They believe that the EFs are not acting to restrict the uses of water, but rather that the demand for water is acting to restrict the EFs, which does not permit the recovery of the fluvial ecosystems. For this reason, they demand a stricter revision of the WPI (Mezger *et al.*, 2019).

Recently, this issue has sparked a political discussion at the national level. In 2019, two environmental NGOs obtained a favourable ruling from the Supreme Court that called for a review of the EFs of the Tagus Basin. This judicial decision alarmed the users of the Tagus-Segura water transfer, who depend on the availability of resources from the Tagus to irrigate 130 000 hectares in the southeast of the country. Finally, in the autumn of 2021, MET increased the EFs by 29% in a stretch of the Tagus River that is key to the transfer, which presumably will cause a reduction in the yearly water volume transferred. This has generated a bitter controversy between the central government, three regional governments, the agrarian lobby and the environmental NGOs.

8.3.4 Water exchanges

In general terms, water cannot be traded in Spain, since it is considered a public good, as established by the Water Law of 1986 and the Constitution. Water rights are strictly attached to land ownership. However, historically, there have been numerous local exceptions to these principles, at the user or collective levels, with different characteristics and institutional frameworks (De Stefano & Hernández-Mora, 2016). These exchanges have historically been tolerated, and in some cases promoted, by the regional governments and the basin authorities, by considering them as flexible local solutions, sustainable and well adapted to their respective contexts. However, this panoply of arrangements occurs with little transparency and without providing knowledge of fundamental aspects such as the volumes exchanged, the prices paid or the impacts on third parties or on the environment. This obviously hinders water governance at local or basin scale.

Since 1999, when the Water Law was reformed, there have been two legal channels for the temporary transfer of the rights to use water (not water itself, since it cannot be traded under its constitutional definition as a public good). These two new instruments, created to ease the allocation of resources and seek greater efficiency, introduced a change in the formal traditional principle of linking water to land (Embid, 2010). The first instrument has allowed the temporary transfer of water right through the Leasing of Water-use Rights (LWR, *Contratos de Cesión de Derechos de Uso de Agua*). LWR are contracts exclusively for temporary cessions, with the exception of non-consumption uses, to temporarily assign their rights between concessionaires or owners of water rights, except private water owners, who are not authorised to use this legal instrument. Water can only be transferred to users of a rank equal to or greater than the transferor, according to the aforementioned priority of uses established by the Water Law. The maximum transferred volumes may be those actually used by the transferor in the last five years and not those that appear in the concession, in order to avoid the transfer of water that had not really been used ('paper water'). The price of the water relinquished is agreed between the two parties, although the administration may impose a maximum value to avoid speculation. The contracts must be authorised by the RBA. The RBA is entitled to not authorise the contract if it negatively affects the basin's resources, third party rights or the environmental flows. Likewise, the RBA is allowed to exercise its overriding right of acquisition. Water transfers that occur between basins should be included in the National Hydrological Plan and were initially required to be authorised by the National government through the Council of Ministers.

The second legal instrument for water exchanges by the Basin Authorities, in agreement with the Council of Ministers, are the Centres for Water-use Rights Exchange (CWRE, *Centros de Intercambio de Derechos de Uso de Agua*). Known informally as water banks, they allow temporary or permanent exchanges of water use rights in cases where there are a series of circumstances, such as extraordinary droughts or declaration of overexploited aquifers. They are organised and managed by the RBAs, which first post the offers of public acquisition of water rights, describing the volume, prices, technical characteristics and duration (temporary or permanent). These centres can not only acquire rights to public water, but also to private water. Subsequently, the centres must carry out offers so that the interested users may acquire these rights. The centres supervise and approve the exchanges once the offers from the parties are received, so these water banks are fully controlled by the public sector.

Once these instruments had been put into practice, several Royal Decrees relaxed the conditions established by the law and widened the scope of these formal markets. These decrees allowed the granting of cession contracts to water users without concessionary rights, allowed the Ministry to authorise contracts without respecting the order of preference established by the Water Law and authorised the signing of cession contracts between users from different river basin demarcations. Later, the Law of Environmental Assessment (Law 21/2013) established the Leasing of Water Rights contracts to be approved at a lower administrative level: by the General Directorate of Water (MET) instead of the Council of Ministers.

The two instruments have been used almost exclusively in times of drought. Between 2001 and 2011, LWR had only exchanged 156.35 Mm^3 for intra-basin transfers and 235.39 Mm^3 for inter-basin transfers. The water banks (CWRE) have only functioned during the drought of 2005–2008, in which the volumes acquired by the administration were 198.34 Mm^3. More than 80% of this volume of transactions was acquired by basin organisations for environmental purposes, mainly to solve the impact of drought on rivers and wetlands. Even in the driest years in the basins, where more transactions have taken place, the total volume of water managed by both types of markets has been less than 5% of total water use (Palomo-Hierro & Gómez-Limón, 2013). Since 2011, very few exchanges have been made.

Water markets seem to hardly have room for growth in this socially and politically consolidated institutional resource allocation system. Nevertheless, they have generated public debate between those in favour of using these economic instruments to introduce more flexibility in the allocation of water resources (Gómez-Limón & Calatrava, 2016) and those who perceive them as a commodification of water that can enable speculation and the hoarding of water rights (Hernández-Mora & Del Moral, 2015). The two groups of authors together provide a balanced view of the Spanish markets, and coincide in highlighting the lack of transparency of these operations, the lack of competition in the markets and the limited development of these exchanges two decades after their legal regulation.

8.4 CONCLUSIONS

Water rights in Spain are assigned through a concessional system, and water allocation is managed throughout the hydrological year by the RBAs depending on water availability, social and environmental priorities and the system operating rules. It is, in short, an institutional approach to water allocation, under state control, but partially open to participation and negotiation with users. Negotiations are based on available information and are generally substantiated on technical criteria derived from decision support systems. These are mechanisms that, in general terms, have produced good results in recent water planning and drought management.

However, this water right allocation model has some weaknesses, mostly derived from the irrigation promotion policies developed throughout the 20th century, which have led to the overallocation of water rights and groundwater overexploitation in some basins, generating structural deficits, dramatic environmental impacts and significant social and territorial tensions. The fair and efficient allocation of resources is heavily burdened by the inherited unsustainable water management model, which in some regions has generated a culture of water demand being strongly rooted in social and political structures. That rivers do not reach the sea is seen as desirable and a source of pride in some regions and economic sectors in which the management model established by the WFD has not yet taken hold.

REFERENCES

Andreu J., Ferrer-Polo J., Perez M. A., Solera A. and Paredes-Arquiola J. (2013). Drought planning and management in the Júcar River Basin, Spain. In: Drought in Arid and Semi-Arid Regions, K. Schwabe, J. Albiac, J. D. Connor, R. M. Hassan and L. M. González (eds.), Springer, Dordrecht, The Netherlands, pp. 237–249.

Calera A., Garrido-Rubio J., Belmonte M., Arellano I., Fraile L., Campos I. and Osann A. (2017). Remote sensing-based water accounting to support governance for groundwater management for irrigation in La Mancha Oriental Aquifer, Spain. *WIT Transactions on Ecology and the Environment*, **220**, 119–126, https://doi.org/10.2495/WRM170121

Carmona M., Máñez M., Andreu J., Pulido-Velazquez M., Haro-Monteagudo D., Lopez-Nicolas A. and Cremades R. (2017). Assessing the effectiveness of multi-sector partnerships to manage droughts: the case of the Jucar River Basin. *Earth's Future*, **5**(7), 750–770, https://doi.org/10.1002/2017EF000545

Castaño S., Sanz D. and Gómez-Alday J. J. (2009). Methodology for quantifying groundwater abstractions for agriculture via remote sensing and GIS. *Water Resources Management*, **24**, 795–814, https://doi.org/10.1007/s11269-009-9473-7

Closas A., Molle F. and Hernández-Mora N. (2017). Sticks and carrots to manage groundwater over-abstraction in La Mancha Spain. *Agricultural Water Management*, **194**,113–124, https://doi.org/10.1016/j.agwat.2017.08.024

De Stefano L. and Hernández-Mora N. (2016). Los mercados informales de aguas en españa: una visión de conjunto (informal water markets in Spain: an overview). In: Los Mercados de Agua en España: Presente y Perspectivas, J. A. Gómez-Limón and J. Calatrava (eds.), Cajamar Caja Rural, Almería, pp. 95–121. English version available from: https://www.researchgate.net/publication/320739827_Informal_water_markets_in_Spain_an_overview (accessed 04 June 2021)

De Stefano L. and Lopez-Gunn E. (2012). Unauthorized groundwater use: institutional, social and ethical considerations. *Water Policy*, **14**, 147–160, https://doi.org/10.2166/wp.2012.101

Díaz-Paniagua C. and Aragones D. (2015). Permanent and temporary ponds in Doñana National Park (SW Spain) are threatened by desiccation, *Limnetica*, **34**(2), 407–424.

Embid A. (2010). The evolution of water law and policy in Spain. *Water Resources Development*, **18**(2), 261–283, https://doi.org/10.1080/07900620220135094

Fornés J. M., de la Hera A. and Llamas M. R. (2005). La propiedad de las aguas subterráneas en España: la situación del Registro y del Catálogo (The ownership of groundwater in Spain: the status of the Registry and Catalog). *Ingeniería del Agua*, **12**(2), 125–136, https://doi.org/10.4995/ia.2005.2556

Fornés J. M., López-Gunn E. and Villarroya F. (2021). Water in Spain: paradigm changes in water policy. *Hydrological Sciences Journal*, **66**(7), 1113–1123, https://doi.org/10.1080/02626667.2021.1918697

García-Mollá M., Sanchis-Ibor C., Avellà L., Albiac J., Isidoro D. and Lecina S. (2019). Spain. In: Irrigation in the Mediterranean: Technologies Institutions and Policies, F. Molle, C. Sanchis-Ibor and L. Avella (eds.), Springer, New York City, USA, pp. 89–121.

Grafton R. Q., Williams J., Perry C. J., Molle F., Ringler C., Steduto P. and Allen R. G. (2018). The paradox of irrigation efficiency. *Science*, **361**(6404), 748–750. https://doi.org/10.1126/science.aat9314

Gómez-Limón J. A. and Calatrava J. (coords.).(2016). Los mercados de agua en España: presente y perspectivas [Water markets in Spain: present and perspectives], Cajamar Caja Rural, Almería.

Hernández-Mora N. and De Stefano L. (2013). Los mercados informales de aguas en España: una primera aproximación (Informal water markets in Spain: a first approach). XVIII Jornadas de Derecho de Aguas, Zaragoza, Spain.

Llamas M. R. (1975). Non–economic motivations in groundwater use: hydro-schizophrenia. *Ground Water*, **13**(3), 296–300.

López-Gunn E. (2009). Agua para todos: A new regionalist hydraulic paradigm in Spain. *Water Alternatives*, **2(3)**, 370–394, https://www.water-alternatives.org/index.php/allabs/66-a2-3-5/file.

López-Gunn E. and Rica M. (2012). La participación activa de los usuarios: la co-gestión como forma de gobernanza del agua subterránea (active participation of users: co-management as a form of groundwater governance). In: E. López-Gunn and M. Rica (coord.) Gestión colectiva del agua subterránea en España, Madrid, Fundación Botín, pp. 15–36.

López-Gunn E., Dumont A. and Villarroya F. (2012). Tablas de daimiel national park and groundwater conflicts. In: Water Agriculture and the Environment in Spain: Can we Square the Circle? L. De Stefano and R. Llamas (eds.), Fundación Botín, Madrid, pp. 259–218.

Magdaleno F. (2017). Experimental floods: a new era for Spanish and Mediterranean rivers? *Environmental Science & Policy*, **75**(C), 10–18. https://doi.org/10.1016/j.envsci.2017.05.011

Mezger G., De Stefano L. and González del Tánago M. (2019). Assessing the establishment and implementation of environmental flows in Spain. *Environmental Management*, **64**, 721–735. https://doi.org/10.1007/s00 267-019-01222-2

Moreu J. L. (2002). Los problemas de la legislación sobre aguas subterráneas en españa: posibles soluciones (The problems of groundwater legislation in Spain: possible solutions). In: Régimen Jurídico de las Aguas Subterráneas, S. Del Saz, J. M. Fornés and M. R. Llamas (eds.), Fundación Marcelino Botín and Ediciones Mundi-Prensa, Madrid, Spain, pp. 1–46.

Palomo-Hierro S. and Gómez-Limón J. A. (2013). El papel de los mercados como instrumento para la reasignación del agua en España (The role of markets as an instrument for the reallocation of water in Spain). *Agua Y Territorio*, **2**, 78–92.

Sanchis-Ibor C., García-Mollá M., Avellà L. and Carles-Genovés J. (2011). Reaching the limits of water resources mobilization: irrigation development in the segura river basin Spain. *Water Alternatives*, **4**(3), 259–278.

Sanz D., Calera A., Castaño S. and Gómez-Alday J. (2015). Knowledge participation and transparency in groundwater management. *Water Policy*, **18**(1), 111–125. https://doi.org/102166/wp2015024

doi: 10.2166/9781789062786_0105

Chapter 9

Managing a common resource in agriculture: an overview of the French nested water allocation system

Josselin Rouillard[1,2] *and Jean-Daniel Rinaudo*[2,3]

[1]*Ecologic Institute, Pfalzburger Str. 43-44, 10717 Berlin, Germany*
[2]*G-EAU (UMR 183)*
[3]*BRGM, Université de Montpellier, Montpellier, France*

ABSTRACT

Water resources in France are the focus of much social and political attention with recurring conflicts between agriculture and environmental organisations. The last 30 years have seen a major transition from open to regulated access to water resources, which has required a deep transformation of the regulatory framework, the development of new planning procedures at different nested levels, the establishment of new organisations, the development of hydrological, hydrogeological and environmental knowledge and significant social change. Nowadays, the French allocation regime has distinct characteristics, giving priority in allocation to the environment and relying on permits that can be modified or cancelled by the State without compensation. A move towards co-management has nevertheless occurred where authorities, users and stakeholders jointly define allocation rules. Co-management is deployed at different nested levels, from river basin district to catchment level and agricultural user communities. In particular, the establishment of agricultural users' organisations (OUGCs) is an innovative attempt to organise reallocation without relying on market mechanisms. The French model still faces major challenges in the future, due to imperfect allocation institutions and increased scarcity and droughts driven by climate change which will further question allocations between water uses and the environment.

Keywords: Agro-food value chains, collective management, common pool resource, reallocations, water user organisations

9.1 INTRODUCTION

France is generally endowed with generous water resources and has not yet faced the extreme cases of water scarcity, droughts, and depletion of aquifers experienced in other countries. However, varying climates are present in the country, including a large Mediterranean ecoregion and dry summers across much of the country. With significant growth in water extractions in the second half of the 20th century, particularly from agriculture which represents 80% of net water use in the dry season, aquatic ecosystems and groundwater-dependent ecosystems have increasingly been impacted by abstraction pressures. In the future, France is likely to be exposed to longer and more severe droughts, as well as growing water scarcity in many regions. According to the study Explore 2070 (MEDD, 2015), average river flow could decline by 50 to 70% in most rivers by 2065, while groundwater recharge could reduce by 55%. In that context, current allocations will increasingly exceed future available water resources, worsening conflicts between agriculture, public water supply and the environment.

Prescriptions under three water laws (adopted in 1964, 1992 and 2006) provide French authorities with significant powers to regulate water use. Authorisations to withdraw water are made through a national permitting regime established in 1964 and, since 1992, users are required to meter their extractions and restrictions on water use applied during droughts. Another important feature of the French model formalised in the 2006 Water Law is its nested and user-based allocation system where users, under the supervision of the State, collectively negotiate who gets water, how much and when (for historical overviews, see Erdlenbruch *et al.* 2013; Rinaudo, 2020).

This chapter presents the legal and institutional framework regulating water allocation in France, including the changes made to the nature of water rights, and the system of collective licences controlling water use in agriculture. Additional operational dimensions of the allocation system are also presented, including approaches taken to define environmental limits and allocation caps, rules for allocating water between agricultural users and rules reducing those allocations to sustainable levels, and mechanisms for monitoring and enforcement. The interaction of the allocation regime with other existing instruments is also explored. The chapter concludes with the key features of the French approach and its overall performance.

9.2 LEGAL AND POLICY BACKGROUND

9.2.1 The nature of water rights

Although the main rivers have been under royal control since the middle ages, non-navigable rivers and 'closed' water (i.e. groundwater, ponds dug by the landowner, springs generating minor flows, and stored rain) were left to landowners' appropriation as long as the water stayed on their land. Access to those resources was progressively restricted, starting with the Civil Code of 1804 and culminating with the 1992 Water Law which stated, in its first article, that all surface water and groundwater is the 'common patrimony of the Nation'. As a result, France now recognises water as a common good, subject theoretically only to rights of use. This provides a strong legal basis for greater oversight by the State over water use (Conseil Etat, 2010), first through a nested system of water management planning instruments and second, through a comprehensive registration and permitting regime.

9.2.2 Water management planning instruments

The current French water allocation regime operates within a nested system of targets and rules, established from the European to the catchment and aquifer levels. Under the European Union Water Framework Directive (WFD), adopted in 2000, Member States were requested to achieve a set of ecological, chemical, and quantitative status conditions in all their surface water and groundwater bodies. To achieve these targets, River Basin Management Plans (RBMPs) are required to be adopted at river basin district levels, which, in mainland France, follow the boundaries of the surface hydrological basins of the six largest rivers.

The preparation of these six RBMPs (SDAGEs in French) is delegated to River Basin Committees (RBCs), representing state administration, regional and local governments, users and civil society. Water agencies support RBCs by preparing and implementing RBMPs and their associated programme of measures, in line with the strategy defined by the RBCs. They are also in charge of collecting and redistributing financial resources in the form of water extraction and pollution fees and subsidies to fund measures in the plans. However, they are not in charge of the operational management of water infrastructure, nor are they in charge of delivering permits to extract water. Water agencies are public administrations linked to the Ministry of Environment and Sustainable Development.

An additional planning level exists in France, at the sub-basin level, organised along hydrological boundaries such as catchments and regional aquifers. A local water commission (LWC) composed of local councils, state administration, and user and civil society representatives is in charge of developing the sub-basin plans (called SAGEs). They refine objectives, strategies and measures laid out in SDAGEs (Rinaudo *et al.*, 2020). As for SDAGEs, these plans are legally binding and the State must ensure coherence of other public policies with the requirements of the SAGE.

9.2.3 The permitting regime

In France, access to water is regulated by several permits: (i) an 'infrastructure' permit, for example for drilling a well, constructing a weir in the river or installing a pumping station; (ii) a pumping permit that defines the capacity of the extraction equipment; and (iii) an 'extraction' permit to withdraw specified quantities of water.

These permits are issued either through a registration process or an authorisation procedure depending on the pumping capacity of the installed equipment and its location inside or outside a 'restricted area'. Restricted areas are catchments that have an imbalance between supply and demand (in France, 'Zone de Répartition des Eaux', hereafter ZRE) (see Table 9.1). The difference between the registration and authorisation procedures only relates to administrative requirements (e.g. whether an environmental impact assessment should be provided with the application). Outside restricted areas, permits are very easy to obtain, provided the projected pumping capacity does not represent a threat to third parties or aquatic ecosystems. In restricted areas, an impact assessment study is required. In both cases, the State can refuse any application it deems incompatible with water management rules.

Pumping and extraction permits are required to be compatible with the SDAGE and SAGE, including specific pumping requirements based on their potential impacts on ecologically sensitive areas, drinking water protected areas, and restricted areas. Users are required to install meters and keep a record of their withdrawals. Permits specify not only a maximum permissible flow rate, but also an annual volume (sometimes expressed in monthly or seasonal time steps) that cannot be exceeded.

The extraction permit is simultaneously tied to the abstraction point and a user. Thus, it cannot be transferred to another abstraction point owned by the same user. The extraction permit can be transferred to a new user in the event of a change of ownership of the abstraction point. However, this is not automatic, and the State can oppose the transfer.

Although infrastructure and pumping permits have no validity limit, they can be modified or cancelled without compensation by the State if it is demonstrated that extraction causes significant

Table 9.1 Declaration and permitting requirements.

	Significance of extraction (volume based on pumping capacity use over one year)	Administrative procedure for pumping permit	Administrative procedure for abstraction permit
Outside water restriction areas (ZRE)	Annual extraction <1000 m^3/yr (domestic use)	Local council declaration	Not applicable
	Annual extraction between 1000 m^3/yr and 10,000 m^3/yr	Declaration to state	Not applicable
	Annual extraction between 10,000 m^3/yr and 200,000 m^3/yr	Declaration to state	Declaration to state
	Annual extraction >200,000 m^3/yr	Application for authorisation (state)	Application for authorisation (state)
In restricted areas ZRE	Pumping capacity exceeding 8 m^3/h	Same procedure as for outside the ZRE	Application for authorisation (state)

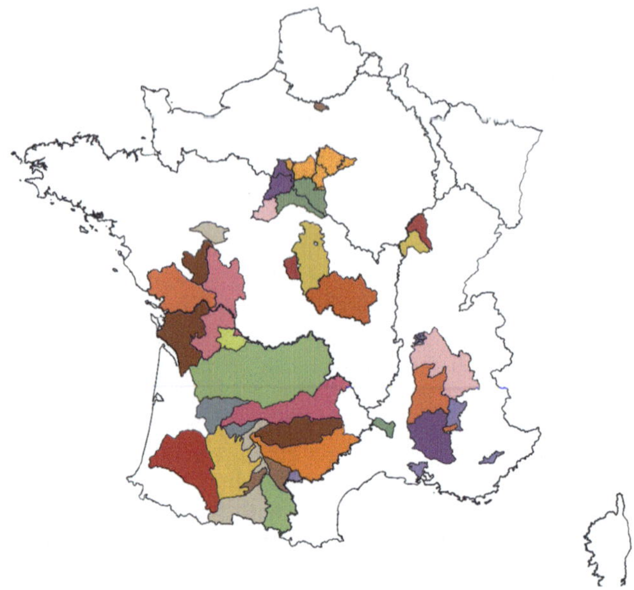

Figure 9.1 Map of river basins (black lines) and OUGCs (coloured units) in France. Data: DREAL in 2018.

environmental impact. Extraction permits are usually renewed annually through a very simple procedure, but the State keeps a possibility to reduce allocated volumes if needed.

Since the 2006 Water Law, a different procedure has applied to agricultural extraction permits withdrawing from 'restricted' catchments and aquifers (ZRE). In this case, all agricultural irrigation extractions (previously under individual extraction permits) are regrouped into a single collective permit valid for up to 15 years. The permit can be revised following revisions of the SDAGE or SAGE, where new hydro-climatic datasets, modelling updates, a better understanding of local conditions, or evolving climatic or socio-economic conditions require a change in the allocated volume.

The permit is issued to a 'Single Collective Management Organisation' ('Organisme Unique de Gestion Collective', hereafter OUGC) created for the purpose of allocating the bulk volume amongst its members (Lafitte *et al.*, 2006). There were 49 OUGCs across France in 2020 (Cinotti *et al.*, 2020, see Figure 9.1). The OUGC is a major institutional innovation established by the 2006 Water Law to facilitate the regulation of agricultural abstraction. The setting of the extraction cap in restricted areas is discussed in Section 9.4 and the functioning of the OUGC in Section 9.5.

9.3 ENVIRONMENTAL BOTTOM LINES

Broadly, three types of requirements have been established to ensure environmental protections for French rivers and aquifers:

- Minimum flows set below each major infrastructure impairing the natural flow of rivers or diverting water
- Target flows and target aquifer levels for major river nodes defined in SDAGEs or SAGEs, and aiming at meeting good quantitative, qualitative and ecological conditions
- Alert and emergency flows and aquifer levels set to restrict water use in the event of a drought

Each requirement is examined below.

9.3.1 Minimum flows

The regulatory framework requires the setting of minimum biological flows guaranteeing the life, reproduction and circulation of water species downstream of infrastructures affecting river flow. These minimum biological flows are servitudes on the operators of the infrastructures, and are gradually adopted as they only apply to new authorisations, renewal of existing authorisations, or existing ones upon request of the State. They are established based on studies focused on local hydrological statistics and considering the linkages between hydraulic and ecological conditions (RF, 2011). In all cases, minimum flows cannot be set below 1/10 of average natural annual flow, or 1/20 for rivers with an average natural annual flow above 80 m^3/s. The average flow rate should be based on all the years for which data are available, with a strict minimum of 5 years, and should recreate an estimated natural flow removing the impact of extraction, discharges and water transfers. The 1/20 also applies as a minimum servitude for infrastructure used to produce peak-time electric production.

9.3.2 Management targets

An additional set of river flow targets are established for major river nodes in SDAGEs and SAGEs. They represent objectives guiding operational management decisions. Called 'target low flows' (in French, 'Débit Objectif d'Etiage' or DOE), they represent the monthly average flow above which authorities consider that downstream water demand can be satisfied without impacting good ecological status under the WFD. Similar to minimum flows (see above), target low flows must be set so they guarantee the life, reproduction and circulation of water species. Target flows can vary between seasons.

Target low flows are set in a nested manner, at the most downstream point of each hydrological sub-unit of the river basin, that is individual catchments, sub-catchments and other management units. Target groundwater piezometric levels are also set for aquifers connected to surface water bodies, to avoid a drop in aquifer levels impairing the achievement of minimum biological flows. The flow targets are considered achieved if it is observed a posteriori that the lowest 10-days average flow (or aquifer level) was maintained above 80% of its value. Flow targets must be met on average 8 years out of every 10. These target low flows are used to calculate the sustainable extraction cap (see below).

9.3.3 Alert and crisis flows

In addition to minimum biological flows and target low flows, French water actors use 'alert' and 'crisis' flows (i.e. Débit d'Alerte and Débit de Crise) below which restrictions on water extractions and uses apply so that essential water uses and the environment are prioritised in the event of droughts:

- 'Alert' level is the average daily flow that indicates that water demand for all water uses downstream may not be met without impacting the aquatic environment. First restrictions on non-priority uses apply.
- 'Crisis' low flow is the average daily flow below which top-priority uses (e.g. essential drinking water provision for humans and animals, and good functioning of freshwater species) are endangered. Non-priority uses are not allowed for the extraction of water.

A 'vigilance' level is also set before the 'alert' level and a 'reinforced alert' level is set before the 'crisis' level in order to smooth the implementation of the alert level (some restrictions) to a crisis situation (full restrictions).

Specific restrictions on water uses apply at each level. An equivalent system based on groundwater levels applies to unconfined aquifers. These targets are set considering the interaction between surface and groundwater, based on studies conducted during the planning process (SDAGE or SAGE).

9.4 DEFINING THE ALLOCABLE RESOURCE POOL

Since the 2006 Water Act, Sustainable Extraction Limits (SELs) must be imposed in restricted catchments and aquifers (i.e. ZRE). SELs are defined in local management plans (SAGEs) by members of the Local Water Commission (Section 9.2.2), based on hydrological or hydrogeological studies. SEL studies are directed by a steering committee appointed by the LWC and include all stakeholders potentially affected by the SEL.

The SEL is legally defined as the volume of water that can be extracted without impairing the environmental objectives of the European Water Framework Directive, that is good ecological, chemical and quantitative status of water bodies. Operationally, the SEL is set to ensure that the low flow targets adopted in the basin plans (Section 9.3) can be met in 8 years out of 10. If the basin is fully allocated within the SEL, allocations will still need to be curtailed on average 2 years out of 10 (drought years). In other words, the SEL is the quantity of water that can be withdrawn with 80% reliability.

The SEL takes the form of a volumetric cap on the water that can be extracted annually from restricted catchments and aquifers, usually sub-divided seasonally (winter/summer), and sometimes monthly and even weekly during intensive use periods. The SEL is specified in volume for each management unit of a catchment and aquifer. Management units can be sub-catchments, parts of a sub-catchment or different aquifers (connected or not connected to surface water).

The SEL is based on an assessment of the catchment and aquifer water balance. Methodologies for assessing the SEL vary greatly, ranging from simple statistical analysis to sophisticated integrated surface-groundwater models. Reporting on an analysis of a sample of those studies, Arnaud (2020) warns that the setting of SELs in France is not always scientifically robust and that SELs have sometimes been overestimated due to various reasons, from insufficient knowledge to political pressure from water users. However, he shows that SELs have often been 10 to 20% below current water use levels, and up to 50% in some cases, posing major challenges to authorities charged with reducing allocations to water users.

Once the SEL is set, the LWC is in charge of allocating water to each main use of the sub-basin, and the OUGC is in charge of allocating water between individual agricultural users. The rules directing the sharing of the SEL are discussed in the next section.

9.5 ALLOCATION RULES

In catchments characterised by water scarcity, water is now allocated using a volumetric management approach, complemented by a system of use restrictions during drought years. These two allocation mechanisms, which were progressively implemented between 1992 and 2010, are presented below, followed by a specific focus on allocations for agriculture within the OUGC.

9.5.1 Volumetric allocations between sectors

Once the SEL is set, the State and stakeholders (in LWCs) negotiate over allocations between sectors (drinking water supply, industry, agriculture). In France, public water supply is given priority over other economic sectors and receives an allocation corresponding to the users' needs. These needs are estimated by taking into account performance standards in terms of water losses and efficiency. The remaining volume of water is shared between industry and agriculture. Industrial water use is usually given priority over agricultural water use, although it depends on the type of industry and the efficiency of use in each industry. Therefore, much of the reduction needed to balance supply and demand in any given catchment or aquifer is usually imposed on agricultural uses.

It is worth noting that, given the use priority system in place, high-priority users have little incentive to limit their claims and the room for negotiation with low-priority users might seem very limited. However, since stakeholders live in the same territory and may have other interests in common and other issues to negotiate, they may be open to compromise.

Once defined, the volumetric allocation granted to agriculture is provided as a bulk permit to the OUGC. According to the 2006 Water Law, the OUGC is responsible for allocating water between irrigators in their operating areas within the limit set by the bulk permit. The OUGC is charged with defining which irrigators are authorised to access water, how much and when.

9.5.2 Allocating and reallocating water in agriculture

The 2006 Water Laws leave considerable leeway to agricultural user organisations (OUGCs) to define their own internal allocation rules. OUGCs took advantage of this and crafted a variety of rules to allocate water between existing users and for including prospective irrigators (Rouillard & Rinaudo, 2020). They are examined in turn below.

9.5.2.1 Allocation between existing claimants

In most cases, OUGCs have used some form of grandfathering, allocating the bulk volume proportionally to past withdrawals (e.g. an average or maximum use over a reference period) or proportionally to the authorised flow rate specified in the original individual pumping permit (especially where accurate data on volumes extracted in previous years by each farmer were not available). This establishes a stable allocation for each irrigator, providing greater security for existing claimants and protecting the value of irrigated land.

However, this has resulted in many cases of overallocation, and OUGCs have had to find strategies to ramp down individual allocations to align with the SEL. To smoothen the process of reducing individual allocations, the first bulk permit usually allows the OUGC to allocate more water than their share of the SEL during the first years. The OUGC is then required to ramp down allocations to the SEL for agriculture over a period of generally 3 to 5 years.

Rouillard and Rinaudo (2020) observed three strategies for ramping down individual allocations so that cumulatively they do not exceed the SEL for agriculture after the permitted period:

- Application of the commonly known 'use it or lose it' approach, whereby an individual allocation is reduced if the reported withdrawals do not match the requested volume. Most OUGCs currently follow this approach despite its potential for disincentivising efficient water use.
- Application of a uniform reduction on all allocations, often with a lower ceiling to protect small allocations.
- Maintaining individual allocations at their initial levels and applying an annual reduction coefficient that reflects resource availability.

Overall, many OUGCs have so far managed to reduce allocations, but few have managed to significantly reduce water extractions. The ramp-down rules implemented mainly eliminated dormant allocations, that is volumes that were systematically allocated but generally not used.

Irrigators are not allowed to transfer volumes between themselves without authorisation from the OUGC. However, OUGCs have established procedures for temporarily reallocating water between users. During the irrigation season, when the OUGC realises one or several irrigators will not fully use their allocation, it may decide to transfer it to other users. This transfer is strictly managed by the OUGC, according to internal rules, in order to avoid the emergence of informal water markets. The internal rule can for instance state that unused volume will be reallocated in priority to cattle breeders or to small farmers. Such rules are approved by OUGC members during plenary assemblies.

9.5.2.2 Incorporating prospective claimants

Farmers who were not previously irrigating had very limited opportunities to access the resource, at least in the first years, as OUGCs have usually grandfathered historical uses. To address this issue, OUGCs have crafted very different rules:

- A first and common approach has been to systematically reject all new requests. In these cases, prospective irrigators can only obtain irrigation water by buying currently irrigated land and obtaining approval for the transfer of the allocation from the OUGC. No legal ground linking allocations with property rights exists in France. However, authorities and OUGCs have so far permitted their simultaneous transfer during land transactions.
- A second approach consists of partially satisfying all new requests, even in fully allocated catchments and aquifers, which implies a reduction of the share allocated to historical users.

- A third approach has been to build an allocation 'reserve' by retaining a share (up to 20%) of the allocation being transferred through land transactions. This reserve is progressively used to endow new irrigators to become permanent claimants.

To select amongst prospective users, OUGCs have developed a 'waiting list' where claimants are ranked according to various criteria which are OUGC specific. Priority can be given to young farmers, to those requesting small allocations (equity), to economically vulnerable farms (social objective), to those growing high-added-value crops or generating employment in the value chain (economic objective), or to efficient irrigation techniques or farmers with ecologically friendly practices (e.g. organic farming).

9.5.3 Drought restrictions

Drought restrictions apply when river flows and aquifer levels reach alert and crisis levels. In effect, drought restrictions are applied frequently in France due to stricter environmental requirements under the WFD, more frequent meteorological droughts and still widespread cases of fully and overallocated river and groundwater basins. Restrictions have been applied almost every year in certain western catchments.

Restrictions apply to users independently of prior appropriation – a notion which does not exist in the French legal framework. Instead, water uses are grouped into pre-defined priority classes: a higher class water use (including drinking water supply and cooling of nuclear power plants) prevails over environmental needs, which, in turn, prevail over sectoral needs such as agriculture or energy. The priority classes are enshrined in the 1992 Water Law.

As a general rule, the following restriction levels are progressively implemented when drought conditions are present:

- *Public water supply:* limitation (and ultimately prohibition) of watering gardens, lawns and green spaces, car washing, and pool filling
- *Agriculture:* prohibition to irrigate 1 day per week, several days per week or at certain times, and ultimately total ban of irrigation
- *Industry:* specific measures imposed on the most water-intensive units requiring a gradual reduction of activity

The State oversees the enforcement of restrictions, but stakeholders have the opportunity to negotiate the phasing of restrictions in Drought Management Committees (DMCs). DMCs are composed of the same actors as the LWCs (i.e. the State together with representatives of agriculture, industry, environment, fishing, and councils).

9.6 COMPLIANCE AND ENFORCEMENT

Despite the partial devolution of allocation responsibilities to LWCs and OUGCs, the State remains responsible for checking compliance of individual agricultural users with the requirements of their infrastructure and pumping permits as well as compliance with the individual allocation granted by OUGCs. OUGCs have a legal duty to collect information on water use from their members and to prepare an annual report comparing extracted volumes with allocations at individual levels – theoretically identifying offenders. However, they have no inspection powers to enter private properties and monitor volumetric meters, implying they have to rely on declarative data.

Notwithstanding this limitation, OUGCs have significantly improved the knowledge of extraction points and volumes withdrawn, which were historically poorly monitored by the State (Montginoul *et al.*, 2020). Farmers were indeed more willing to share information with an organisation that serves their interests and over which they have some control, than with a government agency they did not trust. This willingness increased as farmers realised that OUGCs were able to optimise water allocation, with obvious benefits for themselves, provided they had accurate information of extraction points and volumes.

The State also remains in charge of enforcing fines when individual irrigators do not comply with their allocation. However, some OUGCs implemented sanctions on farmers failing to report water use information since these prerogatives were clearly laid down as their responsibility in the 2006 Water Law. Others apply volumetric penalties (i.e. a reduction of next year's allocation) to farmers exceeding their annual allocation. OUGCs are likely to move in that direction in the coming years, progressively playing a greater role in enforcement and compliance, (see Cinotti *et al.* 2020), but this will require further legal clarification.

9.7 THE BROADER POLICY INSTRUMENT MIX

From the mid-1990s onwards, the progressive tightening of water allocation procedures has posed major challenges to public utilities, industries and farms, who had already invested significantly to achieve more efficient water use. Much of the challenge in reducing total allocations to SEL levels has so far lied with the agricultural sector as the current system prioritises drinking water, nuclear electricity production and industrial water needs. Several studies have highlighted that additional reductions of agricultural allocation, depending on how they are implemented, could significantly impact farm businesses and agro-food value chains (Danel, 2011; Lejars *et al.* 2012).

Remarkable progress has been achieved by the agricultural sector to adapt to the new constraints. For example, water productivity has increased by 30% on cereal crops in 20 years, thanks to reductions in water loss conveyance, improved irrigation piloting, genetic selection and better rotational choices. Farmers have also worked on increasing their security of supply by investing in small-scale reservoirs to store winter flows for use during the summer period, when most restrictions on direct pumping in rivers and shallow groundwater apply (Douez *et al.*, 2020).

Investments in water storage have proved particularly conflictual, leading to violent protests with environmental non-governmental organisations (NGOs) and, in many cases, ending in courts (Rouillard, 2020). To manage these conflicts, the State is currently establishing a new procedure (called Projet de Territoire de Gestion de l'Eau, i.e. PTGE) which brings together stakeholders to jointly define, through negotiation, the conditions for developing storage reservoirs. According to ministerial decrees (RF, 2015, 2019), the PTGE must aim for a balanced quantitative management of water resources, considering the impacts of climate change and taking into account the chemical and ecological status of water bodies, notably by promoting the development of agro-ecological systems and crop diversification to reduce pollution pressures.

The PTGE must also feature a variety of measures applied to all sectors (agriculture, drinking water, and industry) (Bisch *et al.*, 2018). For agriculture, measures should include not only increased water use efficiency and infrastructure modernisation, but also changes in crop production and rotation to reduce water demand and the development of agro-food value chains that incentivise the right production patterns according to local water resources. Overall water savings should take priority before water storage, transfers or reuse are considered. Where water storage is built, it should be in line with and contribute to broader rural development goals.

9.8 CONCLUSION

The development of the French water allocation regime described in this chapter has taken more than three decades. Transitioning out of open access to regulated water use has required a deep transformation of the regulatory framework, the development of new planning procedures at different nested levels, the establishment of new organisations, the development of hydrological, hydrogeological and environmental knowledge and significant social change.

In comparison to allocation policies deployed in other countries, the French approach presents the following specific features:

- The environment is clearly given first priority in allocation: water that can be allocated to consumptive use corresponds to resources available after environmental flows are met. SELs are, at least in theory, set solely based on ecological considerations, without accounting for social and economic dimensions.
- It relies on a system of permits that can be modified or cancelled by the State without compensation. Furthermore, these permits cannot be exchanged directly by individual users.
- The allocation regime relies on a co-management system, where the State and representative of users jointly define the rules, based on partial negotiation. Co-management is deployed at different nested levels, from river basin district to catchment level and agricultural user communities. The authority over allocation is hence shared by many actors at different governance levels, matching the definition of a polycentric model.
- The establishment of agricultural users organisations (OUGCs) is an innovative attempt to organise reallocation without relying on market mechanisms. It aims to reconcile economic efficiency, social justice and environmental sustainability, while favouring the development of solutions adapted to local conditions – hence more likely to be accepted and enforced.

The economic and social impact of this ongoing reform remains difficult to assess, considering the little temporal hindsight we have and the lack of data. A number of factors could turn the reform away from success:

- In some basins, there might be strong resistance to implementing reallocation rules within OUGCs, in particular where historical users endowed with large individual allocations control OUGC boards. There, grandfathering might become the rule and water reallocation be entirely linked to land transactions, water use rights being de facto sold with the land.
- More generally, providing OUGCs considerable leeway to develop allocation rules also creates a risk that influential farmers serving at the board could favour their own interests. There is a need for higher authorities to establish appropriate safeguards to keep rent-seeking behaviours under control.
- Collective management of water allocation can only be successful if users clearly see this approach as beneficial. OUGCs will thus need to develop a range of services, along with allocation, that can compensate for the losses incurred by some users.
- The distribution of enforcement responsibilities needs to be clarified between the State and OUGC, so that the latter can progressively play a greater role in monitoring, controlling and sanctioning. This also requires a shift in technology, with systematic use of telemetry (e.g. smart meters), remote sensing and other ICTs (smartphone applications, etc.).

REFERENCES

Arnaud L. (2020). Conceptual approaches, methods and models used to assess abstraction limits for unconfined aquifers in France. In: Sustainable Groundwater Management, Global Issues in Water Policy, J. D. Rinaudo, C. Holley, S. Barnett and M. Montginoul (eds.), Springer, Cham, Vol. **24**, pp. 211–227.

Bisch P. E., Hubert L., Mailleau C., Denier-Pasquier F. and Servant L. (2018). Cellule d'expertise relative à la gestion quantitative de l'eau pour faire face aux épisodes de sècheresses [Expert group on quantitative water management to face droughts], Ministère de la Transition Ecologique et Solidaire et Ministère de l'Agriculture et de l'Alimentation, Paris, France, p. 128.

Cinotti B., Galtier B. and Granger Y. (2020). Bilan du dispositif des organismes uniques de gestion collective (OUGC) des prélèvements d'eau pour l'irrigation [Expert report assessing the OUGC system], Mission CGEDD n°13017-01-CGAAER n°19089, 88 p.

Conseil Etat (2010). L'eau et son droit [Water and its right]. Rapport annuel du Conseil d'état, La documentation francaise (ed.), p. 584.

Danel J. B. (2011). Conséquences sur les filières agricoles et agroalimentaires de l'atteinte des objectifs quantitatifs de la Directive cadre sur l'eau et du SDAGE dans le bassin Adour-Garonne. [Consequences on agricultural and agrofood value chains of reaching quantitative water targets under the EU WFD and RBMP Adour Garonne]. Conseil général de l'alimentation, agriculture et des espaces ruraux, Paris, France, p. 38.

Douez O., Du Peuty J. E., Leperq D. and Montginoul M. (2020). Developing substitution resources as compensation for reduced groundwater entitlements: the case of the poitou marshes. In: Sustainable Groundwater Management, Global Issues in Water Policy, J. D. Rinaudo, C. Holley, S. Barnett and M. Montginoul (eds.), Springer, Cham, Vol. **24**, pp. 33–353.

Erdlenbruch K., Loubier S., Montginoul M., Morardet S. and Lefebvre M. (2013). La gestion du manque d'eau structurel et des sécheresses en France [Managing water scarcity and droughts in France]. *Sciences Eaux et Territoires*, **11**, 78–85, https://doi.org/10.3917/set.011.0078

Lafitte J. J., Devos P. and Portet P. (2006). Les Organismes Unique de Gestion Collective, Rapport no006313-01 n°006313-01, CGAER-CGEDD, Paris.

Lejars C., Fusillier J. L., Bouarfa S., Brunel G., Rucheton G., Girard X. and Golaz, F. (2012). Impacts de restrictions en eau d'irrigation sur les exploitations et les filières agricoles en Beauce [Impact of water restrictions on farms and agricultural value chains in Beauce]. *Agronomie, Environnement et Sociétés*, **2**(2), 139–154.

MEDD (Ministère de l'Ecologie, du Développement Durable et de l'Energie) (2015). Synthèse du projet Explore 2070 [Synthesis of project Explore 2070], Ed. Ministère de l'Ecologie, du Développement Durable et de l'Energie.

Montginoul M., Rinaudo J. D. and Alcouffe C. (2020). Compliance and enforcement: the achille heel of French water policy. In: Sustainable Groundwater Management, Global Issues in Water Policy, J. D. Rinaudo, C. Holley, S. Barnett and M. Montginoul (eds.), Springer, Cham, Vol. **24**, pp. 435–459.

RF (2011). Circulaire du 5 juillet 2011 relative à l'application de l'Article L214-18 du code de l'environnement sur les debits réservés à maintenir en cours d'eau. [Circulaire relative to the reserved flows to be maintained in rivers].

RF (République Française) (2015). Instruction du Gouvernement du 4 juin 2015 relative au financement par les agences de l'eau des retenues de substitution [Instruction of the Government relative to the financing by water agencies of compensatory storage].

RF (République Française) (2019). Instruction du Gouvernement du 7 mai 2019 relative au projet de territoire pour la gestion de l'eau [Instruction of the Government on the territorial project for managing water].

Rinaudo J. D. (2020). Groundwater policy in France: from private to collective management. In: Sustainable Groundwater Management, Global Issues in Water Policy, J. D. Rinaudo, C. Holley, S. Barnett and M. Montginoul (eds.), Springer, Cham, Vol. **24**, pp. 47–65.

Rinaudo J. D., Marchet P. and Billault P. (2020). Groundwater management planning at the river basin district level: comparative analysis of the Adour-Garonne and Loire-Bretagne river basins. In: Sustainable Groundwater Management, J. D. Rinaudo, C. Holley, M. Montginoul and S. Barnett (eds.), Springer, Cham, pp. 67–91.

Rouillard J. (2020). Tracing the Impact of Agricultural Policies on Irrigation Water Demand and Groundwater Extraction in France. In: Sustainable Groundwater Management: a Comparative Analysis of French and Australian Policy and Implications to Other Countries, J. D. Rinaudo, C. Holley, M. Montginoul and S. Barnett (eds.), Springer, Cham, Switzerland, pp. 461–479.

Rouillard J. and Rinaudo J. D. (2020). From state to user-based water allocations: An empirical analysis of institutions developed by agricultural user associations in France. *Agricultural Water Management*, **239**, 106269. https://doi.org/10.1016/j.agwat.2020.106269

doi: 10.2166/9781789062786_0117

Chapter 10

Turkey's water allocation regime under institutional change

M. Yasir Ak[1], Burcin Demirbilek[2] and David Benson[1]

[1]University of Exeter, Department of Politics and International Relations, Penryn, Cornwall, UK
[2]Faculty of Economics and Administrative Sciences, Department of Political Science and Public Administration, Çankırı Karatekin University, Çankırı, Turkey

ABSTRACT

Due to multiple drivers many countries are experiencing significant water-related risks, particularly to agriculture, making sustainable management of water resources critical for national development. One of the most susceptible countries to these risks is Turkey, where over-abstraction of surface and groundwater resources has occurred in agricultural regions, leading to severe environmental, social and economic impacts. In response to such risks, national water governance is currently undergoing a significant transformation through the implementation of river basin planning in the form of the EU Water Framework Directive alongside institutional innovations for water allocation. In this chapter, we therefore illustrate how this institutional change is occurring and then assess the effectiveness of this new, evolving water allocation regime in the agricultural sector in terms of its sustainability. By examining institutional change in two specific river basin case studies, Konya Closed Basin and Küçük Menderes, this chapter shows that it is not resulting in sustainable use of water resources.

Keywords: Institutional innovation, river basin planning, Turkey, water allocation

10.1 INTRODUCTION

Turkey is experiencing water-related risks, particularly in agricultural regions where over-abstraction of water resources is impacting the environment and food production (Ak *et al.*, 2019). Turkey is located in a semi-arid climate zone and faces challenges in improving water quality, increasing the amount of usable water and ensuring the sustainability of water protection and usage. The annual average annual precipitation in Turkey was 574 mm between 1981 and 2017 (Ministry of Agriculture and Forestry, 2019). Annual water consumption reached 54 billion (10^9) m^3 in 2018 (Ministry of Development, 2018). Of this water, 40.0 billion m^3 (74%) is used for irrigation, 7 billion m^3 (13%) for drinking and 7 billion m^3 (13%) for industrial water needs (ibid.). These consumption rates are unsustainable due to increasing demand. For example, the annual amount of usable water per capita in Turkey is currently around 1519 m^3 but due to population growth and climate change it is expected that by 2040, it will be approximately 1120 m^3 (Körbalta, 2019) – significantly below current levels. Agricultural sectors may also face extreme water deficits.

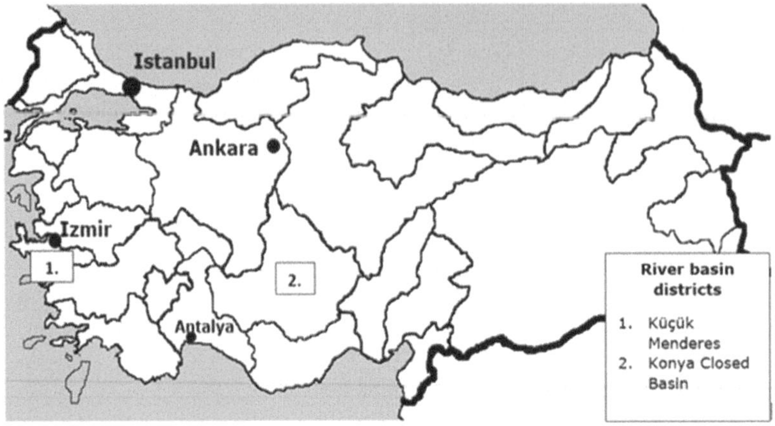

Figure 10.1 River Basin Districts in Turkey, including the Konya Closed Basin and Küçük Menderes Basin.

In response to such water security risks, Turkey is transforming its water allocation regime through institutional innovation. Initially adopted as part of Turkey's European Union (EU) accession process, the country is now implementing the Water Framework Directive, leading to the establishment of new institutional structures and processes for river basin management planning (Demirbilek & Benson, 2019). Pre-existing institutions – both organizational and legal – for water allocation, some dating back to the 1960s, are now being partially integrated within this emerging framework (ibid.). Under the 2014 National River Basin Strategy, the government General Directorate of Water Management (GDWM (SYGM being the Turkish acronym)) is authorized to determine sectoral water allocations in conjunction with river basin management plans (Ak *et al.*, 2019). However, 'institutional incoherence' (Benson & Lorenzoni, 2017) with other government agencies, river basin planning processes and water allocation in basins is an evident problem, raising questions concerning the effectiveness of this regime. The implications for local-scale sustainability in water-scarce agricultural areas are of particular concern.

This chapter aims at understanding the effectiveness of Turkey's emergent water allocation regime in the agricultural sector in terms of its overall sustainability. Here, we adopt a conception of sustainability that includes environmental, economic and social dimensions (Baker, 2016). Initially the chapter will provide an overview of the broad institutional framework of Turkish water management, covering the legal and policy background, water rights and access to water, plus declaration and permitting requirements. Technical-administrative aspects of the regime, including defining environmental objectives, allocation rules and monitoring and compliance, are then described. Related policy instruments are discussed regarding drought measures, economic instruments and provision of public information. Throughout the chapter, evidence from two river basins, the Konya Closed Basin[1] and Küçük Menderes[2], are used to illustrate the current performance of the allocation system.

[1]The Konya Closed Basin is one of 25 river basins in Turkey (see Figure 10.1) and is located in the central Anatolia region. The basin is often called the 'breadbasket' of Turkey due to the dominance of agricultural land use. Agriculture also provides the main income in this river basin (Ribamap, 2018). However, such activity is heavily dependent on groundwater due to its location in a semi-arid climate zone (Ribamap, 2017). As such, 90 percent of basin water abstraction is conducted by the agricultural sector (Berke *et al.*, 2014).

[2]The Küçük Menderes Basin is home to around 3.5 million people, covers the Aegean sea coast of Turkey, including the cities of Izmir and Aydin, and encompasses one of the most significant agricultural and tourism areas nationally (SYGM, 2019). Around 33 percent of the land area is classed as agriculturally productive, of which 43 percent is irrigated (Gülersoy *et al.*, 2015). The main crops produced in the basin are water-intensive potatoes, watermelons, tomatoes and fruits plus cereal crops and olives.

10.2 THE OVERARCHING INSTITUTIONAL FRAMEWORK

10.2.1 Legal and policy background

The main institutional framework for water allocation in Turkey dates back to the 1960s. Prior to this period, surface water allocations were largely determined by local and municipal authorities while groundwater was largely left to landowners' appropriation. A growing centralization of powers for water management occurred after the National Groundwater Law (1960) was enacted, consisting of provisions for protecting groundwater, usage designations and registration requirements for abstraction. The State Hydraulic Works (or DSI using the Turkish acronym), a government agency that is now part of the Ministry of Agriculture and Forestry, became the main implementing body. Article 4 of the Law states that the number, locations and depths of wells within a groundwater operation area should be determined by the DSI and licensed by them. Additionally, the Law mandates that anyone can use groundwaters on their land but must apply for a usage certificate from the DSI, meaning that the state effectively retains water ownership rights (see below)[3].

Other important legal changes related to water allocations have occurred as a result of Turkey's EU accession process. Although this process has effectively been shelved, it has nonetheless led to an ongoing Europeanization of Turkish water law through the implementation of the EU environmental acquis (Demirbilek & Benson, 2018). Multiple by-laws and national laws have been enacted, particularly to support the adoption of river basin management planning under the Water Framework Directive (ibid.). In total, 25 river basins have been established nationally; including the Konya Closed Basin and Küçük Menderes Basin (Figure 10.1). Another important related measure for agricultural water allocations is the by-law on the Control of Water Use and Reduction of Losses in Agricultural Irrigation Activities (16/02/2017, No. 29981). This regulation aims at ensuring irrigation water efficiency (Article 1). In addition, it gives priority to surface water resources in irrigation, preventing water users from taking more water than the amount foreseen in the irrigation water distribution planning according to actual needs, plus using appropriate modern irrigation methods that save water and prevent water loss leaks. However, there is no comprehensive national water law that can aggregate all these laws together, including the Groundwater Law and river basin management planning by-laws, and solve current water management issues (Bulut & Birben, 2019). Therefore, a Draft Water Law, still under Parliamentary consideration, should be enacted to provide a comprehensive legal framework.

However, a national policy does exist for integrating river basin planning and water allocations. The 2014 National River Basin Strategy identifies the need to coordinate water allocations with river basin planning, and sets the goal for water allocation plans to be completed by 2020 for all basins in Turkey, while the short-term target was five basins by 2015 (OSIB, 2014). To achieve this goal, the Strategy obliges government agencies to generate the necessary data for sectoral water allocation and establish a water allocation board (ibid.). Additionally, the National Water Planning policy states that competition for water resources between economic sectors highlights the importance of government departmental coordination as each sector has different regulatory institutions (Ministry of Agriculture and Forestry, 2019).

10.2.2 The nature of water rights

Under Turkish law, ownership of water has changed from historical private water rights to state control. According to Article 679 of the repealed Turkish Civil Code (1926, No. 743), groundwater was considered as spring water and belonged to the landowner[4]. In addition, according to Articles 718 and 756, both surface water resources and groundwater were the property of the landowner from whose land they originate. However, since the 1960s a trend towards state ownership has occurred. The Groundwater Law (23/11/1960, No. 138) made such waters subject to the provisions and disposal

[3]The Groundwater Law (1960, No. 10688): https://www.mevzuat.gov.tr/MevzuatMetin/1.4.167.pdf.

[4]https://www.mevzuat.gov.tr/MevzuatMetin/5.3.743.pdf

of the State (Başpınar, 2016). According to Article 1(1), groundwater was included in state-owned general waters thereby removing the property rights of the landowner (ibid.). Article 756(3) of the Civil Code (No. 4721) was also amended so that groundwater became one of the waters subject to public interest, with the provision added that owning land does not result in owning groundwater (ibid.). In parallel, Article 3(1) of the Draft Water Law, still not enacted, mandates that water resources are under the control and disposal of the state, regardless of the owner or user of the land on which they are located. However, the owner and user of the land still has the right to benefit from such water for their drinking needs (Article 3(2) of the Draft Water Law)[5].

As opposed to water ownership rights, water use rights in agriculture have undergone a limited shift back from state control to privatization in recent decades. Under Law No. 6172 on Irrigation Unions 2011, the government (i.e. the DSI) began to withdraw from governing irrigation, which is the largest water-consuming sector in Turkey, and transferred control of facilities to irrigation unions (Aydoğdu *et al.*, 2015; Şengül, 2013). Irrigation unions consequently manage irrigation facilities under their control and receive payments regarding water provision, infrastructure maintenance and management expenses, with tariffs determined by the DSI. Additionally, irrigation unions collect an irrigator participation fee, water use service charge, and fines that are applied. Not only are the unions responsible for managing some water bodies but also for cooperating with the Agriculture and Rural Affairs Ministry department regarding deciding crop patterns depending on water quantity[6]. While 5 million hectares are irrigated in Turkey, according to DSI data 23 percent is administrated by irrigation unions (Sarıtaş *et al.*, 2001).

10.2.3 Controlling access to water and collaborative decision-making

In Turkey, controls on accessing water are multi-level and complex. The authorized institution responsible for water permitting and charging in basins is still primarily the DSI, through the issuance of water use certificates. Irrigation unions manage water allocations in irrigated areas, under DSI oversight. However, the General Directorate of Water Management (GDWM) is, since the adoption of the National River Basin Strategy, authorized to determine sectoral water allocations in the preparation of river basin management plans, in coordination with different stakeholders within planning processes (Ak *et al.*, 2019). But this inter-actor split in responsibilities has caused coordination problems. River basin planning, coordinated by the GWDM, is still evolving and lacks the full participation of the DSI plus agricultural actors such as irrigation unions. Representatives of the latter are allowed to participate in Basin Management Committees and Provincial Water Management Coordination Committees but their role is merely consultative. Water allocation plans, moreover, have yet to be completed for all river basins. The target set for 2020 was not achieved, with only five basin plans established.

10.3 DEFINING THE AVAILABLE RESOURCE POOL

The water allocation plans calculate environmental needs on which to base management targets. Different methods for calculating 'minimum environmental flow needs' have been used in the plans. One of the most widely used is the Tennant method. This method calculates the 10 or 20 percent of available surface water depending on good or poor ecological status. In all sectoral water allocation plans, the Tennant poor ecological condition was used, which means considering only 10 percent of yearly available surface water resources (SYGM, 2017, 2018a). A WEAP modeling system (Water Evaluation and Planning) may also be used to assess the existing water bodies and calculate secured

[5]https://www.tmmmb.org.tr/images/GORUSLERIMIZ/SU_KANUNU_TMMMB_GORUSLERI_KASIM12.pdf

[6]https://www.mevzuat.gov.tr/MevzuatMetin/1.5.6172.pdf

water resource reserves based on historical data. The planning process incorporates these values and makes allocations based on existing and sustainable thresholds, with the aim not to go beyond sustainable levels of resources (SYGM, 2017, 2018a, 2018b; TOB, 2019). Unfortunately, in many cases, these sustainable levels are exceeded in drought periods as allocation plans should also account for sectoral needs during these events (see next section on allocation rules).

Konya's annual water availability is 4.3 billion m^3 but annual water use is 6.5 billion m^3 (Berke *et al.*, 2014). A deficit of nearly 2 billion m^3 then results in lower groundwater levels each year in the basin as users, mainly farmers, over-abstract from these sources. This leads to decreasing groundwater levels and an increasing number of sinkhole occurrences (Tapur & Bozyiğit, 2015). Because of these sinkholes, some farmers in the surrounding area have given up on agriculture and left their fields fallow, while these vast sinkholes create hazards for people and animals (Bozyiğit & Tapur, 2009). As a result, farmers have decided to migrate to urban areas such as Konya or Kayapinar (ibid). Given these problems, a fairer water allocation regime is needed for the sake of a sustainable ecosystem but also for the continuance of some agricultural practices. In addition, future inter-basin surface water transfer has been assessed, in this case, water conveyance from the River Euphrates to the Konya Closed Basin, and via the Blue Tunnel from the Eastern Mediterranean Basin (SYGM, 2018a).

Such environmental issues are replicated in other Turkish river basins. In the Küçük Menderes Basin, several measures have been adopted to reduce water use and the resulting pressure on water needs for ecology. A vast amount of treated wastewater, 125 hm^3 per year, is added back to the water cycle in the Küçük Menderes Basin, putting it amongst the most advanced watersheds for recycling and reusing water resources in Turkey (SYGM, 2019). A minimum environmental flow was calculated for the basin using the Tennant method. Here, 20 percent of surface water for March, April and May was classed as 'good ecology', however, for the rest of the year, 10 percent of surface water was categorized as 'poor' (ibid). Environmental impacts of water abstraction for agricultural activity are often severe. Pressure on surface waters and groundwaters is increasing, with more than 10 000 wells sunk in the basin (ibid.).

10.4 ALLOCATION AND REALLOCATION RULES

10.4.1 Approach for allocating water between sectors

In the past, allocation was the responsibility of municipalities even dating back to the Ottoman period. Despite the Groundwater Law, it was not prioritized by national government agencies in Turkey until very recently (Demirbilek & Benson, 2018). With the implementation of the Water Framework Directive model, as described above, the General Directorate of Water Management became the responsible government department for strategic water allocation planning in river basins (SYGM, 2017). Allocation and reallocation rules are gradually being developed through the related water allocation plans that seek to reconcile different sector demands with environmental needs in each basin. Most water allocation plans encompass five main sectors: drinking, environmental needs, agriculture, energy, and industry. However, some plans include mining, livestock, geothermal energy, trading and even the tourism sector (SYGM, 2017, 2018b; TOB, 2019).

Water is currently allocated by the DSI and other institutions using the following approach. The DSI has a direct responsibility to implement the planning objectives for water allocation established by the GDWM. The DSI directs water allocation in basins but also other institutions responsible for water provision. For instance, municipalities and special provincial administrations are the institutions responsible for providing domestic drinking water (SYGM, 2018b; TOB, 2019). Municipalities collect revenue through service charges and use it for operational costs (SYGM, 2018a). In addition to these institutions, irrigation unions, cooperatives and the agricultural reform general directorate are included in agricultural water allocation action planning (ibid.). These irrigation unions are not-for-profit institutions that charge for water use. The income is used to invest in irrigation facilities for improving standards.[7]

Currently, water allocations are being calculated based on current and future use but ultimately reflect economic priorities despite recognizing environmental factors. For instance, in the Konya Closed Basin Water Allocation Plan, sectors have been prioritized based on historical use data and income generation. Here, the agriculture sector was prioritized due to its economic value (TOB, 2019).

Water transfer between basins is also included in the allocation plan. The economic added value for each sector was first assessed and the sectoral benefits of water allocation for optimal use were determined (SYGM, 2018a). For instance, industrial use was considered economically more important than any other sector in the Seyhan Basin, so allocation planning prioritized industry needs above those of the other four sectors in the basin (SYGM, 2017). However, the Akarçay Basin allocates the majority of water to the agriculture sector, around 220 hm^3/year, while other sectors are allocated around 49 hm^3/year (2016 is the reference year in the planning process) (SYGM, 2018a).

In the future, sectoral volumetric caps will be set by the water allocation plans. The existing water resources potential was identified while future water potential for individual basins in the case of drought (low-medium-intense) and climatic changes were classified (TOB, 2019). For instance, an intense drought was forecasted for Konya between 2019 and 2021, and the sectoral water allocation plan shows that some of the sub-basins' supply and demand ratio for the agriculture sector would decrease to 80 percent or less (ibid.). Reuse of treated wastewater and water transfer from other basins were discussed and advised (ibid.).

10.4.2 Economic and social performance of the current approach

Economic efficiency of water allocation is desirable for policymakers and water authorities. Here, water allocation could be considered efficient where it supports high-value economic activity, but this is not the case in Turkey. As stated above, 90 percent of water abstracted in Konya is used by agriculture. However, this sector is one of the lowest income-generating sectors per metre of cubic water usage: only 0.45 Turkish liras (tl) per metre of cubic water consumption (SYGM, 2018a). In contrast, the biggest economic generation sectors are energy (1054 tl/m^3), mining (464 tl/m^3) and commerce (553 tl/m^3) (ibid.). Given the amount of water consumption in each sector, the economic efficiency of water allocation nationally is low. There are plans to halve water consumption in the agriculture sector from around 4000 to 2444 hm^3 by 2040 and the expectation is for economic value to double (ibid.). But it would still be far from the economic value created by water consumption in the energy sector.

That said, the economic significance of agriculture varies between basins and the two cases considered in this chapter illustrate the strategic importance of allocating water to agriculture. For instance, around 5 percent of Turkey's agricultural income is produced in the Konya Basin: 15 billion Turkish liras in 2017 (Yildirim *et al.*, 2018). Additionally, nationally critical crops grown in Konya include wheat, barley, seed production, sugar beet, beans and carrots. The basin is also significant for cattle and milk production (ibid.). Agricultural employment share is more than 70 percent of the total in the basin (SYGM, 2015). In the Küçük Menderes Basin, the average agricultural share of economic production is 18 percent, with water used to irrigate high-value agricultural products including mandarin oranges, chestnuts, cherries, peaches, olives and figs (SYGM, 2018b). Agricultural sector employment rose from 20% in 2010 to 25% in 2015 (ibid.). That said, agriculture is still ranked third from last in economic value generated per cubic metre of water (SYGM, 2019). The drinking water sector is the biggest economic producer followed by the mining and livestock sectors (ibid.). Overall, water allocation in the basin appears economically inefficient given that 70 percent of the water is used in the agriculture sector (SYGM, 2018b, 2019).

[7]https://www.mevzuat.gov.tr/MevzuatMetin/1.5.6172.pdf

One of the most important dimensions of integrated water resource management is social equity (Pena, 2011). In Turkey, there is pressure to allocate water to economically significant sectors such as agriculture, meaning little is left for other sectors of society. This will worsen in the future. Research shows that 96 percent of current agricultural sector water needs in the Küçük Menderes Basin can be met in minor drought conditions while other sectors' needs can still be fully met in the same conditions (SYGM, 2019). However, by the 2040s, the supply/demand ratio is projected to go down to 77 percent (ibid.), which could cause conflicts between industrial, domestic and farming interests.

10.5 MONITORING AND COMPLIANCE

Despite commitments to control water abstraction in river basin sectoral plans, as described above only five allocation plans have been produced. Coordination between the WFD process and water allocation planning is therefore limited, preliminary and still evolving. The monitoring of water levels is also historically underdeveloped: the DSI has only six groundwater monitoring stations for the entire Konya Closed Basin (Orhan, 2021). Therefore, real-time monitoring stations and early warning systems should be expanded to include water and groundwater levels alongside water quality. It is therefore necessary to establish a National Monitoring Network to carry out monitoring holistically (Ministry of Development, 2018).

Regulatory compliance is also weak. For instance, according to Article 14 of the 'Regulation on the Protection of Ground Water against Pollution and Detection', the Ministry of Environment and Urbanization is authorized to conduct inspections regarding the protection of groundwater quality and the DSI is also authorized to conduct inspections regarding quantity-related issues[8]. According to Article 18 of the 'Law on Groundwater' (No. 10688, 1960), the necessary permits must be obtained from the DSI for any activity to supply groundwater (Günhan, 2014). There are three types of permissions to be obtained by the user: 'search permission' to drill wells; 'use permission' for abstracting water; and 'modification permission' for then altering conditions of the permit. Administrative fines may be imposed if activities are carried out without obtaining permits. A well can be closed and a fine imposed on the well driller[9]. Under national regulations, the DSI is also the agency responsible for licensing surface water. Depending on the location and the needs of the individual applicant, municipalities, provincial special administrations, and agriculture and forest provincial directorates can license water use rights, under DSI oversight. Restrictions for water use can be imposed on applicants, while allocations are not made where water resources fall below the environmental minimum flow. That said, the widespread issues with the over-abstraction of groundwaters and surface waters in river basins show that such regulations are poorly enforced, primarily due to institutional incapacity

10.6 THE BROADER POLICY INSTRUMENT MIX

In parallel to this emerging national water allocation regime are established government policy instruments for drought and water quality that impact upon its implementation. That said, there is only limited 'coherence' (Benson & Lorenzoni, 2017) between institutional frameworks and responsibilities, necessitating future coordination in water allocation policy at the basin scale.

10.6.1 Drought policy

Turkish water authorities originally recognized drought and other extreme weather events in the 2014 National River Basin Strategy (OSIB, 2014). A drought strategy and drought monitoring system were included within this overarching policy document (ibid.). More recently, a national drought

[8]The Regulation on the Protection of Groundwater against Pollution and Detection (No. 28257, 2012). https://www.mevzuat.gov.tr/mevzuat?MevzuatNo=16038&MevzuatTur=7&MevzuatTertip=5)

[9]The Groundwater Law (1960, No. 10688). https://www.mevzuat.gov.tr/MevzuatMetin/1.4.167.pdf

management strategy and action plan were ratified in 2017 (OSIB, 2017). This plan also added the previous agricultural drought combating strategy and action plan to its framework (OSIB, 2017). The General Directorate of Water Management set the goal for each of the 25 river basins to establish drought management plans covering the period until 2023 (Duygu, 2015). So far, 11 basins have adopted a drought plan, including the Küçük Menderes and Konya Closed basins.

Because of Turkey's location in the Mediterranean Climate zone, the country already experiences frequent and intensive droughts (Seneviratne *et al.*, 2012). These are likely to increase under climate change, meaning drought management plays a crucial role in Turkey's water management planning. The drought management plans involve three phases; the first phase (before a drought), the second phase (during a drought), and the third phase (after a drought) (OSIB, 2017). In agricultural drought management planning, there is a coordinating board that consists of a monitoring, early warning and forecasting committee, a risk evaluation board, and agricultural drought centers under the auspices of local governors (Ministry of Food, Agriculture and Livestock, 2013). Agricultural drought management plans and the national drought management strategy plan were predominantly agricultural-centered plans since the majority of water is used in the agriculture sector (Ministry of Food, Agriculture and Livestock, 2013; OSIB, 2017). Lastly, one of the crucial goals of the national drought strategy was to place the concept of 'drought management' within the scope of the national water law, to provide greater national coordination (OSIB, 2017).

Each drought management cycle has four steps which are monitoring, assessment, mitigation and responses (SYGM, 2018a). Depending on drought intensity, different methods and restrictions are used to ease drought effects. For instance, when moderate drought conditions are exceeded, restrictions on water irrigation such as compelling night irrigation and rotational water use planning are used (ibid.). During the most intense droughts, secondary product planting is forbidden, while priority is given to agricultural sectors such as fruit farmers (ibid.). Implementation of the levels of drought management planning in each basin also depends on drought severity: the full action plan is only enacted during the most severe droughts (SYGM, 2018a). Some of the restrictive actions are then relaxed after drought intensity declines (ibid.). Additionally, the national drought management strategy and action plan informs the public and requires them to participate in plan implementation, while conducting the tasks previously established in the agricultural drought combating strategy and action plan (OSIB, 2017). Moreover, while the drought management plan, agreed by water agencies in Turkey, focuses on water use limitation and restriction, the agricultural drought combatting action plan adopted by the Ministry of Food, Agriculture and Livestock tries to protect farmers and mitigate the adverse effects of drought on production (OSIB, 2017; SYGM, 2018a). Here, it aims to increase the water retention capacity of soil by promoting organic fertilizer and the use of mulch, thus mitigating drought-related yield loss (Ministry of Food, Agriculture and Livestock, 2013). Lastly, this plan recognizes forage needs and supply mechanisms for the livestock sector and supports grain aid distribution to those in need, to reduce the possibility of famine (ibid.).

Despite drought measures being adopted in the 2014 National River Basin Strategy and the adoption of basin drought management plans, the long-term climate resilience of Turkey's water resources can be questioned. One reason for this is that climate change is predicted to increase drought events (Seneviratne *et al.*, 2012). For instance, the Konya Closed Basin will experience medium and severe drought over the next decade, with water allocation analysis showing that water potential and water allocation cannot satisfy all water needs during significant drought conditions, especially for the agriculture sector (SYGM, 2018a). The supply of water to the agriculture sector could be reduced to 75 percent for some sub-basins in Konya, namely Cumra, Beysehir and Karaman, while the supply ratio is 35 percent for the agriculture sector for the Altintekin sub-basin (ibid.). To address these risks, water transfer and recycled water have been recommended to meet sectoral needs (ibid). To an extent, this will enhance climate resilience. However, water transfer makes the basin dependent on external water bodies and undermines one of the key pillars of water security, namely independence. Transfers also impact the resilience of other basins and may therefore not be sustainable. Other approaches such as crop substitution, and water-saving methods and technologies may be required for future climate adaptation.

Some measures have already been adopted to ensure resilience in the Kücük Menderes Basin. The basin has transferred water from the Gediz basin (SYGM, 2019). While providing a short-term solution to deficits, it puts the long-term resilience of the basin in jeopardy, especially when more water transfer is being advised for severe droughts (ibid.). Not only has water conveyance happened outside the basin but there are also intra-basin water transfers occurring due to sub-basins experiencing water scarcity and loss of water independence (ibid.). This issue involves water security at both local and regional levels, requiring strategic government intervention.

10.6.2 Economic instruments

One important government subsidy related to water allocation is aimed at providing financial support for preventing water loss and leakage, as part of Turkey's national water management strategy (OSIB, 2014). If applied properly, significant water conservation can be achieved, especially in times of water scarcity and drought. Promoting ecologically friendly and organic agriculture, the expansion of pastoral areas and afforestation to increase water conservation are also supported by economic subsidies in Turkey (OSIB, 2014).

10.6.3 Awareness-raising

Raising awareness of water scarcity issues is one important part of efficient water allocation management. In Turkey, this is undertaken through national media and educational programs. Media outlets have been used for increasing public awareness (OSIB, 2017). However, the effectiveness of media information as an instrument of government policy is questionable when it comes to conveying water-related messages to the public, especially farmers.

Educational programs for farmers to increase the recognition of water allocation issues are included in drought management plans (Duygu *et al.*, 2017; Ministry of Food, Agriculture and Livestock, 2013; OSIB, 2017). However, it is difficult to assess the usefulness of these educational programs when they can only be tested during extreme weather events. In the assessment of river basin planning, educational inadequacies amongst key stakeholders were observed as a threat to management schemes (OSIB, 2014). Dealing with these deficiencies requires educational programs for water harvesting, increasing the efficiency of water irrigation techniques and the adoption of water use monitoring methodologies (OSIB, 2017).

10.7 CONCLUSIONS

Despite the abundance of laws on water resources management in Turkey, the case studies show that significant problems exist with water allocation in general and in the agricultural sector specifically. Over-abstraction of water in river basins mainly for irrigation is causing severe environmental externalities and, in Konya, major effects such as sinkhole development (Ak *et al.*, 2019). Economic imperatives to promote agricultural production are, to an extent, driving these impacts, while social equity and long-term climate resilience are also reduced. Sustainability of water resources is therefore poor, suggesting that water allocation should be a national priority concern. Problematically, despite recent institutional innovations around water allocation planning, there is no comprehensive national water law that can aggregate existing water measures (Bulut & Birben, 2019). One consequence is institutional 'incoherence' and a lack of coordination between DSI permitting and GDWM strategic planning in controlling water allocation. The Draft Water Law should therefore be enacted to specify responsibilities but it is still under political consideration. In the meantime, Basin Management Committees should be made the only authorized body at the basin scale for determining all planning and applications related to water resource use and should have legal personality (Ministry of Agriculture and Forestry, 2019). Greater collaboration between committees, government agencies and stakeholders such as irrigation unions and farmers is also required to ensure more sustainable forms of abstraction.

REFERENCES

Ak M. Y., Benson D., Demirbilek B. and Scott K. (2019). Participatory groundwater protection regimes in Turkey: influence, control and institutional incoherence. AGU Annual Conference, San Francisco, USA.

Aydoğdu M. H., Mancı A. R. and Aydoğdu M. (2015). The changes in agricultural water management: water user associations, pricing and privatization process. *Electronic Journal of Social Sciences*, **14**(52), 146–160.

Baker S. (2016). Sustainable Development. Routledge, London, UK.

Başpınar V. (2016). Evaluation of the Turkish civil code, law on groundwater and draft water law provisions in terms of water ownership. *Ankara University Faculty of Law Journal*, **65**(4), 2725–2754.

Benson D. and Lorenzoni I. (2017). Climate change adaptation, flood risks and policy coherence in integrated water resources management in England. *Regional Environmental Change*, **17**(7), 1921–1932, https://doi.org/10.1007/s10113-016-0959-6

Berke M. Ö., Dıvrak B. B. and Sarısoy H. D. (2014). Water Today Report in Konya, WWF-Turkey, Istanbul, Turkey.

Bozyiğit R. and Tapur T. (2009). The effect of groundwater on sinkhole formations in the Konya plain and its surroundings. In: Beyşehir Nature Education: 12 Days Together with Science and Nature, A. Meydan (ed.), 2nd edn, Pegem Academy Publishing, Turkey, pp. 61–80.

Bulut M. and Birben U. (2019). Impact of the EU water framework directive on water resources management in Turkey. *Turkish Journal of Forestry*, **20**(3), 221–233.

Demirbilek B. and Benson D. (2018). Legal europeanisation in three dimensions: water legislation in Turkey. *Journal of Water Law*, **25**(6), 294–307.

Demirbilek B. and Benson D. (2019). Between emulation and assemblage: analysing WFD policy transfer outcomes in Turkey. *Water*, **11**(2), 324, https://doi.org/10.3390/w11020324

Duygu M. B. (2015). Drought Management Plan of Konya Basin. Presented at the 7th World Water Forum, Daegu, South Korea.

Duygu M. B., Kirmencioğlu B. and Aras M. (2017). Planning of Drought Management by Integrating Science and Policy. Presented at the International Water Resources Association (IWRA) Conference, Cancun, Quintana Roo, Mexico.

Gülersoy A. E., Gümüş N., Sonmez M. E. and Gündüzoğlu G. (2015). Relations between the land use and land capability classification in Küçük Menderes River Basin. *Journal of Environmental Biology*, **36**, 17–26.

Günhan Ö. (2014). A Methodology Research for Assessment of the Quality of Groundwater. Expertise thesis, The Ministry of Agriculture and Forestry, Ankara, Turkey.

Körbalta H. (2019). Türkiye'de Yerel Su Güvenliği. [local water security in Turkey]. GüVenlik Bilimleri Dergisi, **8**(1), 55–84, https://doi.org/10.28956/gbd.562965

Ministry of Agriculture and Forestry. (2019). National Water Plan (2019–2023). Ministry of Agriculture and Forestry, Ankara, Turkey.

Ministry of Development. (2018). On Birinci Kalkınma Planı [Eleventh Development Plan] (2019–2023). Ministry of Development, Ankara, Turkey.

Ministry of Food, Agriculture and Livestock. (2013). Gıda, Tarım ve Hayvancılık Bakanlığı Türkiye Tarımsal Kuraklıkla Mücadele Stratejisi ve Eylem Planı 2013–2017 [Turkey Strategy and Action Plan for Combating Agricultural Drought]. Ministry of Food, Agriculture and Livestock, Ankara, Turkey.

Orhan O. (2021). Monitoring of land subsidence due to excessive groundwater extraction using small baseline subset technique in konya, Turkey. *Environmental Monitoring and Assessment*, **193**, 174, https://doi.org/10.1007/s10661-021-08962-x

OSIB. (2014). Ulusal Havza Yönetim Stratejisi (2014–2023) [National Basin Management Strategy]. Ministry of Forestry and Water Management, Ankara, Turkey.

OSIB. (2017). National Drought Management Strategy Document and Action Plan (2017–2023). Ministry of Forestry and Water Management, Ankara, Turkey.

Pena H. (2011). Social Equity and Integrated Water Resources Management. Global Water Partnership, Stockholm, Sweden.

Ribamap. (2017). Executive Summary, Draft Article 5 Report for Konya Closed Basin. General Directorate of Water Management, Ankara, Turkey.

Ribamap. (2018). Technical Assistance for the Conversion of River Basin Action Plans Into River Basin Management Plans. General Directorate of Water Management, Ankara, Turkey.

Sarıtaş H., Çınar M. and Çelik A. (2001). Sulama birlikleri ve sulama eğitimi. [irrigation unions and irrigation training]. *Ministry of Agriculture and Rural Affairs Journal of Turkish Agriculture*, **137**, 17–18.

Seneviratne S. I., Nicholls N., Easterling D., Goodess C. M., Kanae S., Kossin J., Luo Y., Marengo J., McInnes K., Rahimi M., Reichstein M., Sorteberg A., Vera C. and Zhang X. (2012). Changes in climate extremes and their impacts on the natural physical environment. In: Managing the Risks of Extreme Events and Disasters to Advance Climate Change Adaptation, C. B. Field, V. Barros, T. F. Stocker and Q. Dahe (eds.), Cambridge University Press, Cambridge, UK, pp. 109–230.

Şengül M. (2013). Türkiye'nin Su Politikası ve Köylülerin Öfkesi: Politikalar ve Sorunlar: Küreselden Yerele [Turkey's Water Policy and Villagers' Concerns: Management of Water Resources: Policies and Issues: From Global to Local]. Nevsehir Universitesi, Nevsehir, pp. 29–41.

SYGM. (2015). Konya Havzası Kuraklık Yönetim Planı [Konya Basin Drought Management Plan]. General Directorate of Water Management, Ankara, Turkey.

SYGM. (2017). Seyhan Havzası Sektörel Su Tahsis Planı [BasinSectoral Water Allocation Plan]. General Directorate of Water Management, Ankara, Turkey.

SYGM. (2018a). Konya Havzası Sektörel Su Tahsis Planı Hazırlanması Projesi [Basin Sectoral Water Allocation Plan Preparation Project]. General Directorate of Water Management, Ankara, Turkey.

SYGM. (2018b). Küçük Menderes Havzası Kuraklık Yönetim Planı [Küçük Menderes Basin Drought Management Plan]. General Directorate of Water Management, Ankara, Turkey.

SYGM. (2019). Küçük Menderes Havzası Sektörel Su Tahsisi Eylem Planı [Küçük Menderes Basin Sectoral Water Allocation Action Plan]. General Directorate of Water Management, Ankara, Turkey.

Tapur T. and Bozyiğit R. (2015). Konya İlinde güncel obruk oluşumları [current sinkhole formations in konya]. *Marmara Coğrafya Dergisi*, **31**, 415, https://doi.org/10.14781/mcd.81669

TOB. (2019). Konya Kapalı Havzası Sektörel Su Tahsisi Eylem Planı [Konya Closed Basin Sectoral Water Allocation Action Plan]. Pub. L. No. 96301635-010.06.02- E.1824734. TOB, Ankara, Turkey.

Yildirim A. I., Demir S. K. and Yayla U. (2018). Konya Tarımı Bilgi Notu [Konya Agriculture Information Note]. Ministry of Agriculture and Forestry, Ankara, Turkey.

doi: 10.2166/9781789062786_0129

Chapter 11

Water allocation in Aotearoa New Zealand: societal values and ecological bottom lines

Edward Challies[1], Stephen Fragaszy[2] and Josselin Rouillard[3]

[1]*University of Canterbury, Christchurch, New Zealand*
[2]*Independent consultant; at the time of writing, Ministry for the Environment, Wellington, New Zealand*
[3]*Ecologic Institute, Berlin, Germany*

ABSTRACT

Water allocation is an increasingly prominent policy issue in Aotearoa New Zealand (NZ), where regulation has largely failed to secure sustainable management of water resources over the past three decades. Although there is abundant water in NZ, the cumulative effects of abstractions and diversions, alongside diffuse pollution from agriculture and other urban and rural land uses, have led to highly degraded and depleted water resources in some locations. This has had significant social and ecological impacts. As a result, governmental planning and decision-making around water allocation (and land-use and development more widely) are increasingly driven by the imperatives to maintain 'environmental flows' and safeguard community values. In the NZ context, the Government has special obligations to partner with Māori (Indigenous New Zealanders) in all aspects of environmental management. This task must be informed by principles and values from *Te Ao Māori* (the Māori world), meaningfully involve Māori in governance and management, and recognise Māori rights and interests in water. Local government (regional councils), which are responsible for defining allocation rules, must ensure rules serve broader freshwater management objectives that are developed through engagement with Māori and wider communities, and which safeguard the health and wellbeing of waterbodies, associated ecosystems, and people.

Keywords: Allocation, Climate change, Collaborative governance, Environmental flows, Freshwater management, New Zealand, Te Mana o Te Wai, Water quality, Water rights

11.1 INTRODUCTION

Aotearoa New Zealand (NZ) is a water-abundant country, albeit with regional as well as intra- and inter-annual variability. While water scarcity has not been a widespread issue historically, the country's drier eastern regions suffer from periodic droughts. Per-capita water demand is increasing, and water underpins social, economic and cultural wellbeing. Water – conceptualised as a wide socio-ecological system including associated ecosystems – is therefore valued in multiple and often competing ways.

Water is of vital importance to Māori (Indigenous New Zealanders)[1], who once managed all water according to Māori laws and customs (*tikanga*) (Macpherson, 2019; Stewart-Harawira, 2020). However, access to water and the water environment for Māori has been systematically eroded through colonial resource management institutions and practices. Aside from hydropower generation, which is generally deemed a non-consumptive use[2], irrigation accounts for the greatest share of water abstraction. Of 12.9 billion (10^9) m^3 total annual allocation[3] for consumptive uses, irrigation accounted for 7.5 billion m^3, or 58% (Booker & Henderson, 2019). Irrigation has expanded considerably over the past two decades, and in some relatively dry regions, demand for new permits now exceeds sustainable supply (OECD, 2017). At the same time, in many regions, more water is allocated than is actually used, suggesting scope for improved efficiency through reallocation.

Today, as local government authorities (specifically regional councils) grapple with over-allocation and negative impacts of land-use intensification on water quality and ecosystem health, instream ecological and indigenous cultural values increasingly guide policy and management. Ongoing reforms over the past decade have largely focused on water *quality*, due to the serious impacts of nutrients, bacteria, and sediment, primarily from farming activities. While water *quantity* has long been an issue for certain catchments and regions, allocation is now emerging as a pressing national-level policy problem as the pervasiveness of over-allocation is recognised, and as questions of Māori rights and interests in water resources come to the fore in legal and policy debates.

This chapter outlines the policy framework for water allocation in NZ, including key laws and policies governing freshwater management. We discuss management approaches, allocation and reallocation rules, monitoring and compliance processes, and interactions with the wider water and environmental policy system. We briefly consider the environmental and social performance of the current allocation system, and reflect on future challenges in the face of ongoing policy reform and social and environmental change.

11.2 THE OVERARCHING POLICY FRAMEWORK

11.2.1 Constitutional and legal framework

The Treaty of Waitangi (signed in 1840) establishes the relationship between the Crown (embodied by the government) and Māori, and is thus a core constitutional document for NZ. Since 1840, water resources have been governed by a mixture of common law and statute. These colonial legal frameworks have, however, largely excluded Māori from decision-making and management (Macpherson, 2019). Common law has generally enabled widespread use of water, but through statute – currently the *Resource Management Act 1991* (RMA) – the Crown has limited some common law rights, including the rights to divert, take, or use natural water, and to discharge into it (Ruru, 2010). As discussed further below, fundamental questions of water rights and ownership are salient at present. Likewise, Māori conceptual ontologies (see Harmsworth & Awatere (2013) for an applicable and cogent overview) are increasingly prominent in statute, policy, and binding regional plans.

11.2.2 RMA and regional councils

The RMA is the primary legislation governing water management in NZ. It articulates an overarching principle of 'sustainable management of natural and physical resources'. The RMA empowers and mandates regional councils to control and manage surface water and groundwater, including in

[1] Indeed, a Māori word commonly used as a translation for 'wellbeing', *waiora*, is a compound word that transliterates as *'wai'* = water and *'ora'* = alive (O'Connell *et al.*, 2018).

[2] Non-consumptive given most schemes return water to the river system downstream. However, hydro schemes do significantly impact river flow regimes, character and instream values including ecology.

[3] In this context 'allocation' refers to permitted water use rights issued by a regulatory authority under law.

Figure 11.1 Map of New Zealand regions.

relation to flows, water quality, and effects on ecosystems. In addition, councils must control land-use and discharges that will affect water quality and quantity. Regional councils therefore have an integrated management function over natural resources within their boundaries (Figure 11.1), which in almost all cases align with catchment boundaries. One critically important point is that in NZ, consideration of water quantity management is inseparable from consideration of ecosystem health and water quality matters[4].

[4] See Clapcott *et al.* (2018) for discussion of the ecosystem health framework.

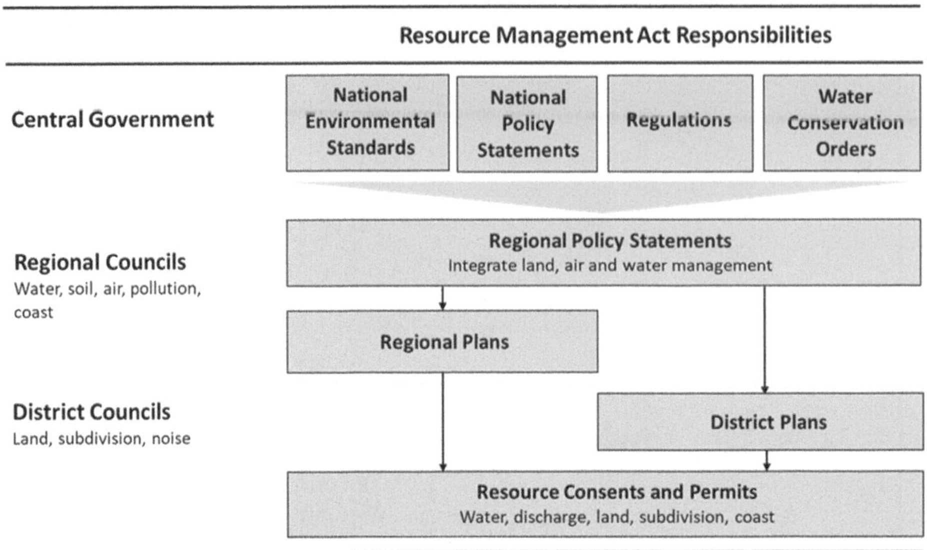

Figure 11.2 Hierarchy of statutory planning instruments and processes under the RMA.

The RMA requires councils to conduct environmental management (Environment Foundation, 2021):

- as guided by a set of core principles that include the principles of the Treaty of Waitangi and sustainable management;
- as directed by national policy statements and standards, which have become the primary vehicle for central government to influence water management in recent years; and
- through effects-based approaches, where rules are predicated on managing the effects of activities rather than the activities themselves.

Regional councils undertake this function by preparing policy statements and plans through participatory processes that are binding once operative. These highly litigious processes produce statutory documents containing policies, rules and management methods related to the use of water and land. For activities governed by a rules regime, such as taking water, individuals must obtain a 'resource consent', a permit which can specify conditions under which the activity is allowed (or not). Figure 11.2 depicts the hierarchy of resource management from the RMA 'downwards' through the chain of governance to consents.

The National Policy Statement for Freshwater Management (NPS-FM) was first introduced in 2011 and amended during each electoral cycle since (2014, 2017, 2020). It plays a principal role in structuring water management by regional councils, and stipulates how councils must achieve the purpose and principles of the RMA as they relate to freshwater. This includes, among other things:

- providing for *Te Mana o te Wai*[5] and
- establishing a process to define geographic units of water management, set future outcomes based on national values, make rules to achieve those outcomes, and evaluate success in doing so.

As the 'fundamental concept' underpinning the NPS-FM, and therefore detailed water governance prescriptions, *Te Mana o te Wai* is defined in the policy (s1.3(1)) as:

[5]Detailed description of this concept is beyond the scope of this chapter, but the reader is referred to Te Aho (2019) for a description of *Te Mana o te Wai*, and Salmond (2014) for a wider discussion of hydrosocial relationships in NZ.

...a concept that refers to the fundamental importance of water and recognises that protecting the health of freshwater protects the health and well-being of the wider environment. It protects the mauri [life-force] of the wai [water]. Te Mana o te Wai *is about restoring and preserving the balance between the water, the wider environment, and the community.*

In relation to water allocation, the NPS-FM imposes a range of requirements on councils. These include:

- Principles to which councils must adhere, such as integrated management of water and land on a whole-of-catchment basis (policy 3), consideration of climate change (policy 4), and efficient allocation and use of water (policy 11);
- Specific obligations, such as to: avoid and phase out over-allocation[6] (policy 11), establish a water accounting framework (policy 13), and set low-flow regimes in relation to ecosystem health (policy 5); and
- Procedures by which councils must undertake planning and rule-making, including most substantively through a National Objectives Framework (NOF).

11.2.3 Controlling access to water

The RMA requires that water 'takes' (i.e. withdrawals) – beyond reasonable use for domestic needs, animal drinking water, or firefighting – be expressly allowed by a national environmental standard, a rule in a regional plan, or a resource consent (s14(3)). Often, a separate 'water use' consent is needed for specific activities, such as irrigation, which may also require 'nutrient discharge' consents to manage nutrient losses to ecosystems. Water takes that are subject to the resource consent process require an assessment of environmental effects (AEE), and are conditional on steps to avoid, remedy or mitigate adverse effects. Specific conditions vary depending on local context, but typically stipulate a maximum rate of take, and a maximum annual or seasonal cumulative take. Water permits can be issued for up to 35 years, although shorter durations are increasingly common as councils seek greater management flexibility.

The practice in NZ has been to replace exercised permits, upon expiry, with a new permit with similar conditions to the original. However, at the time of renewal, the authority does have the opportunity to revise conditions to bring use into line with current rules, targets and limits. Also, the RMA (s128) enables the review of consent conditions in the case of changes to a Regional Plan, or new provisions in national environmental or planning standards. The process of reviewing consent conditions during the course of the consent term is, however, not very flexible, and councils do not routinely do it.

A tendency to prioritise renewal of existing consents over issuance of new consents – especially in catchments that are approaching or have surpassed allocation limits – has served to undermine the public-good nature of the resource. The 'first-in, first-served' approach to allocation has been economically inefficient in that water is often not made available for its highest economic use. Under this approach, councils cannot prioritise certain uses over others (although this can be achieved through specific planning rules). There are also very few incentives for those 'first in the queue' to limit their demand or improve efficiency of use. The system may in fact encourage users to apply for more water than is needed as a form of 'insurance', and this will limit access for subsequent applicants.

The approach is also inequitable in places. Particularly in over-allocated areas, prospective users are routinely prevented from accessing allocation for new uses. In many cases this works against emerging agri-businesses (including Māori businesses), and entrenches existing or long-standing water user interests (Ministry for Primary Industries, 2021).

[6] Where permitted takes exceed available volume of water.

11.2.4 The nature of water rights

Under the common law regime, 'natural water (i.e., water flowing on or under land and not confined to any artificial receptacle) cannot be owned by any person' (Bennion, 2017). Common law has also generally been permissive of the use of groundwater and surface water by landowners for domestic and production purposes. However, as mentioned above, legislation has limited these rights in important ways since the passing of the *Water and Soil Conservation Act 1967*, which 'nationalised rights to use natural water' (ibid.)[7].

Freshwater in NZ is therefore held on trust by the Crown on behalf of the public, and essentially managed as a public good. However, this position is contested by Māori on the basis that the Crown never lawfully extinguished the Māori native title to water (Ruru, 2011). Questions of the nature of water rights are contentious in part because the RMA (s122) declares resource consents to be 'neither real nor personal property' (Warnock & Baker-Galloway, 2015), even though they do confer rights (including exclusive and transferable rights) in water resources that have real economic value (Grinlinton, 2011).

Working through questions and understandings of the nature of water rights in NZ has long been the subject of major contestation. At the best of times, it has also been the source of fruitful society-wide interaction, dialogue, and maturation (Ruru, 2009; Salmond, 2014). These questions are particularly salient now due to recent developments around Māori rights and interests in water. In particular, deliberations of the Waitangi Tribunal[8] over two claims (WAI2357 and WAI2358) about Māori proprietary rights in freshwater and geothermal resources made important determinations and recommendations. Stage 1 of the inquiry (Waitangi Tribunal, 2012) determined that Māori had rights and interests in waterbodies for which the closest English equivalent in 1840 was ownership rights.

In Stage 2, the Tribunal made several recommendations to incorporate the principles of the Treaty of Waitangi[9] (Waitangi Tribunal, 2019), calling for:

- amendments to the RMA and NPS-FM;
- creation of a new freshwater allocation system;
- recognition of Māori proprietary rights and provision of 'proprietary redress'; and
- creation of a national co-governance body, with resourcing to Māori.

Furthermore, Treaty settlements and related laws (such as the *Te Awa Tupua (Whanganui River Claims Settlement) Act 2017*) have widened the legal and policy space related to water management, and substantially expanded the philosophical framework that underpins it. For example, through recognition of 'the inalienable relationship between the *iwi* [extended kinship group] and *hapū* [kinship group] of Whanganui and Te Awa Tupua' (Ngā Tāngata Tiaki o Whanganui, 2021), the settlement restores the rights of the people and the river by re-establishing relational aspects of being. The law (s12) recognises Te Awa Tupua as 'an indivisible and living whole, comprising the Whanganui River from the mountains to the sea, incorporating all its physical and metaphysical elements' with intrinsic values that represent its essence, Tupua te kawa, described in s13 (*Whanganui River Claims Settlement Act, 2017*).

---

[7] Macpherson (2019) observes that the NZ Government began to vest water resources in the Crown as early as 1903 through the *Coal Mines Amendment Act 1903* and the *Water Power Act 1903*.

[8] A permanent commission of inquiry established by law in 1975 to address claims of breaches of the Treaty of Waitangi, as well as existing and proposed legislation. See: https://waitangitribunal.govt.nz/. It is beyond the scope of this chapter, but we note that differences in the wording of the English and Māori language versions of the Treaty are partially the source of subsequent challenges related to water rights (Ruru, 2009).

[9] This Cabinet circular provides guidance for policymakers in applying Treaty of Waitangi principles: https://dpmc.govt.nz/publications/co-19-5-te-tiriti-o-waitangi-treaty-waitangi-guidance-html

The Ministry of Justice noted the difficulty in articulating the regulatory impacts of this new legal framework. Importantly, it is challenging because…existing institutional and statutory frameworks remain in place, but will be influenced by the Te Awa Tupua 'lens' provided through the settlement. That 'lens' will change how decision makers, and others, view and understand the Whanganui River (Office of Treaty Settlements, 2016).

For example, the Tongariro hydropower scheme diverts significant volumes of water out of the Whanganui catchment, and future renewal of consent for these withdrawals may prove controversial. These fundamental issues of water rights are increasingly prominent, with other Māori claims being taken forward[10].

11.3 DEFINING THE AVAILABLE RESOURCE POOL

The available resource pool in a given catchment or 'freshwater management unit' is determined by the regional council, which must work towards achieving the objectives set through processes prescribed by the NPS-FM.

The NPS-FM identifies compulsory 'values' that all regional authorities are required to provide for in freshwater planning and management. These values refer to (1) ecosystem health; (2) human contact; (3) threatened species; and (4) *mahinga kai* (species traditionally gathered and used as food or resources). There are additional, non-compulsory values that authorities must consider (e.g. natural character, drinking water supply, irrigation, hydropower generation), and authorities may identify additional community values.

Figure 11.3 shows how freshwater values are provided for under the NPS-FM. Councils must define measurable and assessable 'objectives' for each value, and select specific 'attributes' to be monitored to track progress towards identified objectives. Mandatory attributes are defined in the NPS-FM (appendices 2A and 2B) – for example phytoplankton, total nitrogen, dissolved oxygen, *Escherichia coli* (*E. coli*) – and regional authorities may use others as well. The NPS-FM defines national bottom

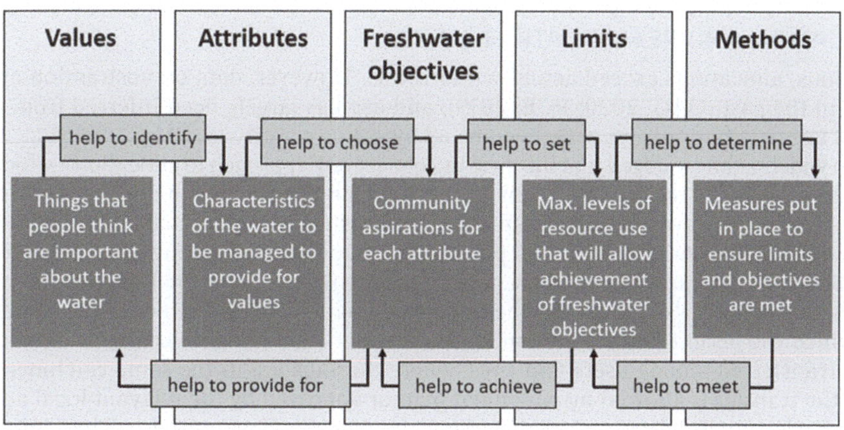

Figure 11.3 Relationships between freshwater values, attributes, objectives, limits and methods. Adapted from Ministry for the Environment (2015).

[10]In late 2020, the representative body of Ngāi Tahu, an *iwi* with authority covering much of the South Island, sought the High Court to recognise its *tino rangatiratanga* ('sovereignty' or 'full authority') over freshwater in its territory.

lines for most listed attributes. Authorities must establish baseline (recent past) states and identify target states for all attributes, and then employ methods to limit resource use in order to achieve those target states.

The methods used by authorities to achieve limits vary. Authorities commonly define minimum flows rather than actual water allocation limits. Where regional councils do apply allocation limits, a variety of methods are used. In allocating surface water, councils may issue blocks or bands of allocations, where for successive bands restrictions are triggered at higher flows. Alternatively, a total cap may be defined, above which no new consents are issued, except where an existing consent expires or is surrendered. Where groundwater and surface water systems are highly interconnected, groundwater allocations are generally tied to surface water limits. Although annual water budget approaches are widely used to quantify groundwater resources, formal setting of allocation limits from groundwater is less common. However, several councils adopt sustainable yield or safe yield approaches and groundwater modelling to set limits.

Waikato Regional Council, for example, specifies minimum flows and 'primary' and 'secondary' allocable flows for surface water bodies by catchment, and differentiated between upland and lowland portions of catchments. Minimum flow is set at 90–95% of the one-in-five-year low flow (Q_5), while primary allocable flow is calculated as the difference between the minimum flow and the Q_5 flow. Secondary allocable flow is a lower reliability allocation, calculated as the difference between 30% of the Q_5 and the primary allocable flow. Aquifers are managed according to a conservative 'management level' until sustainable yields can be established. In Tasman District, allocation limits are defined for surface and groundwater takes across water management zones on the basis of modelling and for the maintenance of minimum flows and environmental bottom lines, as well as security of supply. Limits are expressed as rates (l/second) and, for groundwater zones, also as total annual limits. Graduated reductions are also specified for different rationing steps, with provision for rostering in some zones. As the Waimea Dam water augmentation scheme is under development, transitional allocation arrangements apply differently across affiliated and unaffiliated users, and rules provide for the transfer of allocation to scheme members.

11.4 ALLOCATION AND REALLOCATION RULES

In many regions, allocations exceed actual withdrawals. However, data on abstraction and use have been patchy in the past (OAG, 2020; PCE, 2019), and use has largely been inferred from allocations. Collection of use data has only recently become a key policy priority (see 11.5.1 below).

It has been widely acknowledged that the 'first-in, first-served' approach to allocation – which effectively allows users to perpetually renew water rights (i.e., 'grandparenting') – is sub-optimal and works against efficiency and equity. However, shifting to an alternative system would be challenging, and would likely require clawing back current allocation and potentially compensating existing consent holders.

Transfer of existing permits between users or sites is provided for in the RMA (s136), but is tightly controlled. A permit holder's interest may be transferred with a change in ownership or occupation of the site to which the permit pertains. Permits to use water (excluding diverting or damming waters) may also be transferred among users and sites, where the sites are in the same catchment or aquifer system, and the transfer is allowed by a regional plan or approved by the relevant local authority.

Many regional councils have developed guidance for transfer of water rights. The Canterbury Land and Water Regional Plan (s4.70), for example, enables transfers as a means to achieve 'economic and social outcomes, reduction in water use in over-allocated catchments, improvement in the efficiency of water use, and encouragement of more effective storage and distribution of water'. Transfers must be within the same surface water catchment or groundwater management zone. Transfers within over-allocated catchments or zones may be contingent on a proportion of the allocation being surrendered and retired.

Where water users are organised into collectives such as irrigation schemes, water may be allocated via a global consent, or managed by the collective. The collective entity is accountable to the regional authority for certain management, monitoring and reporting obligations, and is also responsible for ensuring compliance with consent conditions among its membership. Boone and Fragaszy (2018) present two cases of water user groups in the Canterbury and Hawke's Bay regions, which demonstrate how such arrangements can enable greater flexibility and efficiency through collective responses to water shortages or mitigation of adverse effects of land and water use.

Market-based options for the transfer of water rights have long been discussed (Raffensperger & Milke, 2017), and were also advocated by the OECD (2017) in its review of water governance in NZ, but these have, to date, not developed. There are several barriers to implementing market mechanisms for allocation, but one of the most significant in the NZ context is the unresolved nature of Māori proprietary rights and interests in water.

11.5 MONITORING AND COMPLIANCE

11.5.1 Monitoring of water abstraction

Sparse water-use data and limited understanding of effects of localised abstractions have posed challenges in legal proceedings related to water take consents, especially regarding agricultural water use (Boone & Fragaszy, 2018). National regulations issued in 2016 mandated holders of water consents for takes of >5 l/second to install a verified flow measuring device, and report use data annually. The Office of the Auditor General (OAG, 2018), upon reviewing the resulting data, concluded that low data quality limited the effectiveness of the regulations.

In response, amendments to the *Resource Management (Measurement and Reporting of Water Takes) Regulations* in 2020[11] now require most consent holders to record data on water abstraction every 15 minutes, and report to regional councils daily. These reforms, alongside the requirement for development and use of freshwater accounting systems, as required under the NPS-FM, will over time provide a much more nuanced and detailed picture of water use and help to address over-allocation, or free up available water for other users.

11.5.2 Compliance and enforcement

The RMA makes it an offence to contravene the rules relating to water, or any enforcement order, abatement notice, or water shortage direction (e.g. to cease or reduce take) issued by a regional council. The Act specifies penalties (fines or imprisonment) for offending persons or companies.

Regional councils carry out compliance monitoring and enforcement activity. Enforcement officers monitor compliance with resource consent conditions. As authorities lack capacity to monitor all consents, they generally employ a risk-based approach, which focuses on higher-risk consents. Canterbury Regional Council, for example, assesses and grades a sample of consents between fully compliant and significantly non-compliant. In cases of partial non-compliance, a proactive educative approach is taken to support the user to comply, whereas enforcement is reserved for significant and/or ongoing non-compliance. Non-compliance tends to increase in times of water restrictions, and often relates to failures to report use data. There is some concern that penalties are insufficient to deter water takes during restrictions or bans. This may apply in particular to high-value crops that require water at specific periods, where it is perhaps economic to take water in breach of consent conditions and pay the resulting fines.

11.6 THE WIDER POLICY CONTEXT

A major and overarching reform of NZ resource management legislation is currently affecting water policy development. A comprehensive official review (Resource Management Review Panel, 2020) recommended that the RMA be repealed and replaced by three new Acts: a Natural and

[11] See https://www.legislation.govt.nz/regulation/public/2010/0267/latest/DLM3174201.html

Built Environments Act, a Strategic Planning Act, and a Climate Adaptation Act (see New Zealand Government, 2020). Several policy concerns prompted the review including recognition of significant pressures on and decline in the natural environment, poorly managed urban development, and the need to ensure Māori have an effective role consistent with the Treaty of Waitangi.

As the overarching national resource management reform progresses, regional authorities will need to continue to regulate and control freshwater takes and use, in a complex and changing policy and governance context. Several related areas of policy and management will need to be integrated.

11.6.1 Devolved management and collaborative governance

Overall, national policy reforms have resulted in counter-tendencies related to participatory engagement in freshwater planning and management. Whereas the NPS-FM requires broadly participatory and collaborative freshwater planning processes, other national reforms intend to aggregate local authorities' municipal water supply, stormwater, and sewerage infrastructure and governance responsibilities, thus removing them from immediate local control[12].

While it remains to be seen what new resource management legislation will enable or require, regional freshwater planning and management are proceeding with diverse approaches to collaboration and engagement with water users, land owners, *tangata whenua*[13], communities and industry stakeholders. The NPS-FM, and other rules in development (e.g. rules on freshwater farm plans), emphasise close engagement with *tangata whenua* and communities in the formulation of long-term visions for the health and wellbeing of waterways and communities to guide planning (including rule-making) at the local or catchment scale. Increasingly, catchment groups and community environmental groups are integral to the delivery of environmental policy at the local level, and such groups operate with varying degrees of government support (Sinner *et al.*, 2022).

11.6.2 Māori rights and interests

Successive revisions to the NPS-FM have strengthened the obligations on regional authorities to partner with and involve Māori in freshwater decision-making and management. The adoption of *Te Mana o te Wai* as the fundamental guiding concept in the 2020 NPS-FM signals greater prominence for *Te Ao Māori* (the Māori world) principles and concepts to guide resource management and NZ policy generally. The wider resource management reforms also propose stricter requirements on regional authorities to 'give effect to' (rather than merely 'take into account') the Treaty of Waitangi, and to recognise *Te Ao Māori* and *mātauranga* Māori (Māori knowledge) in regional planning and management.

Increasingly, there are calls for a bicultural framing to encompass water allocation and all resource management (Taylor *et al.*, 2021). As noted in Section 11.2.4 above, Māori are also pursuing procedural and proprietary rights and interests through the Waitangi Tribunal process and the court system, and the outcomes of these claims stand to fundamentally shape how water management develops in NZ and perhaps even globally. Already, policy responses to these issues, including the granting of legal personhood such as via the Te Awa Tupua Act 2017 and Te Urewera Act 2014[14], are the subject of intense scholarly interest and activism abroad (e.g. Lambooy *et al.*, 2019; Rodgers, 2017; Westerman, 2019).

11.6.3 Land use change and water quality

Widespread concerns about water quality and impacts on ecology and people have been key drivers of policy reforms. Rapid expansion of irrigation – and particularly irrigated dairy farming – has driven

[12] This is the 'Three Waters Reform' programme: https://www.dia.govt.nz/Three-Waters-Reform-Programme

[13] Meaning: In relation to a particular area, the *iwi*, or *hapu*, that holds *mana whenua* (customary authority) over that area.

[14] See: https://www.legislation.govt.nz/act/public/2014/0051/latest/DLM6183601.html

huge increases in water abstractions, synthetic fertiliser application, and stocking densities. While expansion of irrigation has put pressure on water quantity in some regions, the significant nutrient losses to groundwater and surface waters are even more significant limiting factors on resource use. Indeed, discharge consents are required for particular farming activities in most regions.

In key dairy farming regions, like Canterbury and Southland, it is difficult to see how NPS-FM nutrient requirements can be met without significant land use change in some catchments. New national regulations have also introduced input controls, capping application of synthetic Nitrogen at a rate of 190 kg/ha/yr from 2021 for pastoral land. Regional plans have imposed specific nutrient reductions for particular catchments (e.g. across water management zones in the Canterbury Region – see Jenkins, 2018), and in other areas there are cap and trade schemes (e.g. the Lake Taupo nitrogen trading programme under the Waikato Regional Plan – see Duhon *et al.*, 2015).

11.6.4 Climate change and hazards

As discussed, current reforms aim to integrate wide-ranging environmental policy objectives. The NPS-FM contains the explicit aim that freshwater is 'managed as part of NZ's integrated response to climate change'. In setting environmental flows and limits on resource use, regional councils must have regard to the foreseeable impacts of climate change. These impacts will vary across NZ. Annual average rainfall is projected to increase in the west and south of the country, and to decrease in the east and north, while extremes in terms of rainfall events and dry periods are expected to increase (Collins & Tait, 2016). The occurrence and duration of drought is expected to increase in parts of Northland, Gisborne, Canterbury and Otago (ibid.). Overall, impacts on freshwater resources will be diverse and complex, and it will be difficult for authorities to plan for these in the course of water allocation and water management more broadly. Already, local water restrictions and droughts have severe social and economic impacts for rural and urban water users. For example, Auckland Council's water utility, WaterCare, is drawing 225 million litres per day from the Waikato River in the neighbouring Waikato Region to supply Auckland City in the wake of a sustained drought and water shortages through 2020–2021.

11.7 CONCLUSION

Water allocation is recognised as an issue of national importance in NZ. Allocation by regional councils via resource consents under the RMA has proceeded in a relatively permissive way, insufficiently considering the cumulative environmental impacts of large and small water takes. As a result, many catchments and aquifers are today over-allocated, or face seasonal or periodic water shortages. Furthermore, this approach – along with central government promotion – has enabled significant expansion of irrigation, which has in turn supported ecologically unsustainable intensification of farming in some areas. In this sense, regional councils have largely failed to implement the RMA effectively for the sustainable management of freshwater.

The NPS-FM – first introduced in May 2011 – provided much-needed national direction to regional authorities. It now requires water quality within each region be maintained or improved, and for over-allocation be remedied. To this end, the NPS-FM introduced requirements for regional councils to set freshwater objectives and limits on resource use to achieve those objectives. This approach may help to address the water quantity and quality problems that have developed over the past two decades, but the nature and degree of change required at the catchment scale will be challenging for regional councils and water users. Indeed, regional councils have, in the past, found it difficult to balance development and conservation interests, and the stakes are no lower at the current point in time.

Current reforms, driven by a central government apparently determined to address difficult legacy issues that have degraded the freshwater environment, are ambitious and aim to overhaul NZ's natural resource management system. The success of this far-reaching reform effort, in terms of securing sustainable water allocation, will depend on how implementation efforts address the previously intractable issues discussed, including complex challenges around land-use change. Also, the extent to which regional council processes give effect to the Treaty of Waitangi and address Māori rights and interests in water will certainly have significant implications for water allocation options and for addressing inefficiencies and inequities in the current system.

Finally, the role of regional councils in any future water allocation regime may need to shift. While devolving resource management functions to local government has made sense in many ways, regional councils have also struggled to achieve sustainable management of natural and physical resources. Councils are unevenly resourced and generally do not have the capacity to discharge all of their statutory duties and responsibilities (Kirk *et al.*, 2020). The NZ government is currently reviewing the role of local government. The Future for Local Government Panel (2021) is charged with 'considering how NZ's system of local democracy and governance will need to evolve over the next 30 years'. While the future shape of regional councils is unclear at this stage, they will likely need greater resourcing, better coordination with central government, and greater capacity to adapt and innovate in the area of water allocation systems.

REFERENCES

Bennion T. (2017). Introduction. In: New Zealand Land Law, E. Toomey, T. Bennion and D. Brown (eds.), 3rd edn, Thomson Reuters, Wellington, pp. 1–33.

Booker D. J. and Henderson R. D. (2019). National Water Allocation Statistics for Environmental Reporting; 2018. National Institute of Water & Atmospheric Research, Christchurch.

Boone S. and Fragaszy S. (2018). Emerging scarcity and emerging commons: water management groups and groundwater governance in Aotearoa New Zealand. *Water Alternatives*, **11**(3), 795–823.

Clapcott J., Ataria J., Hepburn C., Hikuroa D., Jackson A.-M., Kirikiri R. and Williams E. (2018). Mātauranga Māori: shaping marine and freshwater futures. *New Zealand Journal of Marine and Freshwater Research*, **52**(4), 457–466, https://doi.org/10.1080/00288330.2018.1539404

Collins D. and Tait A. (2016). Climate change effects on fresh waters. In: Advances in New Zealand Freshwater Science, P. G. Jellyman, T. J. A. Davie, C. P. Pearson and J. S. Harding (eds.), New Zealand Hydrological Society and New Zealand Limnological Society, Wellington, pp. 401–414.

Duhon M., McDonald H. and Kerr S. (2015). Nitrogen Trading in Lake Taupo: An Analysis and Evaluation of an Innovative Water Management Policy. Motu Economic and Public Policy Research, Wellington, New Zealand.

Environment Foundation (2021). Environment Guide. Retrieved 10 December 2021, from https://www.environmentguide.org.nz/

Future for Local Government Panel (2021). Ārewa ake te Kaupapa: Raising the platform – Interim report. Future for Local Government Panel, Wellington.

Grinlinton D. (2011). Evolution, adaptation, and invention: property rights in natural resources in a changing world. In: Property Rights and Sustainability: The Evolution of Property Rights to Meet Ecological Challenges, D. Grinlinton and P. Taylor (eds.), Nijhoff Publishers, Leiden, The Netherlands, pp. 273–304.

Harmsworth G. and Awatere S. (2013). Indigenous māori knowledge and perspectives on ecosystems. In: Ecosystem Services in New Zealand: Conditions and Trends, J. R. Dymond (ed.), Manaaki Whenua Press, Lincoln, New Zealand, pp. 274–286.

Jenkins B. R. (2018). Water Management in New Zealand's Canterbury Region: A Sustainability Framework. Springer, Dordrecht, The Netherlands.

Kirk N., Robson-Williams M., Fenemor A. and Heath N. (2020). Exploring the barriers to freshwater policy implementation in New Zealand. *Australasian Journal of Water Resources*, **24**(2), 91–104, https://doi.org/10.1080/13241583.2020.1800332

Lambooy T., van de Venis J. and Stokkermans C. (2019). A case for granting legal personality to the Dutch part of the Wadden Sea. *Water International*, **44**(6–7), 786–803, https://doi.org/10.1080/02508060.2019.1679925

Macpherson E. J. (2019). Indigenous Water Rights in Law and Regulation: Lessons From Comparative Experience. Cambridge University Press, Cambridge, United Kingdom.

Ministry for Primary Industries (2021). Water Availability and Security in Aotearoa New Zealand: Supporting the sustainability, productivity, and resilience of the food and fibre sector. MPI Information Paper No: 2021/04. Ministry for Primary Industries, Wellington.

Ministry for the Environment (2015). A Guide to the National Policy Statement for Freshwater Management 2014. Ministry for the Environment, Wellington.

New Zealand Government (2020). Natural and Built Environments Bill: Parliamentary Paper on the Exposure Draft. New Zealand Government, Wellington.

Ngā Tāngata Tiaki o Whanganui (2021) Our Story. Retrieved 10 December 2021, from https://www.ngatangatatiaki.co.nz/our-story/

OAG (2018). Monitoring How Water is Used for Irrigation. Office of the Auditor-General (OAG), Wellington.

OAG (2020). Reflecting on Our Work About Water Management. Office of the Auditor General (OAG), Wellington.

O'Connell E., Greenaway T., Moeke T. and McMeeking S. (2018). He Ara Waiora/A Pathway Towards Wellbeing. Discussion Paper 18/11. The Treasury, Wellington.

OECD (2017). OECD Environmental Performance Reviews: New Zealand 2017. OECD Publishing, Paris, https://doi.org/10.1787/9789264268203-en

Office of Treaty Settlements (2016). Regulatory Impact Statement: Te Awa Tupua (Whanganui River) Framework. Office of Treaty Settlements, Wellington.

PCE (2019). Focusing Aotearoa New Zealand's Environmental Reporting System. Parliamentary Commissioner for the Environment (PCE), Wellington.

Raffensperger J. F. and Milke M. W. (2017). Smart Markets for Water Resources: A Manual for Implementation. Springer, Cham, Switzerland.

Resource Management Review Panel (2020). New Directions for Resource Management in New Zealand: Report of the Resource Management Review Panel, June 2020. Resource Management Review Panel, Wellington.

Rodgers C. (2017). A new approach to protecting ecosystems: The Te Awa Tupua (Whanganui River Claims Settlement) Act 2017. *Environmental Law Review*, **19**(4), 266–279, https://doi.org/10.1177/1461452917744909

Ruru J. (2009). The Legal Voice of Māori in Freshwater Governance: A Literature Review. Landcare Research New Zealand, Lincoln.

Ruru J. (2010). Undefined and unresolved: exploring Indigenous rights in Aotearoa New Zealand's freshwater legal regime. *Journal of Water Law*, **20**, 236–242.

Ruru J. (2011). Māori legal rights to water: ownership, management, or just consultation? *Resource Management Theory & Practice*, **7**, 119–135.

Salmond A. (2014). Tears of Rangi: water, power, and people in New Zealand. *Hau: Journal of Ethnographic Theory*, **4**(3), 285–309, https://doi.org/10.14318/hau4.3.017

Sinner J., Tadaki M., McCarthy A., Challies E. and Thomson-Laing J. (2022). Catchment and community environment groups in Aotearoa New Zealand: Goals, activities and needs. Cawthron Report No. 3733, prepared for Ministry for the Environment. Cawthron Institute, Nelson.

Stewart-Harawira M. W. (2020). Troubled waters: Maori values and ethics for freshwater management and New Zealand's fresh water crisis. *WIREs Water*, **7**(5), e1464, https://doi.org/10.1002/wat2.1464

Taylor L. B., Fenemor A., Mihinui R., Sayers T. A., Porou T., Hikuroa D., Harcourt N., White P. and O'Connor M. (2021). Ngā Puna Aroha: towards an indigenous-centred freshwater allocation framework for Aotearoa New Zealand. *Australasian Journal of Water Resources*, **25**(1), 27–39, https://doi.org/10.1080/13241583.2020.1792632

Te Aho L. (2019). *Te Mana o te Wai*: An indigenous perspective on rivers and river management. *River Research and Applications*, **35**(10), 1615–1621, https://doi.org/10.1002/rra.3365

Waitangi Tribunal (2012). The Stage 1 Report on the National Freshwater and Geothermal Resources Claim: WAI2358. Waitangi Tribunal, Wellington.

Waitangi Tribunal (2019). The Stage 2 Report on the National Freshwater and Geothermal Resources Claims: WAI2358. Waitangi Tribunal, Wellington.

Warnock C. and Baker-Galloway M. (2015). Focus on Resource Management Law. LexisNexis, Wellington.

Westerman A. (2019). Should Rivers Have Same Legal Rights As Humans? A Growing Number Of Voices Say Yes. National Public Radio: https://www.npr.org/2019/08/03/740604142/should-rivers-have-same-legal-rights-as-humans-a-growing-number-of-voices-say-ye

doi: 10.2166/9781789062786_0143

Chapter 12

Groundwater allocation in New South Wales, Australia

Joseph H. A. Guillaume[1], Alvar Closas[2] and Andrew McCallum[2]

[1]*Institute for Water Futures and Fenner School of Environment & Society, The Australian National University, Canberra, Australia*
[2]*Water group, Department of Planning and Environment, Parramatta, New South Wales, Australia*

ABSTRACT

New South Wales has more than 100 years of history of water licensing and allocation. This chapter reflects on the approach to water allocation in the current groundwater sharing plans, including general principles and underlying reasoning for application elsewhere. Focus is on groundwater-specific issues for transition from open to regulated access, while embedded within broader water regulations and connections to surface water management. Water allocation is built around water sharing plans that determine extraction limits, with community consultation. Water rights are differentiated in terms of water sources and priority, separated from land ownership, and from time-varying water allocations, subject to available water determinations. Both water entitlements and allocations can be traded, with rules governing impact of trade. Water sharing plans are state-level instruments explicitly connected in applicable regions to the Commonwealth-level Murray–Darling Basin Plan and associated extraction limits. Compliance is based firstly on metering of water extractions. Future prospects are also discussed.

Keywords: Australia, Groundwater allocation, New South Wales

12.1 INTRODUCTION

While Australia is known as the driest inhabited continent, it also spans climate zones with substantially different water availability and water use conditions (Head *et al.*, 2014). As water management is a state responsibility, the focus here is on the state of New South Wales (NSW). NSW covers an area of 800 thousand square kilometres and spans subtropical to arid climates. The Great Dividing Range runs north–south along the east coast of Australia. East of the Great Dividing Range, the terrain is steeper and the climate is generally wetter. West of the Great Dividing Range, long rivers flow across flat landscapes, including those in the over 3000-kilometre-long Murray–Darling basin, which is shared with the states of Queensland, Victoria, and South Australia. Around 40% of Australia's agricultural produce comes from the Murray–Darling basin (MDBA, 2020). While NSW has a population of over eight million, nearly two thirds live in the Sydney area.

New South Wales has more than 100 years of history of water licensing and allocation. This chapter focuses on the approach to water allocation in the current groundwater sharing plans and its implementation, within its historical context and connections to surface water and other policy. Groundwater provides on average around 20% of water needs in NSW every year. Groundwater has important traditional and cultural values for Aboriginal communities who have been using and protecting groundwater resources for millennia. Groundwater is also a vital resource for many natural habitats and ecosystems, towns and rural communities as well as economic activities in NSW. Around 300 000 people in over 250 towns in rural NSW depend on groundwater for their supply and close to $800 million in direct value is generated every year from activities using groundwater. Agriculture is the largest user of groundwater, with 50% of the total share per year, followed by other industries with 26% of the share and towns with 8% (The CIE, 2021).

The chapter begins with the overarching institutional framework (Section 12.2), including distribution of responsibilities for water management, principles that apply to both surface and groundwater, and the historical context of current policy. It then describes the definition of the available resource pool and its allocation (Section 12.3 and 12.4), which involves water plans combining long-term extraction limits derived by a variety of methods, time-varying water allocations, and assignment of tradeable water rights. Monitoring and compliance (Section 12.5) spans water use, state of the system, and performance of water plans. Water plans sit within a broader policy mix for land and water management (Section 12.6). The chapter finishes with an overview of strengths and issues for the future (Section 12.7).

12.2 THE OVERARCHING INSTITUTIONAL FRAMEWORK

12.2.1 Water management as a state responsibility with federal engagements

Water management in Australia is a state responsibility within a federal system (Crase *et al.*, 2012). The institutional framework for water management in NSW has developed locally with federal engagements evolving over time, rather than being led by a top-down hierarchical framework. Historically, the First Nations and Aboriginal people of NSW developed rights and a moral obligation to care for water as part of their culture, connecting them to downstream communities, throughout catchments and over connected surface water and groundwater systems. Within the limits of the Australian constitution, a federal role has been actively forged through international issues, common jurisdictional concerns, and the ability of the Commonwealth to grant financial assistance (Turral & Fullagar, 2007). Despite differences between states, there have been common approaches to water management, as expected through common history, continuing interaction, and common development pathways.

Water management has evolved over time, generally with increasing environmental protection. As summarised by Bell and Park (2006), early policy focused on expanding water use with support for large dams and irrigation schemes until the 1980s. By 1994, a shift towards a sustainability framework had begun, with the Council of Australian Governments (COAG) Strategic Framework on Water Reform recognising water policy as a cross-jurisdictional issue to be approached with principles of sustainable development. The COAG Water Reform Framework included cost recovery pricing and implementation of a system of water allocations by the states with provision for environmental water allocations, separation of land and water titles, and trade in water entitlements (Crase *et al.*, 2000). Water agencies were also required to separate regulation, service delivery and water resource management functions. In 1995, these reforms were linked to the National Competition Policy, which provided monetary incentives to states for water reform (Crase *et al.*, 2000).

COAG reinforced the overarching institutional framework through the National Water Initiative (NWI) in 2004. The NWI is a shared commitment by governments (1) to prepare water plans with provision for the environment, (2) to deal with over-allocated or stressed water systems, (3) to introduce registers of water rights and standards for water accounting, (4) to expand the trade in water, (5) to improve pricing for water storage and delivery, and (6) to meet and manage urban water demands (Crase *et al.*, 2012).

Water sources that fall within the Murray–Darling Basin (MDB) have also driven the need to harmonise water management across states. The Basin Plan 2012 defined water resource plan (WRP) areas within the MDB and associated Sustainable Diversion Limits (SDLs). WRPs are prepared by the states, with the Murray–Darling Basin Authority (MDBA) working with the states to ensure the requirements of the Basin Plan are met. In practice in NSW, existing water sharing plans (WSPs) have been reviewed to align them with the SDLs and Basin Plan, with the WRPs reflecting the arrangements in WSPs and describing how stated requirements are met. Furthermore, groundwater SDLs in NSW were not reduced further from the quantified Baseline Diversion Limits (BDLs), instead retaining extraction limits from existing plans, noting that outcomes of reductions through a previous programme were not yet realised (MDBA 2012a, 2012b). The process of federal harmonisation of state water management is still in progress.

12.2.2 Historical groundwater regulation in New south Wales

The NSW government has long had an involvement in groundwater management, initially by encouraging groundwater development in its varied groundwater resources (see Figure 12.1 and Table 12.1). This started with the first water supplies for Sydney, then expanded to the Great Artesian Basin and inland alluvial basins to support agricultural development, with tenders for the construction of artesian bores in the Great Artesian Basin (GAB) (Public Watering Places Act 1884). Groundwater extraction was further promoted through the Artesian Wells Act of 1897 which enabled groups of

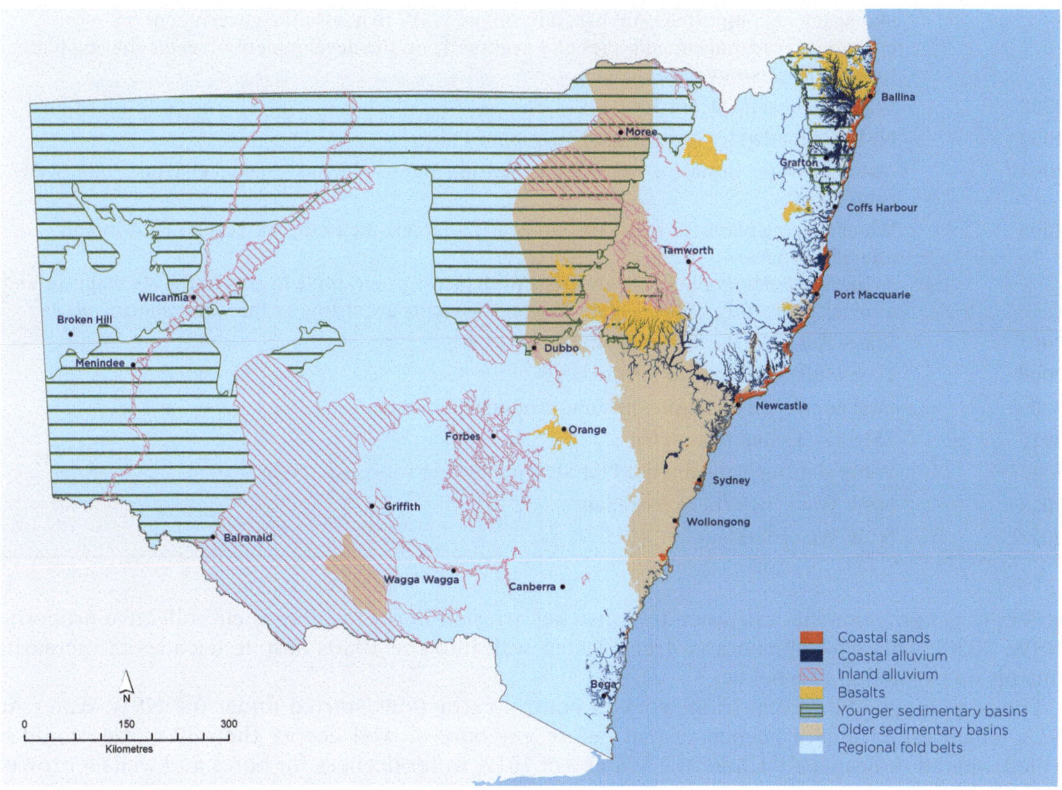

Figure 12.1 Groundwater basins in New South Wales. Source: Department of Planning and Environment.

Table 12.1 Key events and dates for groundwater management in New South Wales (NSW) and Australia.

1884	Public Watering Places Act 1884
1897	Artesian Wells Act
Federation 1901	Water is the responsibility of state and territory governments under the Australian Constitution
1912	Water Act 1912 (NSW) with volumetric licences and tradeable
1914	Murray–Darling Basin Act
1955	Irrigation, Water and rivers foreshores improvement (Amendment) Act 1955 All bores to be licensed
1972 to 1983	Irrigation licences issued on the basis of irrigated area, renewable every five years
1984	Water (Amendment) Act 1984 Conversion from area-based to volumetric allocations and unrestricted licences taking more than 20 ML per year
1986	Water (Amendment) Act 1986 (NSW) Comprehensive volumetric groundwater allocation policies
1994	Council of Australian Governments (COAG) Strategic Framework for Water Reform
1996	Allocation and use of groundwater – A National Framework to Improve Groundwater Management. Adopted by all Australian jurisdictions under the COAG
1997	NSW State Groundwater Policy Framework
1998	NSW Groundwater Quality Protection Policy – Community-based groundwater management committee established in inland NSW to advise the government on groundwater management issues and eventually on the development of water sharing plans for specific groundwater sources
2000	Water Management Act 2000
2002	NSW Groundwater Dependent Ecosystems Policy
2004	National Water Initiative – First water sharing plans for some targeted coastal groundwater sources
2006	Water sharing plans for inland groundwater sources (including the seven major inland alluvial systems) Achieving Sustainable Groundwater Entitlements programme begins in the six major inland alluvials, adjusting to the reduction in entitlements according to the water sharing plans
2007	Water Act 2007 (Commonwealth)
2008	First water sharing plan for the GAB
2009	First controlled allocation in some groundwater systems
2012	Murray–Darling Basin Plan
2016	Water sharing plans finalised for all groundwater sources in NSW
2018	NSW Water Reform Action Plan
2021	NSW Water Strategy

settlers to get government assistance to construct artesian bores to serve their collective properties (NSW, 2020). This developmentalist trend lasted well into the 1980s despite decades of increasing controls on extraction (see below).

The beginnings of a formal framework to control extractions started under the NSW Water Act 1912, which established a licence system before any bore or well deeper than 30 metres could be drilled, altered or deepened. Under the Water Act 1912, water licences for bores and wells for towns, stock and domestic, as well as irrigation users were linked to land and were granted for a fixed term with no restriction on the volume that could be extracted.

The NSW Water Act 1912 was amended in 1955 so that the licence requirements applied to all bores across NSW regardless of their depth. By the 1970s, the expansion of groundwater use began to affect the reliability of supply across parts of inland NSW, so between 1972 and 1983 new irrigation licences began to be issued based on the area to be irrigated. These licences needed to be renewed every five years but did not include volumetric limits on extraction (NSW, 2020).

From 1984, new high-yielding bores and wells were granted a volumetric entitlement and old land area-based licences were progressively converted. The Water (Amendment) Act 1986 introduced comprehensive volumetric groundwater allocation throughout the State and the ability to transfer water entitlements permanently. The 1990s saw embargos on new licences on most coastal, unregulated streams and several groundwater systems (Crase *et al.*, 2000).

12.2.3 Overview of current regulatory framework in NSW

The current regulatory framework in NSW that is the focus of this chapter is the Water Management Act 2000. The object of the Act is 'to provide for the sustainable and integrated management of the water sources of the State for the benefit of both present and future generations'. The water management principles emphasise protection of water sources, water quality, ecosystems, basic landholder rights and features of Aboriginal and cultural heritage or spiritual significance, minimisation of cumulative impacts, maximisation of social and economic benefits to the community, and application of adaptive management. Under the Act, water access licences authorise the taking of water from a water source, including aquifers, within a water sharing plan (WSP) area. WSPs set out rules for allocation of water for 10 years at a time, audited every 5 years and either extended or reviewed. The first WSP was developed in 2004. While previous planning processes involved appointment of water management committees (Bell & Park, 2006), subsequently 'Minister's plans' have involved targeted consultation and public exhibition of rules developed by the state government for a given water source (DPI, 2015). Changes over time have established the Natural Resources Access Regulator (NRAR) as an independent regulator, and the Natural Resources Commission (NRC) as an independent auditor and reviewer. As noted above, within the MDB, WSPs are also currently being aligned with WRPs and the Basin Plan 2012.

WSPs specify the amount of water which is available for extraction from a water source under a long-term average annual extraction limit (LTAAEL, also called 'extraction limit'). WSPs specify restrictions on licences and allocations, including definition of zones within a plan area, rules about carryover of unused water allocations between years and assessing or limiting trade. Distinct rules generally also apply to town water supply and stock and domestic water use. Water for the environment is similarly provided as licensed environmental water (i.e. with access licences and allocations subject to Available Water Determinations, see below) that can be, for example, released from dams to provide environmental flows. However, water is also reserved outside the licence system as 'planned environmental water', which includes water beyond the extraction limit, water not committed by other granted rights, and/or committing to the physical presence of that water (DPI, 2015).

Any licence (entitlement) is associated with a tradeable share of the LTAAEL, along with a tradeable annual water allocation, credited through an Available Water Determination (AWD) (see Section 12.4 for further details). Water licences can be granted with zero shares of water, in which case shares or allocations need to be purchased on the water market. Conditions on licences commonly require, amongst others, metering and keeping a logbook about water taken. Additionally, water supply work approvals are required for construction and use of water bores, and water use approvals are required in the absence of exemptions. The NSW Water Register provides public access to information about water licences, approvals, and trade.

12.3 DEFINING THE AVAILABLE RESOURCE

Each WSP covers multiple water sources and may also split a water source into multiple management zones (DPI, 2015). Boundaries of plans and groundwater sources are defined based on geology, hydraulic connection, and administrative requirements. Water sources associated with the Murray–Darling Basin and Great Artesian Basin are defined to be consistent with their respective management boundaries (DPI, 2015).

The conjunctive use of surface water and groundwater resources is largely left to users. However, depending on the degree of connection between the two resources, the WSPs have rules that do link the management of them. For example, in some areas the annual allocations for groundwater are linked to surface water resource assessments (see below in Section 12.4 on AWDs). In some areas where the systems are highly connected (e.g. on the coast), groundwater is integrated with surface water in the one WSP.

Under the NSW Water Management Act (2000), the WSP must protect each water source, its dependent ecosystems, and basic landholder rights, before making water available for take via a licence. WSPs are the legal instrument to fairly and transparently share water between the environment and consumptive users. Water is reserved for the environment and water is shared between the environment and extractive users on the basis of long-term average annual recharge while recognising existing licensed entitlements. In practice, this has involved reductions from historical usage in stressed systems, setting conservative limits in water sources that are not yet fully committed, and use of groundwater modelling and recharge estimation to inform the LTAAEL (DPI, 2015). In the Murray–Darling Basin, LTAAELs are also demonstrated to comply with SDLs.

In NSW, a proportion of recharge in a given area is assigned for the environment while the remainder is assigned for consumptive use in the form of the LTAAEL. It is worth noting that, while the extraction limit limits the extraction from a given area (a water source), additional rules and regulations restrict extractions at a local scale so that impacts on the resource (in terms of water levels, water quality, and structural integrity) do not unacceptably impact ecosystems or other consumptive users.

Restriction to a proportion of recharge recognises that discharge and groundwater levels are still impacted even if extraction is less than recharge (Pierce *et al.*, 2013). The NWI defines the 'environmentally sustainable level of extraction' as 'the level of water extraction from a particular system which, if exceeded would compromise key environmental assets, or ecosystem functions and the productive base of the resource' (COAG, 2004). Richardson *et al.* (2011) define the term as 'the groundwater extraction regime, measured over a specified planning time frame, that allows acceptable levels of stress and protects dependent economic, social and environmental values.' In this definition, there is flexibility of what is acceptable to different communities. This can evolve over time.

Over the last two decades, various approaches have been used in NSW when determining the extraction limit:

- The first approach applied numerical modelling and was used in most of the highly utilised groundwater systems (e.g. major inland alluvial systems). In this approach, groundwater systems were represented in 3D numerical models (generally MODFLOW) with varying hydraulic properties and boundary conditions combined with time-varying recharge, discharge and extraction patterns. The models were used to compute all inflows, not just recharge, in the water sources: rainfall infiltration, streams and river recharge, floodplain recharge, return flows, as well as 'induced recharge'. These models provided powerful tools for regulators and users to see the impact of different data assumptions and planning scenarios on the resource in the long term. While the models informed water management committee discussions, the final decision was a Ministerial one. Entitlements in many of these systems were then progressively reduced to match the extraction limit (see discussion below).
- The second approach involved using the proportion of estimated recharge or storage, and was applied in systems which are not as utilised (e.g. fractured and porous rocks). Here, the approach has been to estimate recharge to a groundwater source using annual average rainfall, diffuse rainfall recharge modelling (Crosbie *et al.*, 2008), analytical modelling such as the Hill plot method (Sharp, 2016) or Chloride Mass Balance data (Bioregional Assessment Programme,

2015; Crosbie *et al.*, 2010; Scanlon *et al.*, 2002). A proportion of this, based on a risk assessment approach, is then allowed for consumptive extraction (e.g. CSIRO & SKM, 2010a, 2010b, 2010c, 2010d). In a few buried groundwater sources, a small proportion of total storage was set as the extraction limit.

- The third approach, which used the level of usage or entitlements, was chosen for groundwater systems highly connected to surface water systems. In this approach, the level of extraction (either historical or expected in a given planning horizon) or current level of entitlement, plus an estimate for basic landholder rights, was adopted as the extraction limit. The rationale behind this is that capping extraction at existing levels will not lead to further groundwater extraction-induced impacts to surface water features.

12.4 DEFINING ALLOCATION AND REALLOCATION RULES

As seen in the prior section, the extraction limit is a key concept in NSW groundwater management. It defines the size of the resource available for consumptive use. Extractions are not permitted, on average, to exceed this limit.

Conceptually, the licensed shares and annual allocations work within the larger framework of the extraction limit (see Figure 12.2). However, as shown in Table 12.1, rights to access groundwater were granted prior to the introduction of extraction limits. This provided groundwater managers with a challenge for fully committed systems: to either reduce entitlements or allocations. A combination of both approaches was used in NSW.

In the 2000s, WSPs required staged reductions in groundwater use in seven alluvial groundwater systems, which were accompanied by a number of legal attempts to declare plans invalid. A confluence of cooperation of groundwater users across the states, sympathetic federal and NSW Ministers, and threat of further litigation (Schuster *et al.*, 2020) led to a joint federal and state programme known as Achieving Sustainable Groundwater Entitlements (ASGE) from 2005. This programme reduced entitlements based on history of extraction rather than on a pro rata basis, and provided financial assistance based on lost active water, in addition to a community development fund (Schuster *et al.*, 2020). These groundwater sources are also considered SDL resource units under the Basin Plan, which is due to be reviewed by the MDBA in 2026.

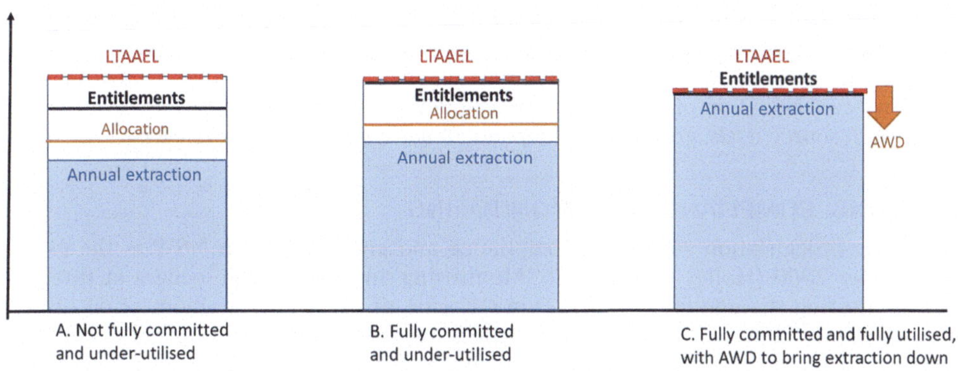

Figure 12.2 Relationship between limits, entitlement shares, allocation and uses. Note: to manage groundwater sustainably, NSW has set up a system to control extraction via the definition of LTAAEL, groundwater entitlements with shares, and allocations combined with metering.

AWDs provide a mechanism for varying the available resource pool each year by crediting the accounts of licensed water users with their allocation through a statutory order. The AWDs are tied to both resource assessment and compliance. For surface water, using the Murrumbidgee Regulated River system as an example, a resource assessment includes calculation of water available in active storage, minimum inflows, volumes remaining in accounts, planned environmental water, and losses (WaterNSW, 2019). Each year, a first assessment and AWD order is made on 1 July, which is then updated twice a month, with the potential to increase water allocations over time.

In most groundwater systems, the AWD mechanism is retained, but has primarily been driven by compliance of a multi-year rolling average extraction value compared against the LTAAEL. If extraction exceeds the limits set in a WSP, a Ministerial decision can be made to reduce allocations in the following years to return average extraction to the limit. This can be made either with an AWD that credits allocations with a percentage of their entitlements, or by reducing the amount of water that can be taken from accounts. This is most likely to occur following a run of dry years during which reliance on groundwater increases, and has typically involved public consultation and Ministerial discretion in determining how much allocations should be reduced. Discretion in the AWD mechanism effectively provides a degree of flexibility recognising that acceptable short-term impacts of extraction are uncertain, even if compliance with long-term sustainable diversion limits is important.

This also aligns with flexibility in how licensed water users can use their allocations over time. While rules differ by WSP, most groundwater systems allow for limited carryover of allocations, that is at least a portion of water credited through an AWD can be retained for later years. This means that groundwater allocations to some extent can be accumulated for use in dry periods, while disincentivising both using up allocations at the end of a year and holding large amounts of unused allocations. In regulated river systems this means that water left in a reservoir will in principle allow for greater AWDs to all users in the next year. In groundwater systems, lower water use reduces impact on the aquifer and the chance of needing to reduce AWDs.

Trade of water shares and annual allocations is the primary mechanism for reallocation of groundwater between users. As noted in describing the overarching institutional framework, entitlements in NSW have been tradable between private diverters since the late 1980s (Department of Agriculture, Water and the Environment, n.d.), an emphasis on market mechanisms was reinforced at national scale in 1994 with the COAG Strategic Framework on Water Reform. Under the Water Management Act 2000, 'dealings' with access licences require consent from the Minister, that is licence holders submit an application for rights or allocations to be assigned (WM Act, 2000). The application form also requires information about the price and reason for trade. Clearing, financial settlement, and the application process may be supported by a water broker. Rules for assessing or limiting trade vary between WSPs, such as defining zones in which inbound trade is not permitted to prevent further concentration of groundwater extraction. Groundwater trades are assessed for local impacts on the resource itself, groundwater dependent ecosystems, and other users.

12.5 METERING, COMPLIANCE AND MONITORING

The main piece of legislation regulating compliance and enforcement in NSW is the *NSW Water Management Act 2000* (Holley *et al.*, 2020). Monitoring and compliance occurs at three levels in NSW: (1) monitoring the aquifer system, (2) monitoring of users and (3) monitoring of plans and evaluation of plans. WaterNSW operates a network of groundwater monitoring bores across the state, including real-time and manual monitoring, and water level and quality, including nested bores monitoring multiple depths. Amongst other uses, the bore network is used to support Annual Reports for groundwater sources, which may include assessment of trends in water levels, assessment of the impact of climate and pumping, and identification of drawdown hotspots.

Water use is generally metered in NSW, either through smart meters with automated telemetry installed or with a logbook for non-metered works. Bore drilling is also regulated with specific licences issued to drillers to carry out bore construction, decommissioning or work on existing bores. Water use is available in the NSW Water Register for a water source as a whole, and is also reported in Annual Reports and the Water Insights web platform, as well as in a dashboard including an assessment of the likelihood of compliance triggers being reached. Annual reports also include water accounts and trade statistics.

The tools and policies supporting groundwater monitoring and compliance in NSW have generally focused on the Murray–Darling Basin whilst the management of coastal groundwater sources remains challenging as monitoring and metering are not as extensive (NRC, 2021).

Since 2018, enforcement is overseen by the independent Natural Resources Access Regulator (NRAR). NRAR uses satellite imagery, site inspections and other data sources in addition to metering data, to ensure that water is lawfully used in NSW and regularly publishes the results of their enforcement actions, which include warnings, penalties, and prosecution.

Along with the creation of NRAR, in 2018 the NSW Government released its Non-urban Water Metering Framework, a key commitment from the 2017 Water Reform Action Plan and the Murray–Darling Basin Compact. The framework aims to improve the standard and coverage of non-urban water meters in NSW, providing water users with clear guidance and support for metering installation and requirements (NSW DPIE, 2021). The roll-out of this framework is being done in stages across the State, with Stage 1 beginning in April 2019 and concerning faulty meters and inactive works and ending with Stage 4 in December 2023 with the Eastern and Coastal regions of NSW (NSW DPIE, 2021).

The Water Management Act (2000) emphasises adaptive management, and therefore monitoring and evaluation of plans. WSPs are intended to be changed every 10 years and reviewed every 5 years, evaluating their effectiveness and user compliance, and revisiting new science and knowledge. Since 2003, the Natural Resources Commission Act has required independent auditing and review of plans and other natural resource management issues by the Natural Resources Commission. Auditing ascertains whether a plan's provisions are being put into practice, that is evaluating that operational arrangements are in place, including delineation of roles and responsibilities across agencies, and performance monitoring arrangements. The review process itself determines whether the implementation of plan provisions has resulted in the expected environmental, social and economic outcomes. This includes a call for public submissions, document review, and technical advice, with consideration of other policies that apply to the region.

The articulation of reliable monitoring and data transfer tasks between the Commonwealth and NSW was further expanded with the Strategic Water Information and Monitoring Plan in 2009. The plan aimed to provide a state-wide view of water monitoring systems and priorities within NSW leading to increased investment for enhanced water data capture and delivery processes to the Bureau of Meteorology.

In addition, within the Murray–Darling Basin, the MDBA is responsible for overseeing compliance of Water Resource Plans (WRPs) with the Basin Plan 2012, that is regulating the regulators. Each of the 11 WRPs includes corresponding WSPs and are intended to ensure that SDLs are not exceeded over time. WRPs are assessed by the MDBA against water resource plan requirements laid out in Chapter 10 of the Basin Plan. In addition to describing compliance with SDLs, requirements for WRPs also cover interception activities, planning for environmental watering, water quality objectives, restrictions on trade, risk management, measuring of water taken and monitoring of the water resource, reviews and amendments of the WRP, the scientific information or models on which the WRP is to be based, planning for extreme events, and Indigenous values and uses. As of 2021, the groundwater WRPs in NSW were being amended prior to re-submission, assessment and accreditation by the MDBA.

12.6 THE BROADER POLICY-REGULATORY MIX

This chapter thus far has focused on water allocations as implemented within the framework of WSPs and the WM Act 2000. This, however, is only one aspect of how water use and corresponding impacts are managed in NSW. Other regulatory mechanisms are available to water managers. Furthermore, to understand water management in NSW in the broadest sense it is important to consider it from a strategic lens. This section of the chapter considers the latter first and then briefly outlines some of the regulatory mechanisms.

In 2021, the NSW government released its first state-wide water strategy (NSW, 2021). Significantly, this strategy puts water on the same footing as other state resources and services. It sets the long-term direction for water management and aims to improve the security, reliability, quality, and resilience of the state's water resource. It sets the necessary priorities and actions to deliver these outcomes (NSW, 2021). As part of the strategy, a renewed emphasis on sustainable groundwater management is included, with Action 3.6 specifically dedicated to developing new opportunities to use and manage groundwater more efficiently, innovatively and sustainably (ibid.).

The NSW Water Strategy is aligned with a suite of place-based water strategies. Two metropolitan and 12 regional strategies provide a roadmap to delivering water security for local communities, enabling economic prosperity, recognising Aboriginal peoples' values and rights, and protecting and enhancing the environment (NSW, 2021a). This is the first time NSW has developed aspirational strategies like this in close consultation with communities. Ultimately, their value will depend on the success of the implementation phase.

Water management in NSW is also integrated with other government initiatives. For example, the NSW Government in 2021 released the Future Ready Regions Strategy (NSW, 2021b). This strategy aims to support community and economic development by building drought resilience and sustainably using natural resources. The first action is to 'Fast track investigations into new groundwater supplies in western NSW'. This project is currently underway. Water is increasingly a consideration in regional planning and enabling economic development.

During droughts, critical human needs become the highest priority under the WM Act 2000. This may involve suspending part or the whole of water sharing plans. This is not done lightly as the intent of the plans is to provide certainty to users about water access. Within the NSW Murray–Darling Basin, the Extreme Events Policy and corresponding Incident Response Guides provide guidance on how droughts, and other extreme events, will be managed. A staged approach, with increasing interventions as the event worsens, is used.

Impacts of development are managed both within and outside WSPs. Within the context of WSPs, supply work and use approvals assess negative impacts of construction or use of a work. WSPs typically include distance limits from other wells, groundwater dependent ecosystems, and surface water bodies. Approvals are also required for flood works and 'controlled activities' that might affect a watercourse. Environmental impact assessments may be required under the NSW Environmental Planning and Assessment Act 1979, Protection of the Environment Operations Act 1997 and at the federal level. At the federal level, the Environment Protection and Biodiversity Conservation Act 1999 (EPBC Act) has included a 'water trigger' since 2013, through which coal seam gas and large coal mining developments require federal assessment and approval if they are likely to have a significant impact on a water resource. The NSW Aquifer Interference Policy (2012) explicitly integrates water use by mining with WSPs, noting that all water taken by aquifer interference activities needs to be accounted for within extraction limits.

Historically, groundwater use has also been closely associated with salinity management. Irrigation, including of rice, historically resulted in high water table levels leading to issues with soil salinisation, and contribution of saline baseflow. This has been tackled by control of application rates and salt interception schemes. Changes in crops and improved irrigation efficiency mean that salinity is not currently a pressing issue. Soil interception schemes continue to operate, involving large-scale pumping to divert saline groundwater and drainage water from rivers. Historically, in multi-aquifer systems like the Lower Murrumbidgee, increased deep groundwater use was also advocated to reduce water table levels, which was a complicating factor in over-allocation of the system (Schuster *et al.*, 2020). Given the significance of saline aquifers in NSW, desalination of brackish water has also been proposed, though this has, to date, primarily been used to reduce salts in river water for drought town water supply.

While water and land use are generally closely coupled, the approach taken in Australia has been to decouple water and land ownership, and encourage market drivers to reallocate scarce water to high-value uses. However, some measures are in place to influence land use and management, in addition to impact management measures, for example relating to salinity. Land use planning includes Local Environmental Plans (LEPs) that guide planning decisions for local government areas, and a policy to maintain land for agricultural industries. Land management has been supported since at least 1989 with the advent of the Landcare programme, which involved establishing state-sponsored local organisations to contribute to rural development, including conservation measures and riparian management. Since 2014 the NSW Local Land Services (LLS) agency has provided integrated customer-focused services to landholders and the community with a focus on outcomes. This replaces separate biosecurity, natural resources management and agricultural advisory services, within which Catchment Management Authorities in particular were previously responsible for regional strategic natural resource management planning in the form of Catchment Action Plans.

Aquifer recharge is not currently actively managed in New South Wales. There is no specific source zone protection. Environmental restoration measures that include benefits for groundwater recharge ('landscape rehydration') are permitted under impact assessment procedures. LTAAELs and SDLs currently assume historical recharge, requiring specific attention to growth in use of surface and groundwater allocations. Diversion limits may need to be revisited as the activation of licences increases, and improvements in irrigation efficiency reduce recharge.

As noted earlier, with the exception of highly connected systems, surface and groundwater are licensed and managed separately, and there is therefore no explicit management of their conjunctive use either. Historically, groundwater use has typically developed more slowly with surface water preferred. Opportunistic annual crops such as rice and cotton did not need to be planted in drought years. At the same time, groundwater could be used as a reserve resource during drought. Recent trends have, however, included increased perennial plantings that increase minimum water requirements (Loch *et al.*, 2020) as well groundwater transitioning from a reserve to primary source under circumstances of low surface water flows for example, that is, basin closure is continuing.

Managed aquifer recharge (MAR) has been discussed in NSW at least since 1994, and is used elsewhere in Australia, in both agricultural and mining contexts. A study for MAR for town water supply in Broken Hill by Geoscience Australia suggested this solution to reduce dependence on the Darling River instead of building a pipeline from the Murray River (Lawrie *et al.*, 2012). Additional work to investigate whether MAR is feasible in the identified sites could be developed. As of 2021, a key issue to adoption of MAR in NSW is that there is no specific policy for assigning rights to recharged water, that is there is no mechanism for a user to be credited an allocation based on water they have recharged. WSPs generally include an option to amend this, however, so that allocations could be granted according to general rules, impact assessment processes, or through operational management processes. The NSW Water Strategy Action Plan for 2021/2022 includes a plan to 'analyse options for implementing a managed aquifer recharge framework in NSW for consultation with stakeholders and the community'.

The NSW Government does not only regulate groundwater resources, whether through allocations or other legal instruments. It also disseminates knowledge about the resource and how it is managed so users can make informed decisions. This includes engaging on relevant topics; publishing news, reports, and policies and legislation; and maintaining public databases and portals. Without transparency and open data access, the trust of the community is undermined.

12.7 DISCUSSION: STRENGTHS AND WEAKNESSES

Up until as late as the 1990s, the approach to groundwater management in NSW was focused on finding and using the resource. A shift in ethos occurred with the introduction of *Ecologically Sustainable*

Development principles via the *WM Act 2000*. This enabled the dual outcomes of protecting whilst using the resource. Water Sharing Plans cover all groundwater sources across the state. These explicitly state how big the resource pool is, and specify how this pool is to be shared and allocated amongst all users, starting with the environment while defining specific resource impact criteria.

With the separation of groundwater rights from land rights, trading of shares and allocations was also facilitated. Institutionally, there is a distinction in responsibilities between who makes the rules (DPIE-Water), who implements the rules (e.g. WaterNSW) and who enforces the rules (NRAR). Since its creation in 2017, the role of NRAR has been strengthened over time (Holley *et al.*, 2020; NRAR, 2021a, 2021b). Despite these strengths, the changing nature of our world requires continued adaptation. The challenge remains to ensure groundwater sustainability and adaptation in light of known and new challenges and of existing community and societal expectations.

Climate change will impact groundwater resources as recharge from rainfall will decrease, affecting groundwater availability and the different uses dependent on it (Barron *et al.*, 2011). At the same time, groundwater demand will likely increase as surface water flows diminish as a result of reduced rainfall, potentially undermining ecosystem needs, communities and industry resilience. A lack of knowledge about these processes poses problems for how NSW's groundwater management framework should adapt in the future.

On the ground, compliance and enforcement of extraction rules requires continued attention as the various NRAR campaigns have shown (NRAR, 2021a, 2021b). More information is required to properly manage the resource, with new regional models and a more up-to-date network of monitoring bores. The long-term sustainability and management of fully committed groundwater sources with scientifically informed extraction limits remains an outstanding issue as WSPs are reviewed. Supporting further changes and incentives for groundwater markets in NSW (especially on the coast) would also increase reliability and security by maximising access to the full resource pool.

Re-building trust through effective and purposeful stakeholder engagement and communication is also necessary as the confidence in government activities by the public remains moderate (Woolcott, 2021). Exploring the possibility to establish co-management of groundwater (Molle & Closas, 2019) with local communities for specific planning and management activities could help improve user agency and trust in that area.

Environmental protection could be reinforced by better integrating and operationalising the joint management of surface water, land and groundwater. The knowledge about surface water and groundwater connectivity needs to increase so that groundwater-dependent ecosystems can be better protected. Considering and addressing the various recommendations found in the 2021 review of the National Water Initiative by the Productivity Commission would help deliver sustainability and protection for groundwater resources in NSW.

Increased local management could support an adaptive, collaborative approach with increased trust and fine-scale understanding of local connectivity. Contrary to previous reform driven by drought, a substantial investment is being made in drought resilience and anticipatory planning. The high value of water is pushing investment in efficiency and innovation, including in new technology such as precision agriculture and automated systems. Irrigation modernisation is in turn driving greater data availability, which enables, for example, spatio-temporal estimation of channel seepage, and greater visibility and control over movement of water through irrigation systems. One can imagine a future in which water users take an operational interest in active management of groundwater resources in order to maximise benefits of its use, for example treating it as a storage rather than simply a water source. While institutional subsidiarity is typically seen as desirable, benefits of technological change should not be taken for granted, and achieving a successful transition will require attention to principles of responsible innovation and adaptive governance able to adjust over time and address competing objectives within a system.

In emphasising adaptive management, the policy framework in NSW recognises that reform is not a destination. From supporting extraction to achieve economic growth to introducing a robust framework of extraction limits and a shared approach to allocations, groundwater management in NSW has moved towards achieving ecosystem protection and sustainable use. Significant challenges remain. The journey of reform has been a long one and must continue.

REFERENCES

Barron O. V., Crosbie R. S., Charles S. P., Dawes W. R., Ali R., Evans W. R., Cresswell R., Pollock D., Hodgson G., Currie D., Mpelasoka F., Pickett T., Aryal S., Donn M. and Wurcker B. (2011). Climate change impact on groundwater resources in Australia: summary report. CSIRO Water for a Healthy Country Flagship, Australia.

Bell S. and Park A. (2006). The problematic metagovernance of networks: water reform in New South Wales. *Journal of Public Policy*, **26**(1), 63–83, https://doi.org/10.1017/S0143814X06000432

Bioregional Assessment Programme (2015). CLM -- Groundwater Recharge Estimates -- Chloride Mass Balance technique v01. Bioregional Assessment Derived Dataset.

COAG (2004). Intergovernmental Agreement on a National Water Initiative, from https://webarchive.nla.gov.au/awa/20140126135336/http://nwc.gov.au/__data/assets/pdf_file/0008/24749/Intergovernmental-Agreement-on-a-national-water-initiative.pdf (Accessed Feb 18, 2022)

Crase L., O'Reilly L. and Dollery B. (2000). Water markets as a vehicle for water reform: the case of New South Wales. *Australian Journal of Agricultural and Resource Economics*, **44**, 299–321. https://doi.org/10.1111/1467-8489.00113

Crase L., O'Keefe S. and Kinoshita Y. (2012). Enhancing agrienvironmental outcomes: market-based approaches to water in Australia's Murray-Darling Basin. *Water Resources Research*, **48**, W09536. https://doi.org/10.1029/2012WR012140

Crosbie R. S., McCallum J. L., Walker G. R. and Chiew F. H. S. (2008). Diffuse groundwater recharge modelling across the Murray-Darling Basin. A report to the Australian government from the CSIRO Murray-Darling Basin Sustainable Yields Project. CSIRO, Australia.

Crosbie R., Jolly I., Leaney F., Petheram C. and Wohling D. (2010). Review of Australian Groundwater Recharge Studies. CSIRO: Water for a Healthy Country National Research Flagship.

CSIRO (2008). Groundwater modelling report – Lower Namoi. CSIRO: Water for a Healthy Country National Research Flagship. Report prepared for MDBA. This report modified the NSW Government model for the CSIRO MDB Sustainable Yields Project.

CSIRO & SKM (2010a). Sustainable Extraction Limits Derived from the Recharge Risk Assessment Method – New South Wales (part 1), report to Murray-Darling Basin Authority. This report describes the numerical models.

CSIRO & SKM (2010b). Sustainable Extraction Limits Derived from the Recharge Risk Assessment Method – New South Wales (part 2), report to Murray-Darling Basin Authority.

CSIRO & SKM (2010c). Sustainable Extraction Limits Derived from the Recharge Risk Assessment Method – New South Wales (part 3), report to Murray-Darling Basin Authority.

CSIRO & SKM (2010d). Sustainable extraction limits derived from the recharge risk assessment method, report to Murray-Darling Basin Authority.

Department of Agriculture, Water and the Environment (n.d.). History of Australian water markets. https://www.awe.gov.au/water/policy/markets/history#:~:text=districts%20in%201995.-,New%20South%20Wales,trading%20was%20enabled%20in%201991 (Accessed 4 Apr 2022)

DPI (2015). Macro water sharing plans – the approach for groundwater. https://www.industry.nsw.gov.au/__data/assets/pdf_file/0005/478184/Macro-water-sharing-plants-the-approach-for-groundwater-A-report-to-assist-community-consultation.pdf (Accessed 28 Feb 2022)

Head L., Adams M., McGregor H. V. and Toole S. (2014). Climate change and Australia. *WIREs Climate Change*, **5**, 175–197. https://doi.org/10.1002/wcc.255

Holley C., Mutongwizo T., Pucci S., Castilla-Rho J. and Sinclair D. (2020). Groundwater regulation, compliance and enforcement: insights on regulators, regulated actors and frameworks in New South Wales, Australia. In: Sustainable Groundwater Management. Global Issues in Water Policy, 24, J. D. Rinaudo, C. Holley, S. Barnett and M. Montginoul (eds.), Springer, Cham, Switzerland, pp. 411–433. https://doi.org/10.1007/978-3-030-32766-8_22.

Lawrie K. C., Brodie R. S., Dillon P., Tan K. P., Gibson D., Magee J., Clarke J. D. A., Somerville P., Gow L., Halas L., Apps H. E., Page D., Vanderzalm J., Abraham J., Hostetler S., Christensen N. B., Miotlinski K., Brodie

R. C., Smith M., Schoning G., Barry K. and Levet K. (2012). BHMAR Project: assessment of conjunctive water supply options involving managed aquifer recharge options at Menindee Lakes. *Geoscience Australia Record*, **13**, 499. https://www.ga.gov.au/about/projects/water/broken-hill-managed-aquifer-recharge (Accessed 28 Feb 2022)

Loch A., Adamson D. and Auricht C. (2020). (g)etting to the point: The problem with water risk and uncertainty. *Water Resources and Economics*, **32**, 100154. https://doi.org/10.1016/j.wre.2019.100154

MDBA (2012a). The proposed Groundwater Baseline and Sustainable Diversion Limits: methods report, MDBA publication no: 16/12, Murray-Darling Basin Authority, Canberra. https://www.mdba.gov.au/sites/default/files/archived/proposed/Proposed-BP-GW-BDL-SDL.pdf

MDBA (2012b). Addendum to Groundwater Methods Report. https://www.mdba.gov.au/publications/archived-information/addendum-groundwater (Accessed 28 Feb 2022)

MDBA (2020). Why the Murray-Darling Basin matters. https://www.mdba.gov.au/why-murray-darling-basin-matters (Accessed 28 Feb 2022)

Molle F. and Closas A. (2019). Co-management of groundwater: a review. *WIRES Water*, **7**(1), 1–19. https://doi.org/10.1002/wat2.1394

NRAR (2021a). Compliance Audit Close Out Report Operation Drawdown. NRAR Natural Resources Access Regulator. https://www.dpie.nsw.gov.au/__data/assets/pdf_file/0011/485228/Bore-Extraction-Limits-Audit-Close-Out-Report.pdf (Accessed 28 Feb 2022)

NRAR (2021b). NRAR Strategic Plan 2021–2023. https://www.nrar.nsw.gov.au/about-us/what-we-do/nrar-strategic-plan-2021-23 (Accessed 28 Feb 2022)

NRC (2021). Audit of the implementation of coastal groundwater sharing plans, March 2021. Natural REsources Commission. https://www.nrc.nsw.gov.au/Final%20report%20-%20Audit%20of%20Coastal%20GWSPs.pdf?downloadable=1 (Accessed 28 Feb 2022)

NSW (2020). Groundwater resource description – NSW Great Artesian Basin. New South Wales Department of Planning, Industry and Environment. PUB20/74. https://www.industry.nsw.gov.au/__data/assets/pdf_file/0007/291175/nsw-gab-resource-description-report.pdf (Accessed 28 Feb 2022)

NSW (2021a). NSW Water Strategy. New South Wales Department of Planning, Industry and Environment. PUB20/882. https://water.dpie.nsw.gov.au/__data/assets/pdf_file/0007/409957/nsw-water-strategy.pdf (Accessed 28 Feb 2022)

NSW (2021b). Future Ready Regions Strategy – supporting drought resilient communities and economies. https://www.nsw.gov.au/sites/default/files/2021-06/Future%20Ready%20Regions%20Strategy%20.pdf (Accessed 28 Feb 2022)

NSW DPIE (2021). Non-urban water metering in NSW – what water users need to know. New South Wales Department of Planning, Industry and Environment. PUB20/754. https://water.dpie.nsw.gov.au/__data/assets/pdf_file/0006/320199/non-urban-water-metering-in-NSW-what-water-users-need-to-know.pdf (Accessed 28 Feb 2022)

Pierce S. A., Sharp J. M., Guillaume J. H. A., Mace R. E. and Eaton D. J. (2013). Aquifer yield continuum as a guide and typology for science-based groundwater management. *Hydrogeology Journal*, **21**(2), 331–340. https://doi.org/10.1007/s10040-012-0910-y

Richardson S., Evans R. and Harrington G. (2011). Connecting science and engagement: setting groundwater extraction limits using a stakeholder-led decision-making process. In: Basin Futures. Water Reform in the Murray Darling Basin, D. Connell and Q. Grafton (eds.), ANU Press, Canberrra, Australia, pp. 351–366. http://doi.org/10.22459/BF.05.2011.22

Scanlon B. R., Healy R. W. and Cook P. G. (2002). Choosing appropriate techniques for quantifying groundwater recharge. *Hydrogeology Journal*, **10**, 18–39, https://doi.org/10.1007/s10040-001-0176-2

Schuster K., Kennedy A. and Holley C. (2020). Reducing groundwater entitlements in the lower murrumbidgee groundwater management area. In: Sustainable Groundwater Management. *Global Issues in Water Policy*, **24**, J. D. Rinaudo, C. Holley, S. Barnett and M. Montginoul (eds.), Springer, Cham, Switzerland, pp. 365–384.

Sharp J. (2016). Sustainability of groundwater resources: Conceptual evolution, opportunities, & challenges, Conference paper.

The CIE (2021). Understanding NSW's economic dependency on groundwater. DPIE internal project document.

Turral H. and Fullagar I. (2007). Ch 15 Institutional Directions in Groundwater Management in Australia. https://www.google.com.au/books/edition/The_Agricultural_Groundwater_Revolution/ru_70R74lMAC?hl=en&gbpv=1&dq=groundwater+management+new+south+wales&pg=PA320&printsec=frontcover

WaterNSW (2019). Murrumbidgee River Operations Plan. https://www.waternsw.com.au/__data/assets/pdf_file/0003/146163/Murrumbidgee-Rivers-Operations-Plan-July-2019.pdf (Accessed 28 Feb 2022)

WM Act (2000). NSW Water Management Act (2000). https://legislation.nsw.gov.au/view/html/inforce/current/act-2000-092 (Accessed 28 Feb 2022)

Woolcott (2021). NRAR Community Pulse Survey – Summary Research Report. https://www.nrar.nsw.gov.au/__data/assets/pdf_file/0018/480222/NRAR-community-pulse-survey-2021.pdf (Accessed 28 Feb 2022)

doi: 10.2166/9781789062786_0159

Chapter 13

Water allocation in Brazil: main strategies, learning and challenges

Guilherme F. Marques

Institute of Hydraulics Research, Federal University of Rio Grande do Sul, Porto Alegre, Brazil

ABSTRACT

Brazil is characterized by highly diverse biomes, hydrologic regimes and water uses. The relevance of irrigated agriculture in the county's economy, combined with the significant contribution of hydropower to the electricity matrix and the presence of large metropolitan areas translate to a strong water-food-energy nexus. Water management under this context requires control and allocation in an extensive network of long rivers, operated with seasonal and multi-year carryover capacity. The future development and overall water security in Brazil will bring increasingly competing demands and growing challenges to water allocation, demanding more robust regulations and institutions. This chapter reviews the nature of Brazilian water rights and the current legal and policy framework determining water allocation, followed by the presentation of several innovative initiatives to control water access, based on locally defined allocation rules and local needs, challenges and opportunities. From this rich experience, we draw recommendations on how to move forward in improving water allocation strategy at the national level and adapting water management practices to cope with the challenges ahead.

Keywords: Stakeholder's engagement, water allocation, water resources management, water rights

13.1 THE BRAZILIAN CONTEXT ON AGRICULTURAL WATER USE: SUPPLIES AND COMPETING DEMANDS

Agricultural production in Brazil has supported an active agribusiness sector, contributing to 10% of the Brazilian GDP, 137 USD billion (10^9) in annual revenues and 15% of the industry jobs (FGV, 2016). This is due to several factors, including mechanization, crop and soil research, access to credit and irrigated agriculture.

The irrigated area has increased over 4% annually since the 90s, reaching 8.2 million hectares in 2019 (ANA, 2021a), which represents 12.36% of the total 66.32 million ha crop-planted area (EMBRAPA, 2022). It currently represents 46.2% of the total withdrawals in Brazil estimated at 2098 m^3/s, and 67.2% of the total consumption (1109 m^3/s). Urban supplies and industries are next, with 23.3% and 9.2% of the withdrawals and 8.8% and 9.5% of the consumption, respectively (ANA, 2017b). Ecosystem water demands are still largely unknown and mostly marginally met with minimum flow requirements. Recent studies (Marques & Tilmant, 2018; Dalcin *et al.*, 2022) indicate that restoring key components of flow regimes will require better knowledge of the trade-offs to other sectors, to support negotiation and water allocation solutions.

Despite its relatively small participation in the total crop area, irrigated agriculture plays a key role in terms of food security. Among all crops, the grains that constitute the staple food in Brazil (rice, beans and wheat), when irrigated presented productivity from 1.9 to 3.7 times the productivity of the non-irrigated areas (ANA, 2021a).

Recent estimates point to an *effective irrigation expansion potential* of 13.1 million ha, of which 4.2 million ha is expected by 2040 (ANA, 2021a). This is likely to reinforce tensions with competing uses (e.g. urban) and increase pressure on aquatic ecosystems, which are already suffering from insufficient minimum flows. An increasingly higher number of municipalities across the country is expected to either boost withdrawals from existing supply sources or find entirely new ones.

Rainfall, flow patterns and water availability vary greatly across the country. Around 80% of the available surface water flows in the Amazon basin, where the population is the smallest (ANA, 2017a). In terms of groundwater, 90% of the Brazilian rivers are fed by base flow from aquifers, which maintain surface waters during the dry seasons (ANA, 2017a). While the total groundwater annual average availability is estimated at 13,205 m^3/s, the aquifer systems in sedimentary basins provide the highest storage potential, covering 48% of the Brazilian territory. This resource is increasingly exploited: from 145 thousand registered groundwater supply sources in 2008 (most are groundwater wells) to approximately 2.6 million wells in 2021, extracting 1083.3 m^3/s (ANA, 2021b). Figure 13.1 presents the major Brazilian watersheds and hydrolithology groups.

Water management in this context requires water control and allocation in an extensive network of long rivers, operated with seasonal and multi-year carryover capacity. The future development of irrigated agriculture, power supply, environmental protection and overall water security will bring increasingly competing demands and growing challenges to water allocation, demanding more robust regulations and institutions. Most of those challenges are associated with four main elements.

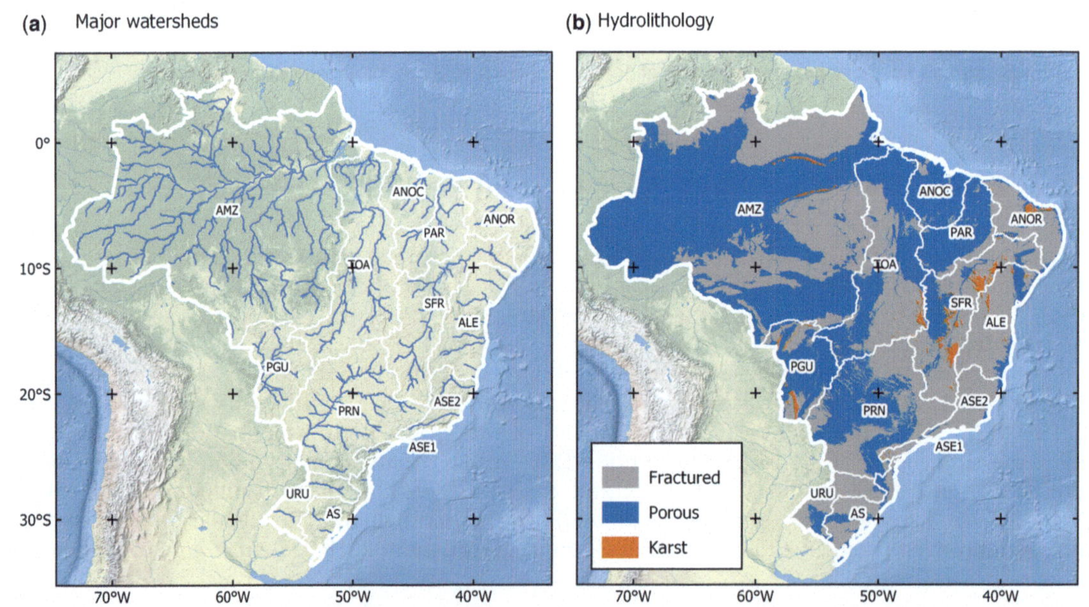

Figure 13.1 Major Brazilian watersheds and hydrolithology groups (source data: Brazilian Geologic Service – CPRM).

The first is the lack of a widespread long-term strategic approach to water allocation mechanisms. Implementation of water allocation policies, with more elaborate criteria for inter- and intra-sectoral water allocation, is the exception rather than the rule (OECD, 2015), as the proper institutional arrangement for this negotiation is insufficient, and the knowledge about the allocation trade-offs is lacking. This is aggravated by the large gap between water management in the state and federal rivers. If water is not properly regulated at the state level, it becomes difficult to account for its availability in the federal rivers. While recent procedures for formal water regulatory agreements (*marcos regulatórios*) and water allocation agreements (*Termos de Alocação de Água*) have improved this aspect, these have been mostly deployed as problem responses.

Second, the negotiation regarding water quantity and quality expected to flow from state to federal rivers is beyond the capability of most current watershed committees, which often have limited implementation capacity due to lacking technical support for such a complex engagement. The determination of boundary 'delivery' agreements (*condições de entrega*) is a fundamental water allocation decision which has yet to be addressed by the Brazilian water resources management framework, except in specific cases responding to local problems.

Third, water allocation must be brought up to sectoral policymaking at nationwide scale, to coordinate large-scale investments (e.g., irrigation, energy and infrastructure).

And fourth, climate change impacts will differ, depending on each region's characteristics and infrastructure. Different coping and adaptation strategies to mitigate the climate impacts in one water use may adversely reflect on others.

This chapter reviews the nature of Brazilian water rights and the current legal and policy framework determining water allocation, followed by the presentation of several initiatives to control water access, based on locally defined allocation rules and accounting for local needs, challenges and opportunities. From this rich experience, we draw recommendations on how to move forward in improving water allocation strategy at the national level and adapting water management practices to cope with the challenges ahead.

13.2 THE NATURE OF BRAZILIAN WATER RIGHTS: PAST AND PRESENT

While past water allocation initiatives in Brazil differ greatly, including informal water markets in the Northeast dating back to 1855 (Campos *et al.*, 2002) and water ditches societies (*associações de vala*) organized by rice farmers dating back to 1927 (Liberato Jr., 2004), the State interfered very little in water allocation until 1934.

State-driven hydropower projects in the 1930s called for formal water regulation to secure reliable supply to the hydropower reservoirs, leading to the 1934 Water Code. This legal milestone formalized the first water allocation system with the public sector controlling an authorization process to grant the right to use water to specific users (Brasil, 1934; Carvalho, 2015). The 1934 Water Code distinguished public waters of common use, which could be property of the Union, the states or municipalities, and private waters, which belonged to the owners of the land on which they were located. It also warrantied access and free use to common waters for basic human needs. Groundwater could be appropriated by the owner on whose land the source was located, provided its use did not interfere with other public and private waters (Brasil, 1934). While the Water Code established fairly advanced concepts for its time, alluding to common water benefits and recognizing the impacts of private uses on others, it was a top-down, sectoral approach, detached from other public policies and focused on preserving the economic potential of the water resources, with environmental considerations mostly absent.

As the Brazilian industry expanded, hydropower systems became interconnected in the 70s and 80s, highlighting the temporal aspects of water allocation (Lopes & Freitas, 2007), and demanding water use and reservoir operation optimization integrated with flood protection. The result was the establishment of minimum flow requirements downstream of federal and state reservoirs and water quality standards with improved regulation of environmental laws in the 80s, which indirectly established environmental water allocation. Given that water scarcity and conflicts were not significant at this time, this was largely uncontested by other users.

The Brazilian Federal Constitution of 1988 provided another milestone, defining water as a common good belonging to the people, to serve public interests (Brasil, 1998). It established the water permits as key instruments to water allocation and management, as well as extinguishing the concept of private waters and the municipal water domains from the 1934 Water Code. After this change, water could be either federal or state domain, and previous private users (e.g., farmers, industries, etc.) had to apply for a permit to maintain access. While significant from a legal standpoint, this change did not bring major contestation from the users. Large water uses (e.g., hydropower) already had formal concessions and maintained access to water. Smaller users applied for permits, which were granted in most cases given the competition for water was not as intense as it is today. However, there were cases where users resorted to legal action to seek compensation.

However, water allocation remained essentially centralized, in a top-down approach controlled by state and federal water authorities. Later, the Brazilian National Water Resources Policy (Federal Law 9433/97) initiated a transition towards more decentralized water management, relying on new organizations, including a National Water Council, State Water Resources Councils, River Basin Committees, State Water Resources Authorities and Water Agencies, drawing inspiration from the French experience and example (Laigneau, 2011). The Federal Law 9433/97 further defined water as a public good. The water permits, along with water charges, watershed plans, water quality standards and water information systems were formally described in the law as water management instruments. An important component in this organizational framework was the formalization of River Basin Committees, with key roles in exercising participative and shared negotiation for water management and conflict resolution (Veiga & Magrini, 2013).

The responsibility for water allocation and management is defined through two different domains: the state and the Union. Public Water Resources Authorities (both at the state and federal levels) are responsible for controlling and registering the water uses permits. Rivers and lakes entirely located within a single state belong to the domain of the corresponding state. Rivers and lakes crossing more than one state, or which form the boundary with another state or country, belong to the domain of the Union. All groundwater is state domain. Waters stored in instream reservoirs built by the federal government are considered federal waters, regardless of the river (state or federal). This shared jurisdiction of state and federal waters is a challenge, as it requires high levels of institutional coordination and decision-making between users and public agencies, often lacking in Brazil (OECD, 2015).

13.3 LEGAL AND POLICY FRAMEWORK TO CONTROL ACCESS TO WATER (GROUND RULES)

13.3.1 Permits and concessions

The formal water allocation process defines water permits and concessions (*'outorgas'* in Portuguese) established by Federal Law 9.433/97. Permits are issued to private users and renewed every 5–10 years (depending on the state). Concessions are issued for public water use (i.e. public utilities and urban water supply systems) and are renewed every 10–12 years. Other concessions, such as for hydropower use, may last up to 30 years. Users submit a withdrawal request to the corresponding Water Resources Authority (state or federal). The request specifies the flowrate and the number of withdrawal hours for each month, followed by the total requested volume each month. Criteria for monthly volumetric caps are based on reference flowrates, which vary across state and federal rivers. The request is usually processed by filling order. Once approved, the permit is capped by the total monthly volume described in the request.

The Water Resources Authority is responsible for maintaining an information system with user and water availability databases to determine if new requests can be accommodated, following general criteria of water availability and request filling order (first come, first served). Water uses considered 'insignificant' (often less than 1 L/s for domestic and animal supply) are exempt from approval, but still require the user to register the use in the database.

While this flow-based allocation procedure is the same throughout the country, specific criteria applied to evaluate the requests are defined by the corresponding River Basin Committee and Water Resources Authority. The River Basin Committee defines the priority water uses, the water quality standards and the insignificant water uses, along with other uses exempt from formal permitting. Priority water uses are defined between sectors only, and often only establish that domestic (human consumption) and small-scale animal consumption come first.

A water permit can be indefinitely suspended, totally or partially and without liability to the public authority, in cases of:

(i) Non-compliance from the user on the water permit terms.
(ii) Water not being used for three consecutive years.
(iii) Water being needed to meet other priorities or under calamities caused by adverse weather conditions.
(iv) Water being needed to prevent or revert severe environmental impact.
(v) Water being needed to maintain navigation conditions.

13.3.2 Issuance of permits for surface water

Withdrawals taking place in a flowing water body are based on a reference flow (flow with a chosen exceedance probability, e.g. Q_{90}), established by the Water Resources Authority, which will be used to calculate the maximum permissible flow rate that can be allocated. The reference flow varies according to the state. Very often, only a fraction of the reference flow is available for allocation so that minimum instream flows can be met.

Finally, other criteria may include an evaluation of the degree of commitment of the waters in the river and the impact of the requested flow relative to the availability (e.g., indicators of the ratio between the user-requested flow and the maximum permissible flow rate or the ratio between all current uses and maximum permissible flow rate). Basins with a very high commitment may go into a 'closure' state, where permits are no longer issued, prompting action for conformity from users.

The Water Resources Authority receives the request and calculates the water budget based on the allocable fraction of the reference flow, subtracting existing permits upstream of the request and checking on existing permits downstream, resulting in the maximum permissible flow rate. The criteria for water permitting are then applied to decide if the permit is approved or denied. For example, a River Basin Committee may determine, in the watershed plan, that new water permits should be restricted if the degree of commitment reaches a critical status. However, very often the water budget is the sole criterion and if the maximum permissible flow rate is larger than the flow requested, the permit is approved. Permits due for renewal take priority over new requests.

If the withdrawal takes place in a surface reservoir with carryover storage, the maximum permissible flow rate is the regulated flow (ANA, 2011), provided the reservoir releases are also able to meet the minimum instream flow to downstream in case of perennial rivers. If the river is intermittent, the allocation can also be volumetric.

13.3.3 Issuance of permits for groundwater

Allocation of groundwater is conducted by the state Water Resources Authority, which defines the criteria to issue pumping and withdrawal permits. User engagement in this process has been essentially reactive and conflict-driven, often taking place in severe cases of interference of pumping in local surface water sources or in other wells. Afonso and Junior (2006) present one example of conflict between irrigators in the state of Minas Gerais in 1991, where small-scale, river-fed, downstream irrigation was affected by intense groundwater pumping upstream. Such cases are managed by the

State with either suspension of some of the water permits or change in the regime of use (e.g., reducing the number of hours of pumping and increasing the interval between pumping operations). Such cases are often started when impacted users formally complain to the State (or the police), which prompts the state Water Resources Authority response. In some cases, a solution is found and users comply, while others end up in court.

To withdraw water from a groundwater source, the user fills out a request with information about the source conditions, including flow measurement during the dry season (performed by the user), hydrogeological characteristics of the well and pumping tests (in the case of wells equipped with pumps), among others. Caps on allowable pumping rates vary depending on the state. Other criteria include analysis of interference in nearby wells, aquifer vulnerability, proximity to other water bodies (e.g., rivers and streams) and environmental protection areas. However, there is no water balance analysis to allocate groundwater, mostly due to lack of information on recharge rates, regional groundwater use, storage capacity, fluxes and surface water interaction, which increases the risk of overallocation. Recent large-scale groundwater study projects are under way to support regulation of allocating ground and surface water between different states.

13.3.4 Regulatory agreements

In regions with very limited water availability and potential water conflicts, the government established regulatory agreements, which define specific water use ruling and are implemented after a negotiation process involving the Water Management Authority and stakeholders. Such ruling may define specific priorities for water permits, minimum efficiency levels for each user sector, adjustment on irrigation schedules, water use restriction based on specific water levels on rivers and reservoirs, maximum allowable flowrates for each user sector in different watershed regions, limits on water storage for new infrastructure, water quality targets and other criteria for water permits (e.g., renewal criteria) (ANA, 2016). The process to reach a regulatory agreement can be initiated by either the users or the Water Management Authority, often in response to water conflicts and crises. It starts with the identification of the water problem (e.g., a conflict), its hydrologic conditions, the users involved, the probable causes (e.g., limited water availability, low efficiency, water quality constraints), the problem time frame (e.g., permanent, cyclical or based on a critical event) and delimitation of the affected water system (ANA, 2017b). Alternatives to solve the conflict are discussed and selected, also considering the watershed plan. A legal document formalizes the multilateral agreement to implement the solution. Finally, monitoring and overseeing of system status are implemented, to re-evaluate the regulatory agreement and make adjustments as conditions change. Examples of alternatives include different scenarios to redistribute the water, capping withdrawals and constraining cropping areas. A recent regulatory agreement in the Brazilian Midwest capped the irrigated acreage in two different states sharing a river and established a minimum flow from one state to the other (ANA, 2010).

13.3.5 Strength and weaknesses of the current permitting regime

Overall, the flow-based allocation system based on water permits provided better control of the water and reduced the possibility of overallocation, which contributed to increasing reliability to existing users compared to the past where regulation was mostly focused on larger scale economic uses (i.e., hydropower).

However, it still suffers from several weaknesses (Spolidorio, 2017):

- No flexibility to accommodate local seasonal hydrology variations.
- Groundwater allocation is not addressed.
- Inadequate during droughts, when water availability may be less than the reference flows.
- More water than needed requested by users.
- Users fail to request cancellation of the water permit when the water use ceases.
- Permits are rarely reviewed or updated, unless under critical events.

By allocating only a flow with high exceedance probability, the chances of a conflict between users with water permits is reduced, at the expense of the flexibility to accommodate new users. Considering the reference flows commonly adopted in Brazil, at least 90% of the time there are flows in the water bodies that are not allocated, which has been long criticized by users.

Finally, water permits and allocations disregard the impact of present decisions on future water demands and availability, having little connection with other development policies and even water management instruments, such as the water resources plans (Dalcin & Marques, 2020). The large number of users, many using water without permits, combined with limited oversight and monitoring by the Water Management Authorities, leads to error in determining available water for new permits, reducing reliability and trust. For Lopes and Freitas (2007), the public authorities still take a lead in managing the water allocation mechanisms. While this increases the guarantees to existing users, it also hinders the development of local control and user participation, increasing the costs of oversight.

13.4 DEFINING WATER ALLOCATION MECHANISMS: MAJOR CATEGORIES AND EXAMPLES

As water problems evolve, the limitations of the simple criteria adopted in the ground rules become more evident, prompting user engagement to propose other mechanisms. These complementary allocation mechanisms, however, do not necessarily replace the ground rules, but rather present solutions to allocate water under more specific conditions or problems (e.g., temporary droughts, permanent scarcity and intermittent water availability, conflict areas). The emerging institutional arrangements vary depending on the region's social capital, local history and challenges. This section presents the complementary allocation mechanisms, illustrated by examples selected based on their innovative approach. The mechanisms are organized as different types, based on the temporal characteristics of their context in Brazil.

13.4.1 Collective water permits

The effective implementation of the ground rules requires a very organized water management framework to register, track and engage users, detect non-compliance and to elaborate and update hydrological studies in the watershed. Not all Brazilian states have a fully functional system, and the most common gaps are incomplete water user databases. As a result, having users operating without permits is not uncommon in some watersheds. To enforce compliance over a river reach, collective water permits are established. Within this mechanism, the permit request lists all users, irrigated area, pumping coordinates and flow rate pumped.

In the example of Rio Grande do Sul state, collective irrigation water permits are requested through the River Basin Committee to the state Water Resources Authority, as in the recent example of a collective permit for rice growers in the Sinos River Basin (DRHS, 2021). The Water Resources Authority verifies the request against the maximum permissible flow rate and, if there is not enough water to fulfill the request, it is returned to the River Basin Committee for user negotiation and adjustment. Once an agreement is made, a new request is submitted and approved.

More recently, an online water permit (http://www.siout.rs.gov.br) system has been implemented in Rio Grande do Sul, integrating a mapping system with hydrology and water users' databases, where each user can individually fill out a request and immediately check his demanded flow against the maximum permissible flow rate at the withdrawal's location (the system performs the water balance considering existing registered permits upstream and downstream). The Water Resources Authority in the state of Rio Grande do Sul is phasing out the collective water permits and transitioning to individual ones through the online system

13.4.2 Long-term, cyclical allocations for reservoir operations

In the challenging climate and hydrological conditions faced in the Brazilian Northeast (NE), water supply depends largely on how the reservoirs are operated, as there are no additional inflows during the dry season (which lasts 9 months). To avoid conflicts and maximize water supply, several institutions are involved in water management and allocation, including a State Secretariat of Water Resources, a State Water Resources Policy (Brasil, 1992) and a state Water Resources Authority (COGERH). This institutional arrangement formalized water allocation procedures that are cyclical, with great attention on how to operate the system and share the released waters every dry season.

Users' participation occurs within the River Basin Committee, responsible for deliberating on criteria for water permits, reservoir releases, execution of new water infrastructure, water charges and how the available resources should be invested in the watersheds. In addition, Reservoir Operation Committees are involved for final validation of reservoir releases, and a users' overseeing committee monitors the implementation of the agreement throughout the dry season.

The general procedure for water allocation is based on evaluation of current storage after the wet season, followed by a series of Reservoir Operations Planning seminars. In the latter, users are presented with simulations of reservoir depletion scenarios and discuss release solutions, negotiating water allocation. Once agreement is reached, reservoir releases are determined for the coming dry season and a formal document is signed by users. During the dry season, reservoir depletion is monitored. If drought conditions are worse than expected, the user committee discusses further changes and water is allocated proportionally to their original shares, with urban/human demands taking priority (Silva *et al.*, 2006; Pinheiro *et al.*, 2011).

This allocation process has been under constant improvement and reservoir releases have been reduced throughout the years (1998–2005), despite growth in population and irrigated areas. According to Silva *et al.* (2006), this reflects increased rational use and public awareness. With smaller releases, there is a higher safety margin and reduced risk in managing the available water stock. Despite this favorable result, there is a need for constant oversight, especially during droughts, when some users may avoid the negotiation process and resort to illegal damming of upstream waters (Pinheiro *et al.*, 2011). In those cases, COGERH has a role to help ensure compliance, and enforcement is carried out by the State.

13.4.3 Short-term, event-based, water allocation mechanisms

Events such as droughts and conflicts trigger adjustments to permit conditions, which can be initiated by users or the Water Resources Authority. In any event, water allocation is negotiated among users and a solution is presented to the Water Resources Authority. The adjustment is temporary, lasting from a few weeks to 18 months. Permit conditions prevail afterwards. In this section we present several examples taking place in different Brazilian states involving allocations to agricultural irrigation.

13.4.3.1 Collective flow rate caps

In the state of Minas Gerais, watersheds with combined demands surpassing the maximum permissible flow rate are declared conflict areas by the state Water Resources Authority. This condition may be detected during the analysis of a water permit, and it may be caused by excess unregistered (illegal) uses. Once a basin is declared a conflict area, the River Basin Committee facilitates user negotiation around the distribution of the available water, so that the maximum allowable flow rate is met. A Local Management Committee is created, with representatives from the affected river reaches. This is a smaller user group drawing inspiration from the French *Commission locale de l'eau* (CLE) to negotiate the withdrawal adjustments. Private consulting is hired by the group to provide technical assistance, and users have equal contribution to the discussion regardless of size or economic power and a full agreement must be reached.

The agreed solution is formalized into a collective water permit request submitted by the River Basin Committee to the Water Resources Authority, which oversees implementation. The state Water Resources Authority is empowered to determine a solution if users cannot reach an agreement. The objective of the negotiated allocation process is to meet economic, social and environmental needs and mitigate occurring conflicts. The proposal must include (SISEMA, 2015):

(i) A technical study of water availability in the area, organized by the Local Management Committee.
(ii) Criteria to define priority withdrawals during severe scarcity.
(iii) Criteria to implement rational water use, considering available technology.
(iv) A schedule to alternate withdrawal pumping.

A second example of this mechanism includes several watersheds in the Federal District (DF) region. The first initiatives took place in the Extrema watershed, when the Water Resources Authority (ADASA) detected violations of minimum instream flows during the 2016 drought. Meanwhile, large-scale irrigators discussed how to minimize and distribute economic losses through water sharing. Users realized that without a mutually agreed solution, they would be unable to secure enough water to irrigate the crops and comply with minimum flows.

ADASA organized users into three groups, based on irrigated area, which would alternate pumping and irrigation schedules in order to maintain the original water permit flows. Users would determine how the schedules would be organized, without ADASA's intervention, resorting to mobile phone message groups to rapidly communicate and adjust pumping schedules. The negotiation and allocation processes were legally formalized in ADASA (2017) with a regulation decree, followed by specific water allocation agreements (*Termo de Alocação de Água)*. By 2017, as hydrological conditions worsened, it became necessary to also reduce the water permit flows and the cropping area. The reduction in irrigated area was based on individual planted areas and determined by users. According to Maniçoba (2021, personal communication), despite some initial resistance from the farmers, a relationship of trust was built along the process, which was key to its effectiveness.

Other initiatives took place in Alto Rio Jardim and Pipiripau watersheds and started back in 2006. The Pipiripau watershed has 71% of its area irrigated, mostly with small-scale agriculture that is more vulnerable to drought, also being responsible for urban water supply downstream. With significant growth of both agricultural and urban areas, water demands surpassed local availability during drought events, resulting in water shortages to urban areas downstream.

Increasing water scarcity led to the issuance of Water Regulations (ADASA, 2006; ANA, 2006; ADASA, 2017) that formalized water allocation regime during droughts. The Water Regulations defined control points in specific river reaches, minimum flows at each control point and water uses exempt from permits. Water Allocation Follow-up Commissions were created for each river basin, with representatives from users, national, state and municipality public power and a state technical agricultural support institution. Each Commission meets monthly, discussing current water availability, proposing actions, and proposing strategies to optimize allocations. Allocation decisions are made by consensus or simple majority, and the results formalized into water allocation agreements (ADASA, 2021). This document informs the current hydrologic state and the flow cap (for each control station), along with the pumping schedule for each user. If users cannot reach an agreement, the state Water Resources Authority intervenes (ADASA, 2020).

As a general rule, agreements require that users adjust withdrawals proportionally to the irrigated area whenever flows reach the minimum limit at the control points. During the 2017 and 2019 dry seasons, users implemented water pumping schedules to avoid concentrated pumping, but also curtailed irrigation withdrawals and reduced the crop area. This maintained the required downstream flows to meet urban demands while minimizing the impact to production. The local water public utility also implemented a water rationing schedule for urban users.

A major improvement over the recent experience by the Extrema watershed was a new real-time supervisory system to monitor individual irrigation systems, built and maintained by the farmers. Full access to the system was given to ADASA and the other users in the watershed. Any off-schedule pumping was detected and adjustment requested. Despite this, very few schedule violations were verified and ADASA's intervention to enforce the schedule was not necessary. For Maniçoba (2019), the acceptance between farmers was very good, and communication was also improved.

The experience in Pipiripau and Alto Rio Jardim further refined the allocation mechanism (ADASA, 2017, 2019), separating the user groups based on crop types (perennial or annual), crop water consumption, management and number of harvests per year. Water balance was improved at the watershed level, which benefited from the new real-time supervisory system.

13.4.3.2 *Individual, compensated caps*

A severe drought in 2001 in the Ceará State (Brazilian NE) triggered the design of a temporary water allocation mechanism which included a form of economic compensation to reduce water demand. The region's 19 000 hectares of irrigated land includes high water demand annual crops (taking up to 55% of the irrigated area and 71% of the total water demand) and perennial crops (taking up 16% of the irrigated area and 12% of the total water demand). According to Oliveira (2008), 70% of the irrigation uses furrow methods, with efficiencies ranging from 60% to 70%, which contributes to a high water demand given that the region is semi-arid.

As the drought event lowered local storage, releases were prioritized to meet human and livestock demands. Only 50% of the total irrigation demand was met. The severe conditions led the National Water Agency (ANA), along with the state government, to propose an emergency irrigation water use plan, the '*Aguas do Vale*'. The main objectives were to (a) improve the water management through higher irrigation water use efficiency, (b) induce more efficient irrigation practices and shift to crop mixes with lower water demand and higher value and (c) bring the water balance to an equilibrium (Oliveira, 2008).

Aguas do Vale also aimed to motivate a crop mix change through economic compensation mechanisms. While annual crops (e.g. rice) consumed 16 670 m^3/ha with an average return (All dollar figures are 2001 values adjusted for inflation to 2020.) of \$0.051/m^3, permanent crops (e.g., melons) demanded 5000 m^3/ha with an average return of \$2.19/m^3 (SRH & SEAGRI, 2001). The plan would target rice planters to reduce demand, allowing the flow to be allocated to higher value/lower water demand perennial crops.

Joining the program was voluntary, and the execution did not involve a direct negotiation between users to determine the flow allocation and compensation value. A rice planter willing to join would fill out a request for a water permit to irrigate his land to full demand, followed by a secondary request to forgo part of the water and fallowing 50% of the planned rice area in exchange for compensation. A requirement was to join a training program on efficient water use and irrigation management. Compensations ranged from \$292.4 to \$438.6/ha to the rice planters, depending on farm size (smaller areas, up to 2 ha, received \$438.6/ha while larger ones, above 100 ha, only received \$292.4/ha). Variables including historical rice prices, yields and average revenues were used to calculate the compensation values. The Federal Government and the Ceará state treasury provided the funds for the compensation mechanisms, which included water charges paid by the water users (OLIVEIRA, 2008). Resources from the water charges in the Jaguaribe River Basin also supported the Plan. After the drought, the program ended, and users returned to their original permit conditions.

The outcome was successful in the short term: the rice planted area was reduced from an original 8370 ha to 637.6 ha in 2001 and 1936 ha in 2002. This significantly reduced the water demand in the period, allowing reservoir releases to be minimized and allocated exclusively to human and livestock uses, while available groundwater was shifted to irrigate higher value perennial crops.

13.5 CONCLUDING REMARKS

The different examples presented in this chapter have common traits and challenges, which are useful to guide water allocation in other contexts. The key instruments to the success of allocation include strong user involvement and ownership, in a 'co-management' approach to organize and discuss the problems before formal involvement of the Water Resources Authority. Formal but flexible rules must also be present to guide the user's behavior, as well as developed and easily accessible monitoring and

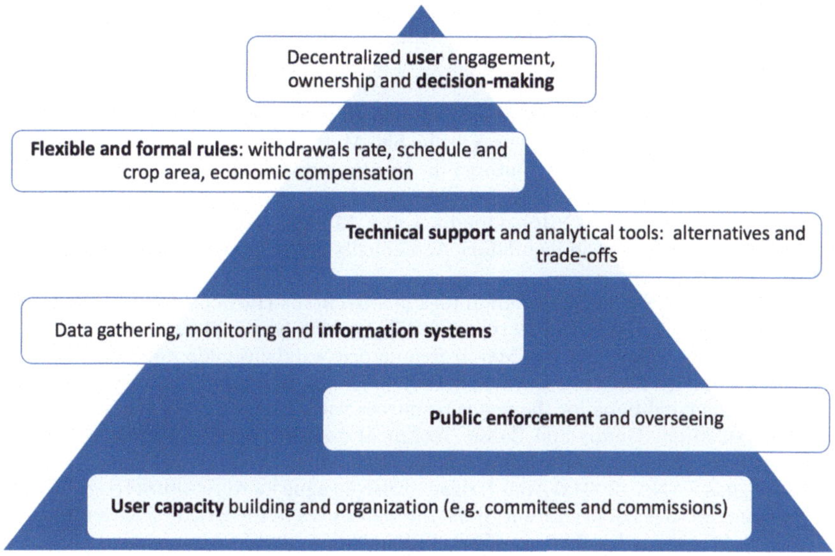

Figure 13.2 Hierarchy of instruments to effective water allocation.

information systems to track conditions and update users frequently, ranging from daily/hourly for short-term drought allocation in small systems, to monthly/yearly for larger ones.

User negotiation and decision-making needs to be supported with analytical (e.g. simulation) tools, to investigate alternative operations and withdrawal schedules, as well as to identify trade-offs between alternatives, including social dimensions. Finally, instruments must be applied in a context of dialogue and transparency among involved parties, as well as proper rule enforcement to build trust. This context requires users' capacity building, so they can understand the actions being implemented and are able to collaborate in the water management process. Those aspects are organized in Figure 13.2.

REFERENCES

ADASA. (2006). Resolução N° 293 [Resolution N°293]. Available at: https://www.adasa.df.gov.br/images/stories/anexos/8Legislacao/Res_ADASA/Resolucao293_2006.pdf (accessed 13 July 2021)

ADASA. (2017). Resolução N° 04. Available at: https://www.adasa.df.gov.br/images/storage/legislacao/resolucoes_adasa/resolucao_adasa_042017_alocacao.pdf (accessed 13 July 2021)

ADASA. (2019). Resolução N° 11. Available at: https://www.adasa.df.gov.br/images/storage/legislacao/resolucoes_adasa/Minuta_Resolucao_Hidrometracao_DODF.pdf (accessed 13 July 2021)

ADASA. (2020). Alocação Negociada Contribuiu Para A Preservação E Usos Múltiplos Da Água No Df [Negotiated water allocation contributes to maintain multiple water uses at DF]. Available at: http://www.adasa.df.gov.br/area-de-imprensa/noticias/1181-alocacao-negociada-contribuiu-para-a-manutencao-do-uso-multiplo-da-agua-durante-a-crise (accessed 13 July 2021)

ADASA. (2021). Termo De Alocação De Água 2021 Bacia Hidrográfica do Ribeirão Pipiripau. [Water allocation regulation 2021 Pipiripau river watershed]. Available at: https://www.adasa.df.gov.br/images/storage/regulacao/2021.zip (accessed 13 July 2021)

Afonso P. C. S. and Junior J. C. (2006). A Questão da Água na Bacia do Riachão – Norte de Minas Gerais: Gestão e Conflitos pelo Uso na Agricultura. II Encontro De Grupos e Pesquisa. [The water issues at Riachão watershed, North of Minas Geraes]. Universidade Federal de Uberlândia.

Agência Nacional de Águas – ANA. (2006). Marco regulatório de procedimentos e critérios de outorga de direito de uso de recursos hídricos na Bacia do Ribeirão Pipiripau, considerando a regularização das intervenções e usos atuais. Regulatory Agreement for procedures and criteria to water permits at the Pipiripau river watershed, considering ajustment to current water uses and interventions]. Available at: https://www.gov.br/ana/pt-br/todos-os-documentos-do-portal/documentos-sre/marcos-regulatorios/resolucao-ana-127-2006.pdf

Agência Nacional de Águas – ANA. (2010). Marco Regulatório do Uso da Água na bacia do São Marcos. Resolução N° 562. [Regulatory Agreement for water uses at the Sao Marcos River Basin, Resolution No 562].

Agência Nacional de Águas – ANA. (2011). Outorga de direito de uso de recursos hídricos. [Water permits for water resources]. *Cadernos de Capacitação em Recursos Hídricos SAG*, **1**(6), 50.

Agência Nacional de Águas – ANA. (2016). Oficina sobre Marcos Regulatórios em Sistemas Hídricos no Semiárido Brasileiro. [Workshop on Regulatory Agreements in the Brazilian semiarid]. Aracaju, SE, 08 e 09 de novembro de 2016. Apresentações.

Agência Nacional de Águas – ANA. (2017a). Conjuntura dos Recursos Hídricos do Brasil: Relatório Pleno 2017. [Water Resources Situation in Brazil: Full report 2017]. Agencia Nacional de Águas, Brasília, Brazil.

Agência Nacional de Águas – ANA. (2017b). Marco Regulatório estabelecendo condições de uso dos recursos hídricos no sistema hídrico Estreito e Cova da Mandioca, nos Estados de Minas Gerais e Bahia. [Regulatory agreement stablishing conditions for the water resources use in the water system of Estreito and Cova da Mandioca, states of Minas Gerais and Bahia. Technical note n° 3/2017/COMAR/SRE]. Nota Técnica n° 3/2017/COMAR/SRE, Documento n°: 00000.009578/2017-88, fevereiro de 2017.

Agência Nacional de Águas – ANA. (2021a) Atlas Irrigação: uso da água na Agricultura Irrigada/Agência Nacional de Águas e Saneamento Básico. [Irrigation Atlas: the water uses in the irrigated agriculture/National Water Agency], 2nd edn, ANA, Brasília, 2021, p. 130 il. ISBN: 978-65-88101-10-0.

Agência Nacional de Águas – ANA. (2021b). Conjuntura dos Recursos Hídricos no Brasil. [Brazilian Water Resources Context]. Available at: https://relatorio-conjuntura-ana-2021.webflow.io/ (accessed 12 May 2022)

Brasil. (1934). Decreto n. 24.643, de 10 de julho de 1934. Decreta o Código de Águas. [Decree nº. 24.643. Water code decree].

Brasil. Lei No. 11.996 de 24 de julho de. (1992). Dispõe sobre a Política Estadual de Recursos Hídricos, institui o Sistema Integrado de Gestão de Recursos Hídricos - SIGERH e dá outras providências. Diário Oficial: 29 de julho. 1992 [Law No 11.996 of July 24th, 1992. Arranges the State Water Resources Policy and creates the Water Resources Integrated system – SIGERH, among other determinations].

Brasil. (1998). Constituição Da República Federativa Do Brasil De 1988. [Constitution of Federative Republic of Brazil]. Available at: http://www.planalto.gov.br/ccivil_03/constituicao/constituicao.htm (accessed 13 July 2021)

Campos J. N. B., Studart T. M. C. and Costa A. M. (2002). Alocação e realocação do direito de uso da água: uma proposta de modelo de mercado limitado no espaço. [Allocation and reallocation of water use rights: proposal of water market model limited in space]. *Revista Brasileira de Recursos Hídricos*, **7**, 5–16.

Carvalho M. M. (2015). A Legislação Sobre A Água No Brasil. In: Água E Cultura: Inventário De Fontes De Água Da Região De Ouro Preto, P. Lemos (ed.), [Brazilian water legislation. In: LEMOS, P. Water and culture: inventory of water sources in the Ouro Preto region], Livraria E Editora Graphar, Ouro Preto, p. 134.

Dalcin A. P. and Marques G. F. (2020). Integrating water management instruments to reconcile a hydro-economic water allocation strategy with other water preferences. *Water Resources Research*, **56**(5), article id. e25558.

Dalcin A.P., Marques G.F., Oliveira D.G. and Tilmant A. (2022). Identifying functional flow regimes and fish response for multiple reservoir operating solutions. *J. Water Resour. Plann. Manage.*, **148**(6), 04022026. https://doi.org/10.1061/(ASCE)WR.1943-5452.0001567

Departamento de Recursos Hídricos e Saneamento Ambiental – Drhs. (2021). Portaria Drhs N° 13/2021. Available at: https://www.sema.rs.gov.br/upload/arquivos/202110/07075842-portaria-drhs-n-13-2021-outorga-coletiva-para-captacoes-diretas-bacia-sinos.pdf (accessed 13 July 2021)

Empresa Brasileira de Pesquisa Agropecuária – EMBRAPA. (2022). Síntese Ocupação e Uso das Terras no Brasil. [Summary of land use in Brazil]. Available at: https://www.embrapa.br/car/sintese (accessed 10 May 2022)

Fundação Getulio Vargas – FGV. (2016). A Indústria de Alimentos no Brasil e na América do sul – Estrutura das Cadeias de Distribuição e Varejo Pulverizado Inserção no Comércio Internacional [[The food industry in Brazil and South America – Structure of distribution chains and retails, Insertion in the International trade, Main market trends and potentialities, Tax and tariff aspects.] Principais Tendências e Potencialidades de Mercado Aspectos Tributários e Tarifários. FGV Projetos, 27.

Laigneau P. (2011). Cobrança pelos usos da água na França. [Water use charges in France]. XIX Simpósio Brasileiro De Recursos Hídricos- Maceió.

Liberato Jr. G. (2004). O caminho das águas: das sociedades de vala ao comitê da bacia. [The way of Waters: from ditch society to watershed committee]. Master degree thesis, Universidade Federal de Santa Catarina, Graduate Program In Sociology And Politics.

Lopes AV. . and Freitas M. A. S. (2007). A alocação de água como instrumento de gestão de recursos hídricos: experiências brasileiras. [Water allocation as water management instrument: Brazilian experiences]. *REGA*, **4**(1), 5–28.

Maniçoba A. M. (2019). Um modelo de gestão participativa: o uso compartilhado de recursos hídricos na Unidade Hidrográfica do Ribeirão Extrema – DF. [A participatory management model: shared water use in the Basin of Ribeirão Extrema – DF.]. Master degree thesis, – Fundação Oswaldo Cruz, Escola Nacional de Saúde Pública Sergio Arouca, Rio de Janeiro, Brazil.

Marques G. F. and Tilmant A. (2018). Cost distribution of environmental flow demands in a large-scale multireservoir system. *J. Water Resour. Plann. Manage*, **144**(6), 04018024.

OECD. (2015). Water Resources Governance in Brazil, OECD Studies on Water. OECD Publishing, Paris, France. https://doi.org/10.1787/9789264238121-en (accessed 13 July 2021)

Oliveira M. C. (2008). Modelos De Alocação E Realocação De Água: Um Estudo De Caso Do Programa 'Águas Do Vale' Nos Rios Jaguaribe E Banabuiú. [Water allocation and reallocation models: case study of the program 'águas do vale' in the Jaguaribe and Banabuiú rivers]. Graduation thesis, Hydraulics and Environmental Engineering, University of Ceará, Brazil.

Pinheiro M. I. T., Campos J. N. B. and Studart T. M. C. (2011). Conflitos por águas e alocação negociada: o caso do vale dos Carás no Ceará. [Water use conflicts and negociated allocation: the study case of vale dos Carás in Ceará]. *RAP – rio de Janeiro*, **45**(6), 1655–1672.

Silva U. P. A., Costa A. M., Lima G. P. B. and Lima B. P. (2006). A Experiência Da Alocação Negociada De Água Nos Vales Do Jaguaribe E Banabuiú. VIII Simpósio de Recursos Hídricos do Nordeste. Gravatá.

Sistema Estadual de Meio Ambiente – SISEMA. (2015). Introdução aos conceitos de área de conflito e processo único de outorga (outorga coletiva). Seminário sobre Ouyorga e Alocaçào de água em bacias críticas. [Introduction to the concepts of conflict area and single water permit process (collective permit). Seminar on Water Permit and Allocation in Critical Basins.] Brasília, Brazil.

Spolidorio P. C. M. (2017). A Alocação Negociada de Água como Estratégia de Regulação Responsiva. [Negociated water allocation as a response and regulation strategy. Journal of Sectoral and Regulatory Law]. *Revista de Direito Setorial e Regulatório*, **3**(1), 183–198.

SRH – Secretaria dos Recursos Hídricos e SEAGRI – Secretaria de Agricultura Irrigada. (2001). Plano de Uso Racional da Água para Irrigação nos vales do Jaguaribe e Banabuiú, Fortaleza: SRH/SEAGRI. [Rational water use plan for irrigation in the Jaguaribe and Banabuiú valleys].

Veiga L. B. E. and Magrini A. (2013). The Brazilian water resources management policy: fifteen years of success and challenges. *Water Resources Management*, **27**, 2287–2302.

doi: 10.2166/9781789062786_0173

Chapter 14

What are we allocating and who decides? Democratising understanding of groundwater and decisions for judicious allocations in India

Uma Aslekar[1], *Dhaval Joshi*[2] *and Himanshu Kulkarni*[1]

[1]*Advanced Center for Water resources Development and Management (ACWADAM), Pune, India*
[2]*School of Geosciences, University of Edinburgh, United Kingdom*

ABSTRACT

India is the largest user of groundwater in the world. With current dependency on the resource, challenges pertaining to management and governance of groundwater in the country have increased tremendously. To address these challenges, various policy and programmatic interventions have been devised and implemented by state and non-state actors. In this chapter, we question the techno-managerial nature of these solutions and propose a socio-hydrogeological approach that integrates the multidisciplinary and decentralised nature of groundwater problems. To illustrate this, we draw on case studies that have emerged as part of our work on strengthening participatory processes of aquifer monitoring and mapping, decentralised groundwater allocation for agricultural decisions and management strategies that evolve through local institutions. Various programmes are underway in the country aimed at improving participation in governance of groundwater resources. If we are to achieve our goals of collectively and sustainably managing this invisible resource, there is a need to adopt approaches that move beyond the techno-managerial paradigm and embody local ways of knowing, using, and managing groundwater.

Keywords: Aquifer, groundwater, management, over-exploitation, participation

14.1 THE CRISIS OF GROUNDWATER DEPLETION AND CONTAMINATION IN INDIA

India has 17.5% of the world's population but only 2.4% of the world's land resources and roughly 4% of the world's freshwater resources (India-WRIS, 2012). The population of India has more than tripled in the last 70 years (Census data[1], various years). Meanwhile, food grain production has increased five-fold (Ministry of Agriculture, 2017). These changes have put enormous pressure on natural resources in different parts of the country, most significantly on water. For a monsoon-dependent country such as India, rainfall plays a significant role in determining the status of water availability each year. While the demand for water has grown across all sectors because of the expansion of agriculture,

[1]www.census2011.co.in

rapid urbanisation and a growing population, there has not been a significant change in the average rainfall over the years[2]

Groundwater has played a pivotal role in fulfilling the increasing demand for water to meet the country's drinking water needs, food security, and economic development. The Green Revolution ushered in an era of achieving food security by making chemicals such as fertilisers and pesticides, high-yielding variety seeds and, most importantly, water available to the farmers that helped the country achieve food self-sufficiency (Singh *et al.*, 2019). Farmers started digging and drilling wells and bore wells to secure water supplies for agriculture. About 84% of the increase in net irrigated area in the last four decades has been sourced from groundwater.

At present, nearly 70% of India's irrigation depends on groundwater (Kulkarni *et al.*, 2015; MoA, 2014) and India is the largest user of groundwater in the world (Shah 2010). As of 2017, the total annual groundwater extraction in India has been assessed as 249 billion (10^9) cubic metres (BCM) out of 393 BCM, which is the total annual extractable groundwater resource in India (CGWB, 2019). Nearly 85% of rural drinking water supplies are dependent on groundwater (NITI Aayog, 2018) and 48% of the water requirement in cities is fulfilled by groundwater (Narain & Pandey, 2012).

Such unprecedented usage of groundwater is unfolding into a crisis with various ramifications. Firstly, expansion of groundwater abstraction in the 1970s and 1980s started to be felt in the 1990s as the shallow water table was slowly depleted. As a result, farmers started exploiting the deeper confined aquifers and further faced negative impacts of resource exhaustion on their lives and livelihoods (Shah 2010; Shah & Kulkarni, 2015). Groundwater exploitation has resulted in a decrease in basic groundwater availability; in some areas, this has meant persistent, inter-annual water scarcity. The Central Groundwater Board (CGWB) is the National Apex Agency entrusted with the responsibilities of providing scientific inputs for management, exploration, monitoring, assessment, augmentation, and

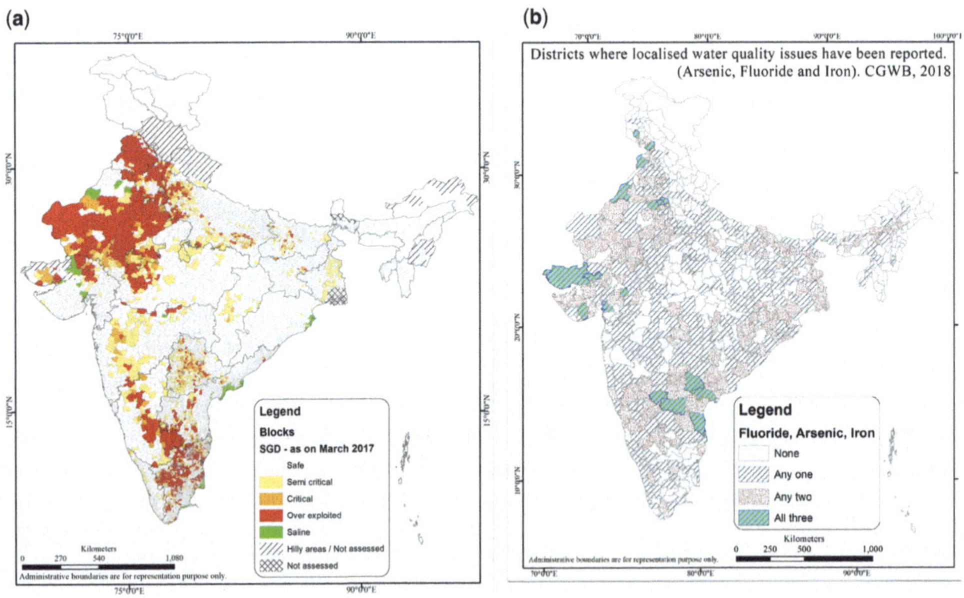

Figure 14.1 (a) Groundwater development in India by district and (b) groundwater quality in shallow aquifers by district (CGWB, 2019).

[2]https://www.tandfonline.com/doi/abs/10.1080/02626667.2010.481373

regulation of the country's groundwater resources. They conduct periodic assessment of replenishable groundwater resources of the country jointly with the State Government agencies concerned. As per the 2020 assessment, 35% of assessment units, that is the sub-district level administrative units (tehsils and blocks in different states of the country), are in various 'stages' of groundwater exploitation. As shown in Figure 14.1a below, the districts are marked in red, amber, and yellow indicating groundwater over-exploited[3], critical[4] and semi-critical[5] districts, respectively.

Secondly, problems of groundwater contamination have emerged, sometimes in association with reduced availability, leading to an acute water quality degradation. According to a 2017 report by Centre for Science and Environment (CSE), half of the country's groundwater is contaminated; 55% of districts are contaminated with high nitrate concentrations, 47% with fluoride, 22% with arsenic, and 13% with high lead. Nearly 60% of India's districts have become vulnerable to groundwater depletion, contamination, or both during the last 10–15 years (Kulkarni & Shankar, 2009), endangering the water security of many regions in the country. Figure 14.1b indicates areas experiencing groundwater quality issues.

Groundwater overextraction in India threatens drinking water supply and food security, impacts base flows, and leads to the drying-up of springs and streams. There is immense competition amongst users to tap both shallow and deeper aquifers, leading to conflicts between users. Consequently, there is an urgent need for improved water management and governance in India.

14.2 EFFORTS TO ADDRESS THE GROUNDWATER CRISIS TO DATE

Facing these challenges, the Indian government is implementing several programmes to tackle groundwater depletion. The most important of these programmes include the Integrated Watershed Management Program, one of India's flagship programmes for many years under the Ministry of Rural Development (GoI, 2008), the Drought Prone Areas Program (DPAP), and the Indo German Watershed Program. These watershed development programmes (collectively referred to as 'Programs') have resulted in a positive change in the land use pattern in most of the regions through improvement in the irrigated area and agricultural production (Palanisami & Kumar, 2009). The focus of these Programs is on supply-side interventions and resource augmentation, but these Programs fail to take into consideration the resource itself. Without an understanding of the aquifers, where groundwater resides, these interventions have severe limitations. The watershed development Programs were implemented in a 'one-size-fits-all' manner without considering the hydrogeological conditions that often govern the location, nature, and impacts of a watershed management programme on groundwater resources (Kulkarni *et al.*, 2000; Vijay Shankar *et al.*, 2011). This is evident through the periodic assessments done by the CGWB over the years. There is no significant improvement in the number of critical and over-exploited blocks in the country. Therefore, the success of watershed development programmes with respect to groundwater management seems to be limited.

In all these programmes, much emphasis was given to managed artificial recharge (MAR) that refers to the augmentation of groundwater resources involving the transfer of surface water to aquifers through human intervention (Athavale, 2003; Gale *et al.*, 2002). Initially, artificial recharge in rural India was primarily implemented through watershed development projects, and therefore designed at that scale. More specific methods such as direct recharge through farm-ponds, open wells, boreholes, and roof-top rainwater harvesting gained popularity as integral components of watershed

[3]Groundwater over-exploited area indicating groundwater extraction exceeding the annually replenishable groundwater recharge.

[4]Critical areas are the areas where the stage of groundwater extraction is between 90% and 100% of annual extractable resources available.

[5]Semi-critical areas are the areas where the stage of groundwater extraction is between 70% and 90% of annual extractable resources available.

development, as well as separate, specific, 'groundwater recharging' initiatives. Some of the village-level case studies that have emerged through this approach include Hiware Bazaar, Ralegan Siddhi and Randullabad (Aslekar *et al.*, 2013; Singh, 2012).

While MAR is an important strategy, it is equally pertinent to focus on developing an understanding of groundwater resources amongst stakeholders that leads to improved decisions about cropping choices and drinking water security. Such an approach was adopted by the Food and Agriculture Organization (FAO)- supported Andhra Pradesh Farmer-Managed Groundwater System (APFMGS) programme in Telangana and Andhra Pradesh, located in central eastern India. The APFAMGS[6] project developed a participatory hydrological monitoring programme to build farmers' capacities by devising protocols for monitoring groundwater levels and developing the understanding of groundwater. During the project period (between 2004 and 2009) 638 village-level Groundwater Monitoring Committees (GMCs) were formed in 7 districts of Andhra Pradesh and Telangana. These committees regularly monitored hydrogeological parameters for developing climate-resilient crop water budgeting on an annual basis. Such has been the impact of the project that it was recognised as an inspiration for the World Bank-supported Atal Jal programme that is currently underway (World Bank, 2018).

The Sardar Patel Sahkari Jal Sanchay Yojana (SPSJSY) project, located in the Saurashtra region of Gujrat in northwestern India, is a promising example of a community-based distributed groundwater recharge system focused on developing interventions including check dams aimed at improving the recharge of groundwater. Inspired by this programme, the Government of Maharashtra implemented the Jal Yukta Shivar (literally, the water abundant farms campaign) programme between 2015 and 2019, aiming to make the state drought free by constructing cement and earthen stop dams and digging farm ponds through community participation and convergence with other programmes. About 254 000 water and soil conservation works have been completed in 16 522 villages with a total cost of 7692 crores, roughly 1100 million US dollars (Bhadbhade *et al.*, 2019). The Program's limited outcomes have been attributed to the lack of groundwater management at local scales.

Although groundwater is a common pool resource, the ownership and access to it is private in nature. The Easement Act of 1882 states that anyone who owns land owns the resources under it, meaning groundwater sources such as dug-wells and bore-wells are private in nature. There is no licensing or permit system for groundwater in India or any of its states. Although recent state-led legislations have tried to identify groundwater sources and levy cess on existing sources beyond a certain depth, there is very little implementation and enforcement of the laws.

Given that nearly 65% of irrigated agriculture in the country is dependent on groundwater and with increasing evidence of improved water-use efficiency using micro-irrigation techniques (Wani *et al.*, 2016), the government is now subsidising the adoption of these technologies by farmers under the Pradhan Mantri Krishi Sichai Yojana (Prime Minister Agriculture Irrigation Scheme). In India, as of 2020–21[7] about 938 193 ha[8] of land have been brought under micro-irrigation. However, it is observed that in the absence of regulation over groundwater abstraction, farmers who acquire micro-irrigation techniques prefer to intensify production rather than conserve water (Birkenholtz, 2017)

Recent water management programmes were backed by the energy reforms. Jyotigram Yojana was launched in the state of Gujarat with the intention to address the problems of growing energy subsidies and groundwater overdraft (Gupta, 2012; Shah & Verma, 2008; Shah *et al.*, 2004) by providing assured 8-hour uninterrupted energy supply to the farmers. However, the programme failed

[6]https://www.fao.org/climate-smart-agriculture-sourcebook/enabling-frameworks/module-c1-capacity-development/c1-case-studies/case-study-c111-the-andhra-pradesh-farmer-managed-groundwater-systems-apfamgs-project/en/

[7]https://pmksy.gov.in/mis/rptAchievement.aspx

[8]Ha – Hectare (1 Ha = 2.5 acres)

to address the issue of groundwater overextraction. All of these programmes have focused on 'supply side interventions' to tackle the groundwater crisis.

14.3 CHALLENGES IN THE CURRENT PARADIGM

Despite some success in improving groundwater management through MAR and groundwater monitoring-based crop water budgeting, studies by different government and non-governmental organisations have highlighted the rather limited impacts of the programmes. Questions have emerged over the technical aspects of the programmes in terms of design, planning, implementation, and their efficacy in resolving the groundwater crisis (Argade & Narayanan, 2019; Bhadbhade *et al.*, 2019; CAG, 2020; Richard-Ferroudji *et al.*, 2018; Shah & Narain, 2019). In 2004, the CGWB started assessing the groundwater situation in the country using the Groundwater Estimation Methodology-1997[9], with subsequent assessments conducted in 2009, 2013 and 2019. During this period, the number of safe blocks, which are blocks where there is no significant long-term decline in the pre- and post-monsoon water levels and where there is groundwater development potential[10] , has gone down from 72% to 63% within a span of 15 years

Legal frameworks play a crucial role in enabling efficient groundwater governance (Mechlem, 2012). Water is governed at the state level in India and, as per the constitution of the country, the responsibility lies with the state departments to enhance, develop, and manage these hydrological systems. Every state in India frames its own rules and regulations based on the Model Legislation Bill proposed by the central government to regulate and control the development and management of groundwater that was first proposed in the 1970s (Cullet, 2019). In their analysis of the 1993 Maharashtra Groundwater Act that aimed to protect drinking water sources, Phansalkar and Kher (2006) highlighted the Act's limited understanding of groundwater resources and poor enforcement by the administration. The Act missed the premise that a single aquifer system often caters for all types of users and uses, where diverse types of uses come into conflict even within a single village that depends on groundwater for meeting agricultural and domestic needs (Kulkarni & Vijay Shankar, 2014). Some scholars argue that the close interdependencies between various aspects of management strategies such as crop choices, energy pricing, and state control often undermine the effectiveness of such regulatory mechanisms (Hoogesteger & Wester, 2017). The right to pump groundwater has been regarded as a landowner's impregnable right. This pre-existing provision has led to a strong social consensus in favour of irrigators, even if that compounds the difficulty in fetching groundwater for drinking water.

Various limitations have been outlined by scholars and practitioners such as the bureaucratic nature of the Programs and their discard of local experiential knowledge favouring 'technical guidelines' (Baviskar, 2004; Samuel *et al.*, 2007). In addition to the often narrowly-focused 'supply-side' nature of such Programs mentioned above, many scholars have also outlined the (mis)conception of the village community as a homogeneous and harmonious entity as an important factor in yielding poor outcomes for the Programs (Agrawal & Gibson, 1999). Although the Program includes establishing a

---

[9]Groundwater Estimation methodology-1997 was the outcome of the recommendations of a 'High Power Committee' constituted by the Ministry of Water Resources, Government of India that includes detailed computational procedures to be followed while applying the groundwater estimation methodology and the assumptions behind the estimation of all components of groundwater assessment. (http://cgwb.gov.in/Documents/GEC97-Detailed_Guidelines.pdf)

[10]Safe block: The groundwater resources are assessed in units, i.e. blocks/talukas/mandals/watersheds. These assessment units are categorised for groundwater development based on two criteria – (a) stage of groundwater development, and (b) long-term pre-and post-monsoon water levels. The long-term groundwater level trends are computed generally for the period of 10 years. (http://cgwb.gov.in)

committee through a formal process of Gram Sabha[11], there are power dynamics at play which result in a capture of such spaces by the villages' social and political elites (Kale, 2011).

14.4 MOVING BEYOND TECHNO-MANAGERIAL SOLUTIONS

The various Programs discussed thus far are often very technical in scope and lack adequate mechanisms to change mindsets and secure the buy-in of communities. The participatory processes included in these Programs tend to limit the participation of communities in understanding their resources. For example, the Government of India's ambitious Jal Jeevan Mission focuses on improving the public drinking water supply by installing tap connections in all rural households by 2024. In doing so, the emphasis lies on creating infrastructure, laying pipes, installing taps and pump-sets. While this is necessary, it is much more important to focus on the institutional capacities and participation of communities to ensure inclusive and efficient local-level planning and management. Evidence suggests that groundwater-based projects that promoted local participation were far more successful than those focused solely on technical interventions (Kerr & Chung, 2002; Palanisami & Kumar, 2009). Hence, it is essential to transform how participation is envisaged in these programmes and move towards more engaging, active, and empowering participation, where communities organise and design strategies to manage and govern their water resources (including groundwater).

The Farm Pond Program promoted by the government of Maharashtra is an example of a supply-side, technical project that resulted in undesirable outcomes that could have been prevented with better, participative planning. This programme aimed to improve access to water for rainfed and drought-prone farmers by creating ponds that harvest rainwater. Despite its promises, its implementation (called '*Magel tyala Shet Tala*' which literally means 'whoever wants shall have a farm pond') has led to water equity and sustainability challenges. Many farmers have resorted to abstracting groundwater from deeper bore-wells and storing it in their farm ponds, which has led to severe evaporation losses and the capture of a common pool resource (Joshi, 2017; Kale, 2017).

Assessing the problems as simply technical or managerial in nature (e.g., lack of data) promotes techno-managerial solutions and practices (Joy *et al.*, 2014; Mehta *et al.*, 2019; Prakash, 2005). Instead, embedding the technical solutions into the challenge of unsustainable water use in a fully participative process would have ensured that (technical) solutions are associated with the right understanding (i.e., water is part of a complex, interconnected system) and mindset (i.e., acknowledging that an action on one part of the system will impact another part of the system). Moving away from the current techno-engineering approach requires recognising the 'governance' dimensions of water management, that is that water use is an embodiment of politics, science, culture, histories and values (Anand, 2011; Mosse, 2003; Orlove, 2002) and requires frameworks and decision-making processes that confront perspectives, understandings, and values.

14.5 TOWARDS A MULTIDISCIPLINARY AND PARTICIPATORY FRAMEWORK

Drawing from our work in this field for nearly a decade in India and other research studies, we propose a multidisciplinary framework that aims to bring together understanding from various disciplines to design groundwater management strategies for improving its allocation amongst various users and uses. We define multidisciplinarity as a practice that aims to integrate multiple voices, beyond the sciences, and values the multiple nature of groundwater beyond the biophysical characteristics, to devise a process that leads to improved governance. Through multiple nature, we intend to recognise diversity of dependencies and contexts within which groundwater–society relationships have emerged

[11]Gram Sabha means a body consisting of all persons whose names are included in the electoral rolls for the panchayat at the village level. The term is defined in the Constitution of India under Article 243(b).

over the last decades. In doing so, the multidisciplinary framework aims to emphasise the role of aquifers rather than wells and the role of communities rather than individuals (Zwarteveen *et al.*, 2021). This will also mean transforming the policy and programmatic landscape through a set of subsidies, interventions, and incentives.

Participatory processes can help enable collective decision-making. Creating engagement spaces in the form of institutional arrangements, such as aquifer federations and watershed committees, that are recognised by state and non-state actors and work closely with the communities is important. Recent changes in groundwater legislation in Maharashtra, for example, promote such decentralised decision-making spaces in the form of a Watershed Water Resources Committee, aquifer federations, and so on. Similar arrangements have been proposed in watershed Programs, such as the Jal Jeevan Mission (wherein the emphasis is on the Village Water and Sanitation Committee constituted by the Gram Panchayat[12] through Gram Sabha).

One of the impediments towards improving groundwater governance and management is the lack of data. Researchers have emphasised the role of groundwater knowledge in improving governance. Collective groundwater monitoring has been attempted in various instances such as the APFAMGS Program in Andhra Pradesh, aquifer management organisations (COTAS or technical groundwater committees in Spanish) in Mexico and groundwater user associations in Spain (Garduno *et al.*, 2009; Lopez-Gunn & Cortina, 2006; Wester *et al.*, 2011). Varady *et al.* (2016) highlight groundwater data availability as a prerequisite for governance strategies. Groundwater is an invisible resource, which is why the understanding of groundwater becomes even more challenging. In India's context, given the country's hydrogeological diversity, it is difficult to implement a common data programme that will document the nuances of different aquifer systems across the country. Another important observation is that given the sheer magnitude of groundwater use in the country and the number of sources (nearly 30 million sources), arriving at representative data in terms of their granularity and frequency is often a cumbersome task. For example, the CGWB currently monitors about 15 640 observation wells

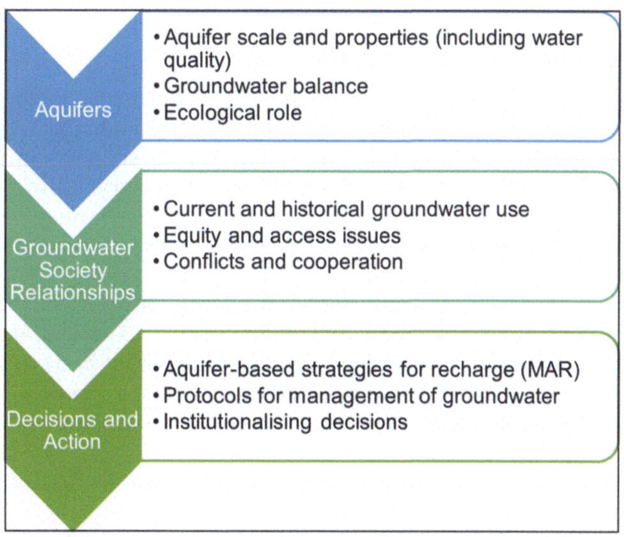

Figure 14.2 Elements of a socio-hydrogeological approach.

[12]Gram Panchayat is the local elected governing council at the village level.

across 5740 units (block, taluka, mandal) which equates to about 2.7 wells per unit. Each of these units roughly consists of about 100 villages, thus the representative figure equates to about 1 well in 35–40 villages. This gives us an idea of the scope of the challenge in terms of groundwater monitoring and associated costs.

Participatory frameworks of data collection, analysis, and interpretation will help address some of the problems discussed above. Such frameworks will also help communities and other stakeholders improve their collective understanding of local groundwater conditions. Engaging community members in data collection and preliminary interpretation shapes pathways to local management strategies.

Going beyond efforts for participatory monitoring of groundwater, our framework promotes groundwater socio-hydrogeology, which can be defined as a set of practices that interact and shape an approach that integrates the societal–groundwater relationships beyond groundwater's biophysical nature. It aims to engage with social inequities, resource distribution and questions of access to arrive at collective ways of addressing governance challenges. Experiences from ACWADAM's[13] work across India have emphasised the need for such an approach that has led to a diversity of practices to understand, manage, and govern groundwater over-abstraction (Figure 14.2).

A socio-hydrogeological approach integrates three key aspects of governance as depicted below:

These three aspects together help develop a robust groundwater management and governance plan at the local level. The next section discusses some of the innovative methods and approaches to understanding the three pillars of socio-hydrology. We shall discuss the method of Participatory Aquifer Mapping (PAQM) that has emerged from our work in rural and urban areas and engages in participatory processes of mapping and managing aquifers. To elaborate groundwater–society relationships, we draw from the experiences of groundwater balance-based crop water budgeting and allocations in rural settings where agriculture is the dominant driver of groundwater use. Lastly, we bring forth our experiences on how this improved understanding leads to decisions and actions around groundwater management and governance. We draw from diverse case studies across different hydrogeological and regional settings.

14.6 PARTICIPATORY MAPPING OF AQUIFERS

Groundwater accumulation and movement is governed by the hydrogeology of the region. The types of rocks and their properties determine the transmissivity and storativity of groundwater in the aquifers. Hydrogeologically, India is a diverse country with six predominant hydrogeological regimes (Figure 14.3), viz. alluvial (unconsolidated), volcanic, crystalline, sedimentary (hard), sedimentary (soft) and mountain systems (Kulkarni & Vijayshankar, 2014). Therefore, an understanding of local groundwater systems and aquifers as a prerequisite for groundwater management programmes becomes pertinent.

Aquifer mapping is a hydrogeological enterprise, predominantly driven by geological sciences. However, the experiences of authors suggest that communities' experiential knowledge regarding water-bearing and water-resisting rock formations forms an important basis to develop this collective understanding. Participatory Aquifer Mapping has evolved as a process where techniques of hydrogeology are coupled with local experiential knowledge to reach a common understanding of the aquifer (Figure 14.4).

The process entails conducting field surveys for observing local geology and collecting lithology samples from sources that are being drilled. A groundwater monitoring network is set up to collect water-level data for at least one hydrological year. It involves identifying sources and then setting

[13] ACWADAM stands for Advanced Center for Water Resources Development and Management. ACWADAM is a Pune based not-for-profit organisation working on the issues pertaining to groundwater and its management.

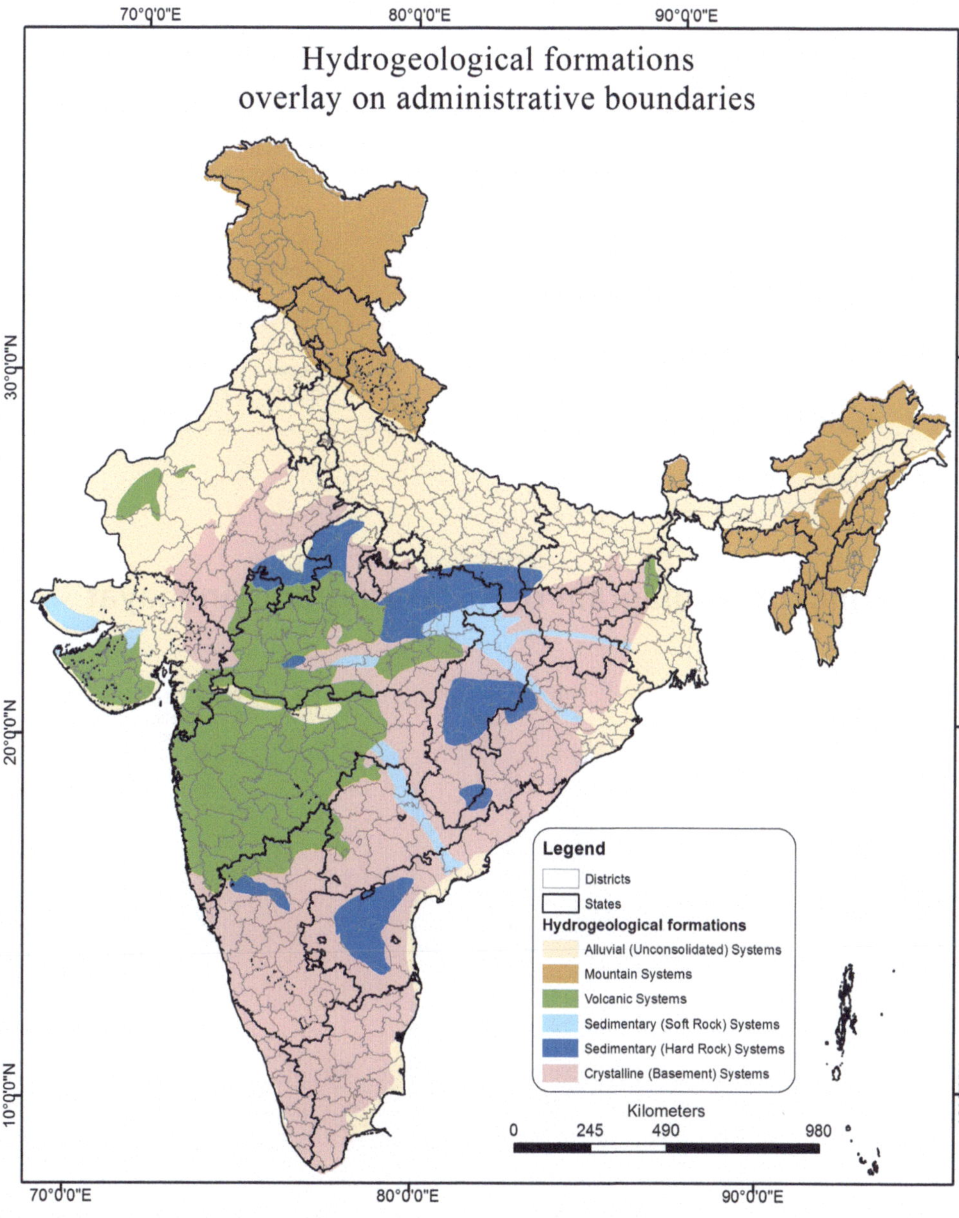

Figure 14.3 Hydrogeological formations in India (after Kulkarni & Vijayshankar, 2014).

a monitoring frequency (monthly, quarterly, seasonal). A questionnaire containing information on source type, use, year of construction, depth and water struck is then used to document responses from community members. These crowdsourced data are then validated through field observations and information from local drillers who have thorough knowledge of the groundwater systems in the area. Based on this collective effort, a conceptual map of aquifer systems present in the area is

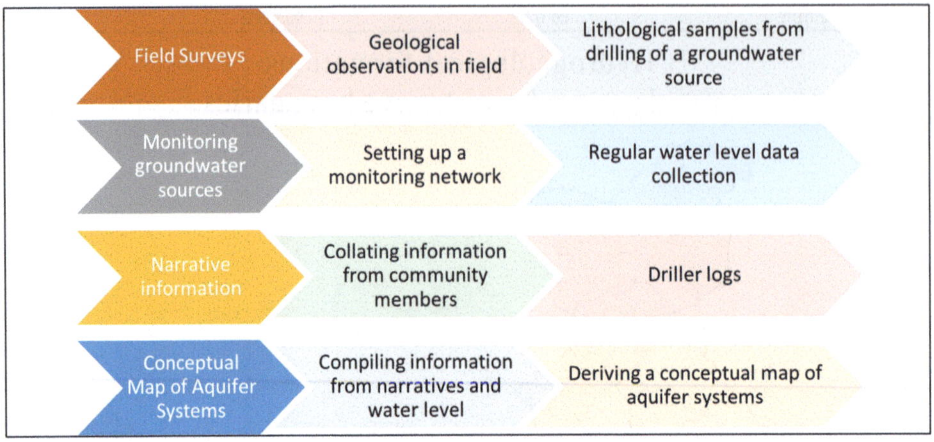

Figure 14.4 Participatory aquifer mapping process.

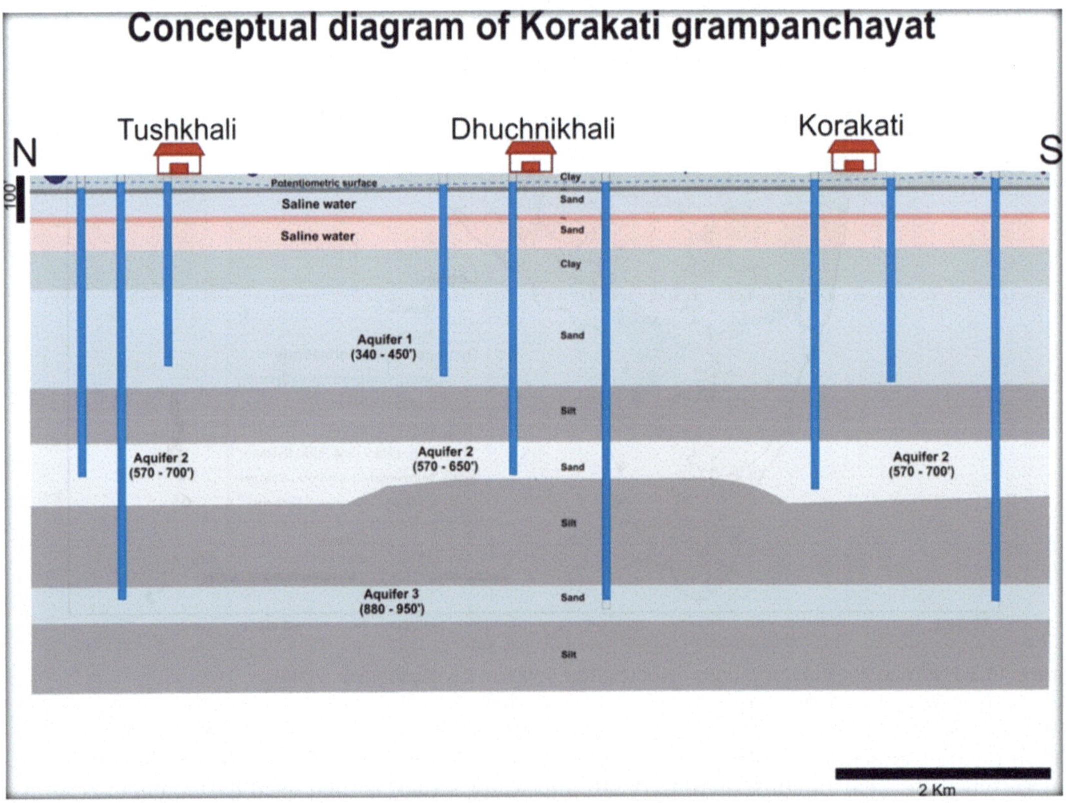

Figure 14.5 Conceptual Aquifer Map of Korakati Gram Panchayat. Source: ACWADAM (2021).

derived. In areas where further validation is required, data can be substantiated using geological techniques such as vertical electrical sounding (VES)[14] resistivity surveys, drilling observation wells, and so on.

Application of the above process was undertaken in Korakati Gram Panchayat of the Sundarbans region of West Bengal. Korakati is a part of the alluvial aquifer settings of the Ganga delta region of Sundarbans. The community depends on groundwater for their livelihoods and drinking water security; thus, mapping and monitoring groundwater becomes crucial for ensuring equity and sustainability. The PAQM process involved inputs from local drillers and community members with experience tapping handpumps and tube-wells at various depths. Figure 14.5 presents the conceptual map that emerged for the area.

The map helped create a common and shared understanding of the resource system which otherwise is understood at an individual level. A shared understanding of resource systems is critical for any positive action for resource governance. A similar exercise was carried out in multiple locations across India to prepare the aquifer maps and conceptual diagrams for different villages. Even an approximation of aquifer boundaries and aquifer characteristics can form a robust platform for launching a community-level, participatory form of groundwater management.

14.7 PARTICIPATORY WATER BUDGETING IN GROUNDWATER-BASED IRRIGATION

As stated earlier, studies have shown that the use of micro-irrigation techniques often leads to the intensification of cropping and irrigation practices (Birkenholtz, 2017). Thus, reliance on micro-irrigation alone as a panacea to tackle the agricultural water use may be misleading. Behavioural change through improved and shared understanding of resources shall enable a change in perception of farmers and their orientation towards water use in agriculture. The experiences of APFAMGS and case studies such as Randullabad have shown the importance of this approach (Aslekar *et al.*, 2013; FAO, 2016; Singh 2012).

Many studies rely on standards of crop water requirements developed by various international and state agencies such as FAO[15], Water and Land Management Institute (WALMI) (2013), and so on. However, such an approach leads to approximations and may not be reflective of the practices adopted by farmers to irrigate their crops. Hence, to arrive at a more accurate understanding of crop water requirements, we implemented an approach involving identifying farmer plots that are cultivating crops grown in the village/area, complemented with documented irrigation cycle data based on farmer diaries and consolidated to arrive at crop water requirements. This allows us to undertake groundwater balance-based crop water budgeting based on local data that are more accurate and more accepted by local communities. In the Waki village in the Dhule district of Maharashtra, derived crop water requirements were compared with the groundwater balance situation to assess total annual recharge and discharge of groundwater along with changes in aquifer storage. An estimate of the groundwater balance enabled us to understand patterns of groundwater recharge, discharge and change in aquifer storage, and estimate the stage of groundwater development in an aquifer/area

14.8 PARTICIPATORY GROUNDWATER MANAGEMENT

Much of the groundwater use is allocated to agriculture in any given village. While winter cropping (known as Rabi season) is the dominant driver, farmers are increasingly resorting to irrigating crops in Kharif season, owing to irregular monsoons. Collectivising the allocation of groundwater in

[14] Vertical electrical sounding (VES) is a geophysical method for investigation of a geological medium. The method is based on the estimation of the electrical conductivity or resistivity of the medium.

[15] See e.g.: http://www.fao.org/docrep/S2022E/s2022e07.htm

rural settings is one of the biggest challenges. Although most communities understand the need for collective action for allocating groundwater appropriately, it is difficult to adopt new practices and change cropping patterns, as this is intricately linked to livelihoods, income security and the larger agricultural economy. However, community-driven, hydrogeology-based participatory groundwater management planning can help shape better local allocations.

In the Baretha village, a set of marginalised and rainfed farmers were identified, who agreed to sharing wells for irrigation. Using the information developed through local-level groundwater monitoring, potential recharge areas were demarcated to identify the location for the new wells. Once the wells were in place, a formal agreement was sought between the farmers for allocating water for each member of the group. The agreement had points referring to shared well ownership, mutual arrangement for irrigation cycles and cropping patterns, and contribution towards maintenance costs.

In the case of Sitaljhiri village, a local governance institution called Bhujal Prabandhan Samiti (Groundwater Management Committee) was constituted by the Gram Panchayat through a Gram Sabha. This institution was entrusted with decisions for managing and recharging groundwater. The Samiti decided to adopt certain norms for crop choices, which included a reduction in wheat cropping area, cultivating Bengal Gram, and adopting an improved wheat variety (locally known as 1544) that required three cycles of irrigation as compared to an existing variety called Lokwan, which needed five cycles. These choices were adopted by some farmers in the village and enabled the judicious use of groundwater. Similarly, in the Randullabad village of the Satara district of Maharashtra, the community has taken a decision to put a complete ban on drilling bore wells in the village to secure groundwater in the villages' confined aquifers. This decision is meticulously followed by the community even after 20 years, reiterating the importance of knowledge, participation, and decentralised groundwater governance (Aslekar *et al.*, 2013).

14.9 CONCLUSIONS

Today, india is the largest user of groundwater in the world, a seemingly remarkable fact that hides nuances of social disparity, iniquitous competition, hidden conflicts, and the tension between traditional and modern practices. The individualised nature of groundwater sourcing, access, and distribution with groundwater usage from millions of wells (and springs in the mountains) means a centralised command and control type of regulation is difficult to implement. Decentralised mapping of aquifers and a combined understanding of groundwater characteristics and social behaviour is necessary if India's groundwater resources are to be allocated collectively. Groundwater governance through principles of managing common pool resources cannot be fully achieved unless collective decision and action is enabled through community participation and institutional decentralisation. The concept of socio-hydrogeology offers a useful instrument to conduct participatory mapping, measurement and monitoring of aquifers in India's varied geography and contexts.

Socio-hydrogeological mapping of aquifers, using people's narratives on well-logs, groundwater pumping, groundwater levels, groundwater quality and the ecosystem changes in an area can help create a three-dimensional understanding of groundwater resources. Such an approach incorporates a collaborative understanding of local groundwater dynamics, paving the way for collective decisions and actions at community levels that help prioritise allocations amongst users and uses. Formal institutions of governance, such as the Gram Panchayats (in rural India) and Municipalities (in urban India) can remain well-informed about the groundwater situation in their respective jurisdictions and about protocols of conservation, Managed Aquifer Recharge and Participatory Aquifer Management. This approach not only incorporates decentralised groundwater governance, but also embodies principles of democratising the management and governance of groundwater, particularly given India's peculiar groundwater situation.

ACKNOWLEDGEMENT

We acknowledge the support from Bharat Rural Livelihoods Foundation for the Program on 'Participatory Groundwater Management in Central Indian Tribal Belt' along with non-governmental organisation partners BAIF, PRASARI and Lupin Foundation. We would like to extend our sincere gratitude to the communities of Korakati Gram Panchayat, Sitaljhiri Gram Panchayat and Waki Gram Panchayat for collaborating on the Program.

REFERENCES

Aayog N. (2018). Composite Water Management Index: A tool for water management.

ACWADAM. (2021). Towards Decentralised Groundwater Governance in the Central Indian Tribal Belt: evidences for participatory processes and inputs for policy. ACWA/Hydro/2021/H112.

Agrawal A. and Gibson C. C. (1999). Enchantment and disenchantment: the role of community in natural resource conservation. *World Development*, **27**(4), 629–649.

Anand N. (2011). Pressure: the polytechnics of water supply in Mumbai. *Cultural Anthropology*, **26**(4), 542–564, https://doi.org/10.1111/j.1548-1360.2011.01111.x

Argade P. and Narayanan N. C. (2019). Undercurrents of participatory groundwater governance in rural jalna, western India. *Water Alternatives*, **12**(3), 869–885.

Aslekar U., Upamanyu A. and Kulkarni H. (2013). Participatory groundwater management in Randullabad. Report on ACWADAM's Action Research Initiative under the PGWM Program with Support from Arghyam Trust, Bengaluru.

Athavale R. N. (2003). Water Harvesting and Sustainable Supply in India. Rawat Publications, Jaipur.

Baviskar A. (2004). Between micro-politics and administrative imperatives: decentralisation and the watershed mission in Madhya Pradesh, India. *The European Journal of Development Research*, **16**(1), 26–40, https://doi.org/10.1080/09578810410001688716

Bhadbhade N., Bhagat S., Joy K. J., Samuel A., Lohakare K. and Adagale R. (2019). Can Jalyukt Shivar Abhiyan prevent drought in Maharashtra? *Economic & Political Weekly*, **54**(25), 13.

Birkenholtz, T. (2017). Assessing India's drip-irrigation boom: efficiency, climate change and groundwater policy. *Water International*, **42**(6), 663–677.

Central Ground Water Board. (2019). National compilation on dynamic ground water resources of India, 2017. Ministry of Water Resources River Development and Ganga Rejuvenation, Govt of India. Available at: http://www.cgwb.gov.in

Comptroller and Auditor General of India (CAG). (2020). Report no. 3 Government of Maharashtra (https://cag.gov.in/cag_old/sites/default/files/audit_report_files/Report_No_3_of_2020_General_and_Social_Sector_and_PSUs_Government_of_Maharashtra.pdf)

Cullet P. (2019). Governing groundwater: fostering participatory and aquifer-based regulation. In: Water Governance: Challenges and Prospects. Springer, Singapore, pp. 117–129.

Department of Agriculture. (2017). Annual Report 2016–17. Department of Agriculture and Framers Welfare, Govt of India, Krishi Bhawan, New Delhi. Available at: http://agricoop.nic.in/sites/default/files/Annual_rpt_201617_E.pdf (accessed 8 February 2021)

FAO. (2016). Global Framework for Action: to Achieve the Vision on Groundwater Governance. Food and Agriculture Organisation, Rome. www.fao.org/3/a-i5705e.pdf

Gale I., Neumann I., Calow R. and Moench dan M. (2002). The effectiveness of Artificial Recharge of groundwater: a review. British Geological Survey. (CR/02/108N) (Unpublished).

Garduño H., Foster S., Raj P. and van Steenbergen F. (2009). Addressing groundwater depletion through community-based management actions in the weathered granitic basement aquifer of drought-prone Andhra Pradesh-India. GW-MATE Case Profile Collection, **19**.

GOI. (2008). Common Guidelines for Watershed Development Projects. Department of Land Resources, Ministry of Rural Development, Government of India, New Delhi.

Gupta A. (2012). Red Tape: Bureaucracy, Structural Violence, and Poverty in India. Duke University Press, North Carolina.

Hoogesteger J. and Wester P. (2017). Regulating groundwater use: the challenges of policy implementation in guanajuato, central Mexico. *Environmental Science & Policy*, **77**, 107–113, https://doi.org/10.1016/j.envsci.2017.08.002

India-WRIS. (2012). River Basin Atlas of India, Ministry of Water Resources, Government of India, RRSCO West, ISRO, Jodhpur.

India - Atal Bhujal Yojana (ABHY) - National Groundwater Management Improvement Program (English). World Bank Group, Washington, D.C. http://documents.worldbank.org/curated/en/697581528428694246/India-Atal-Bhujal-Yojana-ABHY-National-Groundwater-Management-Improvement-Program.

Joshi, D. (2017). Promoting Farm Ponds. *Economic & Political Weekly*, **52**(16), 71.

Joy K. J., Kulkarni S., Roth D. and Zwarteveen M. (2014). Re-politicising water governance: exploring water re-allocations in terms of justice. *Local Environment*, **19**(9), 954–973, https://doi.org/10.1080/13549839.2013.870542

Kale, E. (2011). Social exclusion in watershed development: evidence from the Indo-German Watershed Development Project in Maharashtra. *Law Environmental and Development Journal*, **7**, 95.

Kale E. (2017). Problematic uses and practices of farm ponds in Maharashtra. *Economic & Political Weekly*, **52**(3), 20–22.

Kerr J. and Chung K. (2002). Evaluating watershed management projects. *Water Policy*, **3**(6), 537–554.

Kulkarni H. and Shankar P. V. (2009). Groundwater: towards an aquifer management framework. *Economic and Political Weekly*, 13–17.

Kulkarni H. and Shankar P. V. (2014). Groundwater resources in India: an arena for diverse competition. *Local Environment*, **19**(9), 990–1011, https://doi.org/10.1080/13549839.2014.964192

Kulkarni H., Deolankar S. B., Lalwani A., Joseph B. and Pawar S. (2000). Hydrogeological framework of the Deccan basalt groundwater systems, west-central India. *Hydrogeology Journal*, **8**(4), 368–378, https://doi.org/10.1007/s100400000079

Kulkarni H., Shah M. and Vijay Shankar P. S. (2015). Shaping the contours of groundwater governance in India. *Journal of Hydrology: Regional Studies*, **4**, 172–192. https://doi.org/10.1016/j.ejrh.2014.11.004

Lopez-Gunn E. and Cortina L. M. (2006). Is self-regulation a myth? Case study on Spanish groundwater user associations and the role of higher-level authorities. *Hydrogeology Journal*, **14**(3), 361–379, https://doi.org/10.1007/s10040-005-0014-z

Mechlem K. (2012). Groundwater governance: a global framework for country action-thematic paper 6: Legal and institutional frameworks. GEF: Groundwater Goverance: A Global Framework for Country Action, GEF ID, *3726*.

Mehta L., Srivastava S., Adam H. N., Bose S., Ghosh U. and Kumar V. V. (2019). Climate change and uncertainty from 'above'and 'below': perspectives from India. *Regional Environmental Change*, **19**(6), 1533–1547, https://doi.org/10.1007/s10113-019-01479-7

Ministry of Agriculture, Government of India (2014). Web based land use statistics information system, Various Land Use Statistics Reports, Directorate of Economics and Statistics, Department of Agriculture and Cooperation. http://lus.dacnet.nic.in/dt_lus.aspx.

Mosse D. (2003). The Rule of Water: Statecraft, Ecology and Collective Action in South India. Oxford University Press, New Delhi.

Narain S. and Pandey P. (2012). Excreta Matters: How Urban India is Soaking up Water, Polluting Rivers and Drowning in its Own Waste. Centre for Science and Environment, New Delhi.

Orlove, B. (2002). Lines in the Water: nature and Culture at Lake Titicaca. University of California Press.

Palanisami K. and Kumar D. S. (2009). Impacts of watershed development programs: experiences and evidences from Tamil Nadu. *Agricultural Economics Research Review*, **22**, 387–396.

Phansalkar S. and Kher V. (2006). A decade of Maharashtra groundwater legislation: analysis of the implementation process. *Law, Environment and Development Journal*, **2**, 67.

Prakash, A. (2005). The Dark Zone: groundwater Irrigation, Politics and Social Power in North Gujarat. Wageningen University and Research.

Richard-Ferroudji A., Raghunath T. P. and Venkatasubramanian, G. (2018). Managed aquifer recharge in India: Consensual policy but controversial implementation. *Water Alternatives*, **11**(3), 749–769.

Samuel A., Joy K. J., Paranjape S., Peddi S., Adagale R., Deshpande P. and Kulkarni S. (2007). Watershed development in Maharashtra: Present scenario and issues for restructuring the programme. SOPPECOM, Pune.

Shah T. (2010). Taming the anarchy: Groundwater governance in South Asia. Routledge.

Shah T., Scott C., Kishore A. and Sharma A. (2004). Energy-Irrigation Nexus in South Asia: improving Groundwater Conservation and Power Sector Viability. IWMI, Colombo, Sri Lanka, Vol. **70**.

Shah M. and Kulkarni H. (2015). Urban water systems in India. *Economic and Political Weekly*, **50**(30), 57–69.

Shah S. H. and Narain V. (2019). Re-framing India's 'water crisis': an institutions and entitlements perspective. *Geoforum*, **101**, 76–79, https://doi.org/10.1016/j.geoforum.2019.02.030

Shah T. and Verma S. (2008). Co-management of electricity and groundwater: an assessment of Gujarat's Jyotirgram scheme. *Economic and Political Weekly*, **43**(7), 59–66.

Singh S. (2012). Local Governance and Environment Investments in Hiware Bazar, India. Ecological Economics From the Ground Up. Routledge, London, pp. 191–202.

Singh A., (eds.). (2019). Water Governance: Challenges and Prospects, Springer Water, https://doi.org/10.1007/978-981-13-2700-1_1. Springer Nature, Singapore Pte Ltd.

Varady R. G., Zuniga-Teran A. A., Gerlak A. K. and Megdal S. B. (2016). Modes and approaches of groundwater governance: a survey of lessons learned from selected cases across the globe. *Water*, **8**(10), 417, https://doi.org/10.3390/w8100417

Vijayshankar P. S., Kulkarni H. and Krishnan S. (2011). India's groundwater challenge and the way forward. *Economic and Political Weekly*, **46**(2).

WALMI. (2013). Irrigated Crops – Part I, 4th edn. Water and Land Management Institute Publication no. 44, Aurangabad, Maharashtra.

Wani S. P., Bergvinson D., Raju K. V., Gaur P. M. and Varshney R. K. (2016). Mission India for Transforming Agriculture (MITrA), Research Report IDC-4.

Wester P., Minero R. S. and Hoogesteger J. (2011). Assessment of the development of aquifer management councils (COTAS) for sustainable groundwater management in guanajuato, Mexico. *Hydrogeology Journal*, **19**(4), 889–899, https://doi.org/10.1007/s10040-011-0733-2

Zwarteveen M., Kuper M., Olmos-Herrera C., Dajani M., Kemerink-Seyoum J., Frances C., Beckett L., Lu F., Kulkarni S., Kulkarni H., Aslekar U., Börjeson L., Verzijl A., Guzmán C. D., TeresaOré M., Leonardelli I., Bossenbroek L., Ftouhi H., Chitata T., Hartani T., Saidani A., Johnson M., Peterson A., Bhat S., Bhopal S., Kadiri Z., Deshmukh R., Joshi D., Komakech H., Joseph K., Mlimbila E. and De Bont C. (2021). Transformations to groundwater sustainability: from individuals and pumps to communities and aquifers. *Current Opinion in Environmental Sustainability*, **49**, 88–97.

doi: 10.2166/9781789062786_0189

Chapter 15

Legal frameworks for agricultural water use in Canada: a comparative study of Alberta and Québec

Hugo Tremblay

Associate Professor, Faculty of Law, University of Montreal, Montreal, Canada

ABSTRACT

Canada encompasses vastly different hydrological conditions, agricultural practices, and legal regimes for water rights. In Alberta, extensive agricultural irrigation has developed since the late 1800s in water-supply constrained conditions based on a prior-appropriation regime for water rights. In comparison, the legal framework for agricultural water rights in the relatively water-abundant province of Quebec has more recently evolved into regulated riparianism where most of the crops are still rain-fed. This chapter compares these two provinces to highlight how sub-national jurisdictions on water rights and differences in legal traditions can address diversity in agricultural practice and hydrological regimes.

Keywords: Federalism, prior appropriation, riparianism

15.1 INTRODUCTION

Although Canada is considered water-rich, there are variations in water supply among provinces. Alberta experiences significant water shortages while Quebec has relatively abundant water resources. This is reflected in the historical evolution of water law in the two provinces. Water law in Alberta has adapted to water scarcity by adopting prior appropriation, while riparianism in Quebec has endured in the absence of acute competition for limited water supply. At present, Alberta's comprehensive statutory framework for water allocation establishes detailed provisions governing irrigation and other agricultural uses. By comparison, water allocation in Quebec still relies on common law water rights and remains subsumed in the general framework for environmental protection, attesting to the prevalence of water quality issues, especially in relation to agriculture.

This chapter provides a comparative analysis of the legal frameworks for water allocation in these two provinces to show how important differences in hydrological regimes impact the normative governance of agricultural water uses. The comparison relies on three axes: a presentation of hydrological regimes and agricultural water uses (Section 15.2); a summary of the institutional framework from a historical perspective (Section 15.3); and an analysis of the current allocation regime with an emphasis on agricultural abstractions (Section 15.4).

15.2 BACKGROUND

Canada is often ranked among the countries with the most abundant freshwater supplies (McKitrick *et al.*, 2018). Such rankings are deceptive. The country's vast territory, spanning temperate latitudes up to the Arctic over different climatic and geographical regions, allows for stark variations in water supply. The concentration of Canada's population near its southern border coupled with predominantly northward water flows compound freshwater's uneven availability. These factors result in profound disparities between hydrological regimes and agricultural water uses in Alberta and Quebec.

15.2.1 Hydrological regimes

Alberta holds about 2.2% of Canada's freshwater (Government of Alberta, 2010). It lies in the rain shadow of the Rocky Mountains, and its prairies form part of the driest area of southern Canada. The province has a semi-arid climate, with precipitations that can average less than 50 mm per year in the southeast. More than 80% of Alberta's water supply is found in the northern part of the province, while 80% of the demand lies in the south.

Alberta has seven major watersheds that cover almost all its territory, that straddle provincial or international borders, and that flow northeast from the Rocky Mountains beyond Alberta's borders. Among these watersheds, the North and South Saskatchewan River basins that cover about 200 000 km^2 and most of the southern half of the province, including the major urban centers, have a combined mean annual discharge of 16 440 000 000 m^3. Lakes are scarce in the south, although more abundant in Alberta's northern half.

Climate warming will affect glaciers, snowpacks, and evaporation in Alberta, three factors crucial to the province's hydrologic regime (Schindler & Donahue, 2006). Extreme weather events – floods in particular – are expected to combine with cyclic drought and increasing human activity to cause a crisis in water quantity and quality with far-reaching implications.

In comparison, Quebec has relatively abundant freshwater resources, particularly in the densely populated south of the province subject to a temperate continental climate, where precipitations average 1000 mm per year with rainfall accounting for almost 75% of this total. In total, 22% of Quebec's territory is covered by surface waters of which 36% flow within about 40 significant watersheds towards the Saint-Lawrence River and its estuary (MELCC, 2014). Natural hydrological variability is dampened by extensive groundwater recharge due to snowmelt and spring discharge sustaining low water levels in the summer. Climate change is expected to reduce total precipitations and snowmelt recharge in addition to making rainfall less regular. The resulting heightened variability will be compounded by increased extreme weather events causing more floods and local water shortages.

15.2.2 Agricultural water uses

The difference between hydrological regimes in Alberta and Quebec is most starkly illustrated in relation to agriculture by the relative importance of irrigation in the two provinces, which stand at opposite ends of the Canadian spectrum.

In Alberta, agriculture averages around 60 to 65% of all water consumption, principally sourced from surface waters. In 2018, Alberta accounted for 66% of water used for irrigation in Canada, and the 625 000 hectares of irrigated land in that province correspond to around 70% of the land receiving irrigation in the entire country (Ansieta & Marzook, 2021). Irrigation to supplement natural precipitation for consistent crop yields amounts to 95% of total water allocations to the agricultural sector, the rest being used for livestock production. Sprinkler irrigation is the predominant method in the province.

Irrigation occurs mostly through 13 irrigation districts that cover 525 000 hectares in the southern half of the province (Alberta WaterPortal, 2021). The districts rely on extensive distribution networks composed of 8000 km of conveyance works, 52 storage reservoirs, and 4900 km of drainage canals returning surplus water to sources. Apart from supplying agriculture, irrigation districts also deliver water for domestic use to around 50 municipalities with a total population of over 50 000 residents. Most of the water used by irrigation districts comes from rivers in the South Saskatchewan River Basin including the St. Mary, Bow, Oldman rivers and their major tributaries, with irrigation comprising almost 73% of all water allocated across the entire basin. A smaller supply comes from the Milk River, a tributary of the Missouri–Mississippi river system.

In comparison, agriculture in Quebec is generally rain-fed and localized in the most densely populated corridor around the Saint-Lawrence River where farmland covers 50% of the territory. Only 5% of irrigated farmland in Canada is located in Quebec, with micro-irrigation being used 52% of the time, mostly for fruit crops and small-scale vegetable production and conditioning, including for protection against frost and heat (Statistics Canada, 2014). Agriculture mostly affects water quality through point-source effluents from livestock production and diffuse pollution caused by organic fertilizer and pesticide runoff as well as increased turbidity and sedimentation due to land use changes, encroachment on riparian areas, and extensive drainage operations. With respect to water quantity, agriculture amounts to less than 10% of total withdrawals in the province, with municipal and industrial abstractions accounting respectively for 59% and 27% of that share, but 90% of irrigation water and 80% of livestock watering is consumed without return flows (MELCC, 2020). The impact of water abstraction for farming on aquatic ecosystems is compounded by its timing, as half of the total withdrawal for agriculture occurs during summer low flows.

15.3 OVERARCHING INSTITUTIONAL FRAMEWORK

The management of agricultural water uses is based on broader legal frameworks that reflect fundamental principles of Canadian law and their respective historical evolutions in Alberta and Quebec under different hydrological conditions (Figure 15.1). The Canadian constitutional set-up establishes overarching constraints uninformed by hydrological science or principles for water management (15.3.1), while the private law of land ownership provides the fundamental building blocks for water rights in two different legal traditions: Alberta departed early from the English common law of riparian rights to adapt to water shortages (15.3.2); ancient-régime French law transplanted in Quebec gradually evolved in a riparian regime influenced by relative water abundance (15.3.2).

15.3.1 Constitutional principles

Canada's constitutional set-up affects the legal framework for agricultural water management. Firstly, constitutional primacy requires that laws and regulations be consistent with the Constitution to have any effect (*Constitution Act, 1982*, s.52). Secondly, the Constitution grants proprietary rights over natural resources, including internal waters, to the provinces (*Constitution Act, 1867*, s.109, 117; Gibson, 1969). As a result, Alberta and Quebec have a right-based power of water management, within their provincial boundaries, that is limited by the allocation and exercise of legislative powers (Newman, 2013). Thirdly, the *Constitution Act, 1867*, apportions legislative powers to the federal and provincial governments (see s.91, 92). Both orders of government have the constitutional authority to regulate matters pertaining to agriculture and water management.

Agriculture is subject to concurrent legislative powers, although the constitutional text explicitly grants paramountcy to federal law over provincial legislation when both are incompatible (*Constitution Act, 1867*, s.95). Irrigation may fall within this concurrent legislative power (La Forest, 1973). Yet, it is not fully determinative for agricultural water uses, as its traditionally narrow interpretation concerning the 'encouragement or support of agriculture' may only apply to activities behind the farm gates (Newman, 2013).

Water itself is not enumerated by the Constitution among the heads of legislative powers. Jurisdiction validating provincial or federal law relates to various provincial and federal powers listed in the constitutional text. Laws regarding the conservation, impoundment, supply, and distribution of water for various purposes as well as effluent discharge, generally lie within the provincial jurisdiction

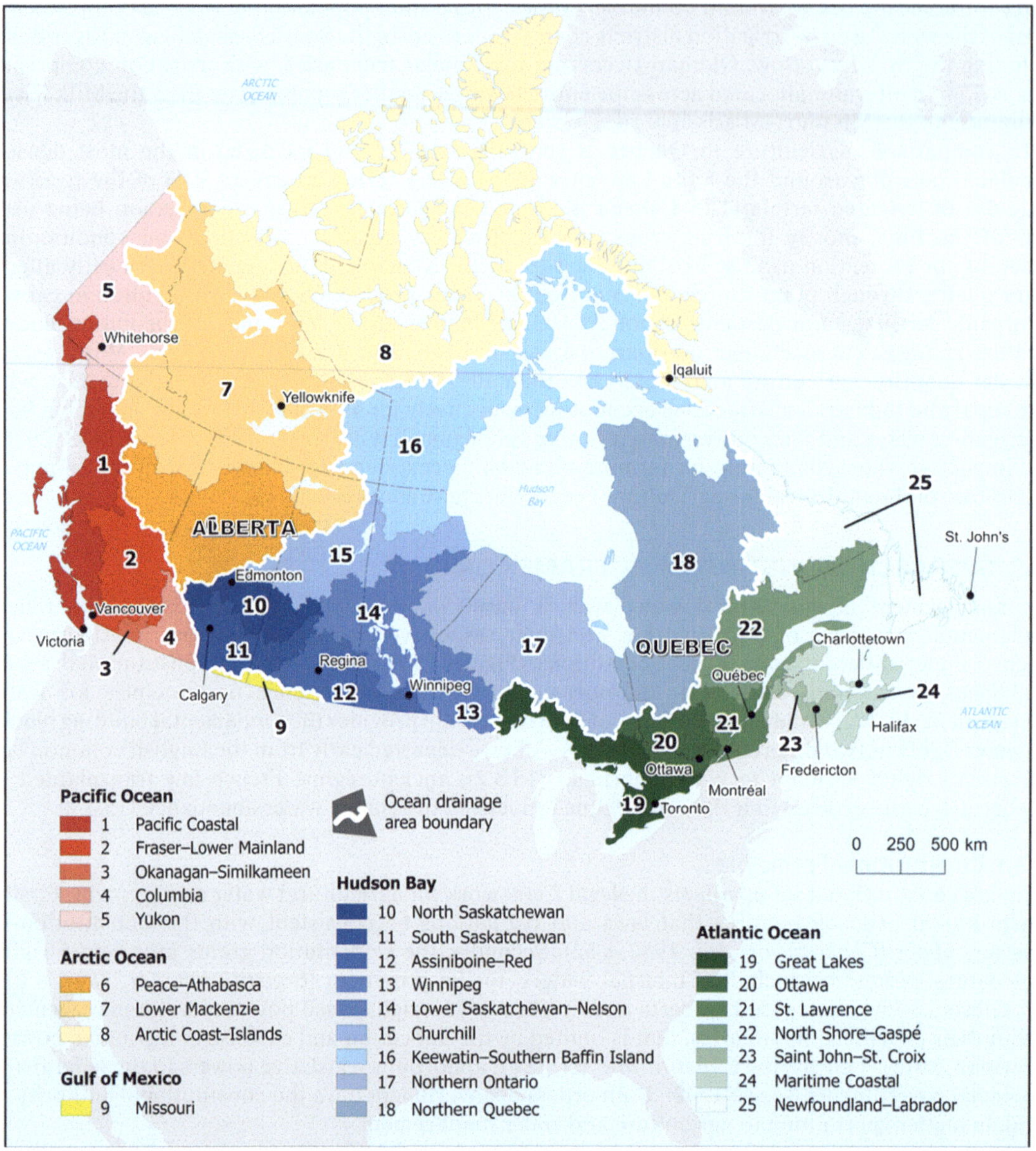

Figure 15.1 Map of drainage basins and provinces in Canada. Source: StatsCan.

based principally on the powers to legislate, respecting property and civil rights, the management and sale of public lands, as well as local matters and local works or undertakings (*Constitution Act, 1867*, s.92(5), (10), (13), (16)). Nevertheless, the federal parliament has extensive legislative jurisdiction over aspects of water management that fall within numerous heads of power, including navigation and shipping, trade and commerce, fisheries, aboriginal lands and rights, criminal law, extra-provincial works and undertakings, international treaties, and a residual general power (*Constitution Act, 1867*, s.91(2), (10), (12), (24), (27), (29)).

This constitutional framework imposes constraints on water management (Alhéritière, 1976). The federal regime fosters legal fragmentation (Saunders & Wenig, 2007). The broadest right-based and legislative powers over water accrue to the provinces only to the extent of their respective territories, which do not correspond to river basin divides. Legislative powers are allocated on a functional basis reflecting a sectoral outlook at odds with integrated management. The federal–provincial division of legislative powers is critical in this respect. Tensions generated by the exclusivity of respective powers, the lack of express jurisdiction over water in the constitutional text, and the inevitable overlaps among operational regulations from both government levels, are mediated through judicial processes involving delays and uncertainties (Kennett, 1991). The application of the doctrine of federal paramountcy developed by the courts can render longstanding provincial regulation inoperative when courts identify conflicts with federal law. These legal factors are compounded by hydrogeomorphology in Alberta and Quebec, where the most important river basins straddle provincial boundaries, thus falling within federal jurisdiction over international and interprovincial rivers (Saunders, 1988).

15.3.2 Prior allocation in Alberta

The riparian rights regime was initially received in Canadian colonial law as part of the existing body of English common law (Burchill, 1948; Getzler, 2004). It can be described as follows:

> *Each riparian owner is entitled to receive the flow of water to his property undiminished in quantity and unimpaired in quality, subject to the right of upstream riparians to abstract sufficient water for their domestic purposes. In addition, upstream riparians are entitled to use water for other than domestic purposes, provided that they do not diminish perceptibly the flow of the stream and thereby interfere with the rights of other riparians. In this way limited upstream development of water for uses such as mills or irrigation can be permitted, but so long as the consumption involved does not cause a noticeable reduction in the downstream flow. Although on the face of it this rule sets a very strict standard which would permit only minor upstream abstractions, in practice it has been mitigated by the application of the principle de minimis non curat lex. All uses of water, whether for domestic or other purposes, are restricted to the riparian tenement and cannot readily be transferred to other land even with the consent of the riparian owner* (Percy, 1977).

Towards the end of the 19th century, major settlement in Alberta followed by lengthy drought in the southern prairies exposed major deficiencies in the riparian regime, potentially hindering the development of this arid region (Percy, 1988). Firstly, land development distant from adequate sources of water was inhibited, as riparianism prevented riparian owners from transferring their water entitlement to non-riparian landowners. With respect to agriculture, this encouraged early farmer settlement on riparian lands that would be fenced-off to prohibit access to range cattle affecting stream flow quantity and quality. Secondly, riparianism impeded major consumptive uses such as large-scale irrigation projects because resultant water flow reductions would illegally interfere with downstream riparian rights. Thirdly, projects requiring secure water supply were complicated by uncertainty in the quantity of water riparian landowners were entitled to. Fourthly, equal claims to the resource among riparian owners meant no prioritization rule could allocate water to more important uses, invariably leaving downstream users to suffer shortages.

These defects, coupled with growing demand for irrigation at a time when dryland farming was still nascent, resulted in a movement to reform riparian common law, culminating in 1894–1895 with the adoption of a statute declaring that the property in, and rights to the use of, all water was vested in the State. That statutory regime was primarily designed for irrigation, but it regulated all water appropriations and remained the basis for water use legislation in Alberta for around a century (Percy, 1996). It authorized water withdrawals and diversions for specific quantities through licenses without maximum durations, it removed common law restrictions on water transfers including to non-riparian owners, and it established a rule for prioritizing water users in ascending order of seniority during shortages based on the principle of 'FITFIR' (first in time, first in right). The resulting prior-allocation regime was inspired by, and indeed very similar to, the American prior-appropriation doctrine. The two systems' most significant differences hinged on the State's ownership of the resources in Alberta, and on seniority in that province being pinned to the moment when license is granted rather than to the date on which water is put to beneficial use (Percy, 1988).

This statutory regime underwent numerous amendments in the following decades (Percy, 1996). Most notably, an order of prioritization was added in 1920 that gave highest priority to water uses for domestic purposes, followed by uses for municipal, industrial, irrigation and other purposes, thus allowing a person to obtain a water license for a preferred use through the canceling of another use for a lower purpose. In 1962, groundwater was brought under the statutory regime to attempt to reduce uncertainties caused by the regime's interaction with vestigial riparian rights based on the rule of capture (Percy, 1987). In 1975, a change in the order of preferential uses gave priority to irrigation over industrial purposes. Still, defects in Alberta's prior-allocation regime emerged over the years despite the amendments: barriers to new water users, entrenchment of consumption patterns, obstacles to rationalizing allocations between sectors, and disincentives to water conservations related to the lack of volumetric pricing for water abstractions (Percy, 1986).

In this context, Alberta's government launched a broad reflection based on wide-ranging consultations that led to a fundamental reform of the long-standing statutory regime for water allocation and transfers with the adoption of the *Water Act* in 1996. The new statutory regime that came into force in 1999 currently constitutes the general framework for agricultural water uses and irrigation described in Section 15.4.1.

15.3.3 Riparianism in Quebec

Originally, the common law regime for water allocation in Quebec was similar to the riparian regime inherited from English law in Alberta (*Miner v Gilmour*, 1858; Schorr, 2015). However, the legal roots of riparianism in Quebec differed, as the doctrine was transplanted from customary law codifying royal edicts in pre-revolutionary France before the English conquered New France in 1760 (Brun, 1969).

This piecemeal, incomplete, and ambiguous regime for water allocation was clarified to some extent by the Seigniorial Court at the abolition of the seigneurial regime in Quebec (Lelièvre & Angers, 1856). In principle, riparian owners were entitled to the 'ordinary' use of water for domestic and agricultural purposes on their lands (excluding major abstractions for sawmills, factories and manufactures), as long as other riparian owners could exert the same rights. This implied that riparian users were bound to leave downstream water flow apparently unaltered, although some degradation in downstream water quality was allowed (Brière, 1975). As was the case in Alberta, some of this regime's features hampered development: restrictions on transfers; uncertainties regarding water quantities subject to riparian rights; an absence of prioritization between types of uses; and hazy corelative interactions between riparian rights.

These constraints on development were removed in 1856 when the *Act to authorize the exploitation of watercourses* was passed to foster 'economic prosperity'. The *Act* granted riparian owners the power to use and exploit watercourses abutting their properties by any undertaking, including canals, dams, dikes, and levees, required to operate mills, factories, manufactures or other machines of any kind. Nuisances or damages caused to other riparian rights, including the flooding of neighboring properties, required compensation, but relief stopping the injurious use of water would only be granted in the absence of compensation. That the entitlement to the 'extraordinary' commercial use of water conferred by the *Act* remained subject to the requirement that riparian users leave downstream water flow unaltered for corelative users, speaks to the relative abundance of water in Quebec.

At the time, an almost complete lack of administrative control on water uses in line with a liberalist attitude further encouraged the development of water resources in the province. Absent water shortages, the numerous ambiguities marring this modified regime – including the riparian rights' qualification at law, the legal status of who held them, the possibility of their transfer to non-riparian users, and the type of purposes for which they could be used given unresolved interactions between the *Act* and common law rights – did not preclude the regime's perpetuation relatively unchanged for more than a century (Lord, 1977). With respect to agriculture, a significant uncertainty pertained to whether irrigation or the use of water to raise crops or cattle was covered by the extended rights for commercial purposes conferred by the *Act*.

Since then, Quebec's riparian regime has been codified in the *Civil Code of Quebec* (see s. 979–982). The general principles structuring this framework continue to reflect the province's relative water abundance. Lower land is subject to receiving water flowing onto it naturally from higher land, a rule that lets the upstream agrarian owner carry out drainage work to facilitate runoff if it does not aggravate the situation of the downstream owner. A riparian owner may, for his needs, make use of surface waters and leave them without substantial change in quality or quantity to allow other riparian owners to exercise the same rights. The common law rights thus provided by the Civil Code are extended by rights to the 'extraordinary' use of water carried over from the *Act to authorize the exploitation of watercourses* in the *Watercourses Act* (Giroux *et al.*, 1997). The current version of these statutory rights now renders possible the expropriation of riparian lands to enable water uses for hydropower generation or water-works systems destined for domestic or industrial purposes. Significantly, the rights to hydraulic forces are essentially reserved by the state and leased to a public corporation providing more than 80% of the electricity generated in Quebec. These rights constitute the basic legal framework for the use of water, now overlain by a detailed administrative regime controlling water withdrawals, including for agricultural purposes as described in Section 15.4.2.

15.4 ADMINISTRATIVE FRAMEWORKS FOR AGRICULTURAL WATER USES

The water allocation regimes that evolved through long historical processes in Alberta and Quebec, have both been thoroughly reformed in recent decades. At present, Alberta's comprehensive statutory framework for water allocation establishes detailed provisions governing irrigation and other agricultural uses (15.4.1). By comparison, water allocation in Quebec still relies on common law water rights and remains subsumed in the general framework for environmental protection, attesting to the prevalence of water quality issues, especially in relation to agriculture (15.4.2).

15.4.1 Alberta

Under the *Water Act* as under predecessor legislation, the property in and the right to the use of water is generally vested in the province. Notwithstanding the common law, an extensive statutory regime relies on this basis to establish a complex water allocation framework abundantly supplemented by administrative directives to address issues related to water shortage between competing uses at the watershed level.

Licenses and registrations are the principal entitlements under the *Water Act*. The license is the primary mechanism to authorize the diversion of a specified volume of water for stipulated purposes prescribed in the *Water (Ministerial) Regulation*, including municipal, agricultural, irrigation and commercial. Notably, licenses may be issued to implement a conservation objective retaining reserved water to ensure streamflow rates or water levels requirements for the protection of a natural water body or its aquatic environment. The Act also carries on as 'deemed licenses' various permits to divert water issued under predecessor legislation. Registrations have a much narrower scope and cover persons on riparian lands who diverted up to 6250 m^3/year of water for the purpose of raising animals or applying pesticides to crops before 1999 and duly registered this diversion prior to 2002. In addition to these entitlements, a catch-all approval process twinned with the *Environmental Protection and Enhancement Act*, generally targets works and activities such as drainage, flood or erosion control, and channel realignment or even water well drilling, that may alter the flow, level, direction, or location of water, whether temporarily or permanently, but these approvals generally do not themselves provide the right to divert or use water for a specific purpose. Some withdrawals are exempted from the *Water Act*'s allocation regime, including diversions listed in the regulations from works such as agricultural dugouts with a capacity under 2500 m^3. Traditional agriculture users eligible to, but without, a registration can continue to divert water. Also, household users on riparian lands may commence and continue the diversion of 1250 m^3/year for the purposes of human consumption, sanitation, fire prevention and watering animals, gardens, lawns and trees without a license or registration.

The management of licenses under the *Water Act* is subject to discretionary power governed by a detailed normative framework. Either directly or further to a preliminary certificate, the regulatory authority may decide to accept or refuse to issue a license considering a number of factors: the diversion's potential or cumulative effects on the aquatic environment, its hydraulic, hydrological and hydrogeological effects, and its effects on household users; other licensees and traditional agriculture users, including water conservation objectives; land suitability for irrigated agriculture and applicable water management plans. If the regulator believes that no further allocation of water should be made in a water management area or another geographical area, applications for licenses may be refused for a specified period. When the regulator decides to issue a preliminary certificate or a license, the water diversion may be subject to appropriate terms and conditions considering the same factors. For example, the volume of water authorized under a license may be affected by transboundary apportionment obligations to Saskatchewan under Schedule A of the intergovernmental *1969 Master Agreement on Apportionment and By-Laws, Rules and Procedures*. By contrast, registrations must be effected without discretionary power when the conditions and regulatory requirements are met. This major difference between registrations and licenses remains at later stages in the legal framework. Whereas registrations have no expiry date and may only be canceled or suspended if there is an emergency or if it is necessary for public safety, licenses must be assigned an expiry date considering an array of factors (25 years by default for agricultural, irrigation or water conservation) and may not be renewed on specified grounds including: public interest; inconsistency with an approved water management plan; meeting a water conservation objective, such as avoiding significant adverse effects on the aquatic environment.

The *Water Act* establishes a priority scheme that reflects the FITFIR principle. Under this scheme, senior entitlement holders can divert their whole specified allocation of water before junior license or registration holders can divert any water during shortages. As a result, junior users bear the risk of flow reduction compared to senior users with secure access to water. In general, licenses are prioritized in consecutive order corresponding to the dates when complete applications for the licenses were received. Deemed licenses retain their original priority. Registrations are assigned priorities that correspond to the first known date of diversion of water, but no earlier than 1894. Exempted household users have priority over licensed and registered diversions but no priority among themselves. Traditional agriculture users without registration have no priority.

In general, water entitlements are location-specific and issued in relation to appurtenant lands. Under the current framework, there are three methods to acquire water rights from an existing rights holder (Bankes, 2006). Firstly, there remains the traditional means of acquiring water rights by purchasing the land to which an existing entitlement is appurtenant. Secondly, an agreement to 'temporarily' assign water under the *Water Act* allows licensees or registered traditional agriculture users to transfer their water diversion to other entitled users for a limited period. In principle, temporary assignments may be used to avert a crisis (e.g., getting water to a crop within hours or days) or might be used on a seasonal basis when snowpack assessments suggest that flows will be inadequate to meet all summer irrigation needs. Temporary assignments are allowed when the transfer is possible without new works to effect them, and when they cause no adverse effects on senior rights, household users, or the aquatic environment. Thirdly, the *Water Act* allows transfers through formal arrangements subject to governmental review and approval by which all or part of licensed diversions may be acquired, either permanently or for a limited amount of time. Transfers require consent from the licensed user and may serve to change the appurtenance of a water license, even if the user remains the same. Transfers are allowed from one waterbody to another, but they must take place in accordance with an approved water management plan for the relevant area. The *Water Act* lists the factors considered by the regulator in deciding whether to approve a transfer, including factors stipulated in an approved water management plan such as the transfer's absence of impairment on water quality or the aquatic environment, and the absence of adverse effect on water apportionment or public health and safety.

Provisions respecting water management plans under the *Water Act* establish a detailed framework for the spatial management of water allocation. The government may approve a water management plan to adopt an integrated approach with respect to water, land and other resources in cooperation and consultation with the public, local authorities, governmental agencies and other jurisdictions. One water management plan approved in August 2006 for the South Saskatchewan River Basin prohibits the issuance of most new licenses or registrations and restricts transfers due to concerns over water shortages. In accordance with the *Framework for water management planning* enabled by the *Water Act*, a more comprehensive water management plan was approved in 2014 for the Battle River Basin that sets a volumetric water allocation limit beyond which new licenses cannot be issued, that enables water transfers subject to a conservation holdback reserving 10% of the transferred diversions to the regulator, and that establishes a conservation objective at 85% of the river's natural flow. Further details in this spatial management framework can materialize through the regulator's discretion to reserve water for conservation and management under the *Water Act*, which resulted in the issuance of the *Oldman River Basin Water Allocation Order*, whereby 11 000 acre-feet of water are allocated within the Oldman River Reservoir Area for various purposes, such as municipal, commercial, and agricultural (other than irrigation). In addition, Alberta's seven major river basins are recognized under the Act, and water transfers are prohibited between them. Likewise, licenses cannot be issued for transferring water from Alberta outside Canada to promote water conservation and the wise allocation and use of water.

Spatial management under the *Water Act* is further developed and complemented by the *Irrigation District Act* (IDA), which provides the legal framework for Alberta's 13 irrigation districts located in the South Saskatchewan Basin. The purpose of these districts is to convey and deliver water through their irrigation works, and to divert and use quantities of water in accordance with licenses under the *Water Act*. The IDA thus provides derivative water rights called 'irrigation acres' that allow greater allocation flexibility within each district. The irrigation acres rights come in a variety of contractual forms that enable the use of irrigation districts' facilities for irrigation, conveyance, and household purposes. Fees for the use of water under these subsidiary rights are established by district bylaws and may vary depending on the type of subsidiary right, the amount of water consumed and the purpose of water use.

15.4.2 Quebec

In Quebec, the extensive regulation of farming's impacts on water quantity and quality is a recent phenomenon. Historically, agricultural water requirements have been met by rainfall and groundwater resurgence during low flows. Water abstractions for agriculture were an order of magnitude smaller than domestic consumption, itself dwarfed by industrial withdrawals amounting to around 90% of total water consumption in the province. Until the 1970s, statutory and administrative controls over agricultural activities in relation to water mostly pertained to drainage works required to cultivate farmlands. Major efforts to reform the legal framework applicable to water resources did not specifically consider the agricultural sector until the end of the 1990s (Ministère de l'environnement, 2003).

Significant administrative controls over farming's effects on water resources initially materialized through the general framework for environmental protection. The *Act to remedy the pollution of water* passed in 1961 to reduce point source pollution and improve wastewater treatment was replaced in 1972 by the *Environment quality Act* (EQA), a comprehensive framework legislation designed to tackle environmental degradation at large, including agricultural contamination affecting water quality. Regulations taken under the EQA, first in 1981 with the *Regulation respecting the prevention of water pollution in livestock operations* replaced in 1997 with the *Regulation respecting the reduction of pollution from agricultural sources*, subjected the undertaking or enlargement of livestock operations to discretionary authorizations and imposed various constraints on liquid manure storage and spreading. Similar measures for pesticide management were also introduced gradually. Yet, many agricultural activities were still exempt from sector-specific regulation.

The current framework encompassing the most important issues for water quantity and quality in the agricultural sector results from the iterative evolution of a fragmented legal landscape over the last two decades. At the beginning of the 2000s, a wide-ranging initiative to improve water management at the provincial level resulted in the formulation of a water policy that identified several strategic orientations, two of which were particularly important for agriculture. On the qualitative side, the reform would implement stronger remedial measures for point source and diffuse agricultural pollution in watersheds where urban and agricultural uses conflicted. On the quantitative side, the integration of water management based on river basins in southern Quebec would be anchored to a new authorization regime applicable to most water withdrawals.

The current administrative framework for water withdrawal management was put in place further to complementary legislative and regulatory initiatives in 2009, 2014 and 2017. Under the EQA and related regulations, the rights to use water (see Section 15.3.3 above) are generally subject to a preliminary authorization regime with tiered controls that vary according to the withdrawal's expected impacts. In the highest bracket, withdrawals such as dams impounding lakes greater than 0.2 km^2, river diversions, large-scale livestock production, and dikes flooding wetlands and bodies of water over an area greater than 0.1 km^2 for the operation of cranberry farms, are subject to an extensive environmental impact assessment and a governmental authorization process that may include public consultations. In the lower bracket, other water withdrawals are subject to a ministerial authorization process. Among the various exclusions to this regime, some of the most significant include withdrawals under 75 m^3/day unless they provide drinking water, and withdrawals through man-made, naturally-fed irrigation ponds. Agricultural withdrawals are also exempt from volumetric charges under the *Regulation respecting the charges payable for the use of water.*

Under this regime, the power to grant an authorization is discretionary. The authorization can be refused in the public interest to ensure adequate protection of the environment, of the health and welfare of human beings, of other living species, or to prevent adverse effects on property. The decision to authorize a withdrawal must be exercised so as to ensure the protection of water resources considering climate change, and must aim to reconcile the needs of aquatic ecosystems with the needs of agriculture, aquaculture, industry, energy production and other human activities. Authorizations can impose discretionary restrictions on withdrawals to monitor, prevent, limit and remedy environmental damage, including through environmental flow protection, or to prevent, limit or remedy interference with other water rights. Withdrawals subjected to ministerial authorizations must be renewed after 10 years and may be canceled. Such authorizations can be transferred with ministerial approval, but the underlying water rights remain bound to appurtenant lands.

The authorization regime ties into the spatial management of water resources. In application of a transboundary agreement signed by all riparian provinces and U.S. states in the North-American Great Lakes Basin, withdrawals from the St. Lawrence River Basin on Quebec's territory cannot be transferred out of it, subject to restrictive exceptions. At the regional and local levels, planning instruments are developed by watershed bodies representing various stakeholders, including farmers, to foster the conservation and efficient use of water. These plans must be taken into account when a withdrawal project is submitted for authorization. Spatial management for the protection of riparian areas under the EQA through the *Protection Policy for Lakeshores, Riverbanks, Littoral Zones and Floodplains* implemented by municipalities, links water quantity and quality issues with prohibitions, restrictions, and authorizations applicable to withdrawals and agricultural or other polluting activities within various perimeters around surface waters.

With respect to water quality, the *Agricultural operations regulation* (AOR) establishes the main sector-specific provisions. The AOR essentially aims at reducing contaminant runoff into waters bodies from livestock raising and crop fertilization through interrelated measures. Firstly, livestock waste such as fecal matter and urine must be stored in watertight facilities and managed to prevent or minimize runoff and rainwater contamination. Secondly, stored livestock waste must be reclaimed or destroyed in conformity with environmental norms. Reclamation is done by spreading over cultivated parcels solely for fertilization purposes in compliance with a certified agro-environmental plan. Norms and standards of practice applicable to spreading in the AOR and the agro-environmental plan determine maximum waste production levels per livestock production facility, and maximum annual loads in phosphorus, nitrogen and potassium per pre-designated cultivated areas depending on the type of culture and soil. Thirdly, cattle grazing and fertilizer spreading is restricted or prohibited in protective perimeters around watercourses, lakes, drainage ditches and riparian areas. Fourthly, the AOR generally prohibits increasing cultivated areas in degraded watersheds where the environment's carrying capacity for phosphorus has been exceeded. The *Water Withdrawal and Protection Regulation* completes this set of measures by imposing spatial prohibitions or restrictions on livestock production and livestock waste storage or spreading in bacteriological and virological protection zones around vulnerable water withdrawals made for human consumption or food processing purposes.

15.5 CONCLUSION

The legal frameworks for agricultural water use in Alberta and Quebec have evolved in parallel trajectories mostly unaffected by the fragmented constitutional context. From similar albeit distinct legal origins, differences in water availability have resulted in the development of prior allocation in Alberta as opposed to regulated riparianism in Quebec.

In Quebec, the abundance of freshwater resources, the prevalence of rain-fed agriculture, the relatively limited availability of cropland, and the predominance of other industries, have contributed to maintaining a static allocation regime with poorly defined entitlements and without sector-specific provisions for agriculture until legislative reforms initiated 25 years ago. By comparison, relative water scarcity in Alberta is reflected in a detailed water allocation regime that has focused on agriculture, and irrigation more specifically, as an engine for colonial settlement and economic development since the end of the 19th century.

As a result, Alberta's extensive legal framework contains specific features designed to address water shortages absent from Quebec law: a clear prioritization mechanism allocating risks caused by supply variability to more recent users; provisions allowing water transfers that increase flexibility and sever entitlements from appurtenant lands, particularly with respect to irrigation; detailed protection for environmental flows and the needs of aquatic ecosystems, including the possibility to issue licenses or reserve water specifically for conservation purposes. Planning in water-stressed areas of southern Alberta where agriculture is extensive, has resulted in localized water allocation regimes that emphasize the need to live within the carrying capacity of the watershed, and the need to improve the health of the aquatic ecosystem. Quebec law provides no equivalent mechanisms, although water quality issues due to agricultural pollution ultimately limit the expansion of farming in highly degraded river basins, thus reflecting the carrying capacity of watersheds.

Despite these differences, the legal frameworks in the two provinces share similarities. Both have evolved into administrative regimes centered and dependent on discretionary power to attain legislative objectives. Both integrate water allocation issues with watershed management involving stakeholders to different extents. Finally, both confer significant advantages to some agricultural uses. Alberta's *Water Act* recognizes historical agricultural practices, exempts registrations from renewal, and entrenches the most significant licenses in terms of volume and priority that were issued in perpetuity under predecessor legislations, many to the province's irrigation districts.

ACKNOWLEDGEMENT

I acknowledge financial support from Genome Canada and Génome Québec trough the ATRAPP project (Algal Blooms, Treatment, Risk Assessment, Prediction and Prevention through Genomics).

REFERENCES

Alberta WaterPortal. (2021). Why Alberta Irrigation Matters. https://albertawater.com/topics/irrigation/ (accessed 31 December 2021)

Alhéritière D. (1976). La Gestion des Eaux en Droit Constitutionnel Canadien (Water Management in Canadian Constitutional Law). Editeur officiel du Quebec, Quebec.

Ansieta P. and Marzook E. (2021). Agricultural irrigation patterns in Canada from 2012 to 2018. Statistics Canada. https://www150.statcan.gc.ca/n1/pub/16-508-x/16-508-x2021001-eng.htm (accessed 31 December 2021)

Bankes N. (2006). The legal framework for acquiring water entitlements from existing users. *Alberta Law Review*, **44**(2), 323–376.

Brière J. (1975). Les droits de l'État, des riverains et du public dans les eaux publiques de l'État du Québec (The Rights of the State, of the Riparians, and of the Public in the Public Waters of the Quebec State). Ministère des Ressources Naturelles, Quebec.

Brun H. (1969). Histoire du droit québécois de l'eau: 1663–1969 (History of Quebec Water Law: 1663–1969). Commission d'étude des problèmes juridiques de l'eau, Quebec.

Burchill C. S. (1948). The origins of Canadian irrigation law. *Canadian Historical Review*, **29**(4), 353–362, https://doi.org/10.3138/CHR-029-04-02

Civil Code of Québec, CQLR c.CCQ-1991.

Constitution Act. (1982). Schedule B to the Canada Act 1982 (UK), 1982, c.11.

Getzler J. (2004). A History of Water Rights at Common Law. Oxford University Press, Oxford.

Gibson D. (1969). The constitutional context of Canadian water planning. *Alberta Law Review*, **7**, 71–92, https://doi.org/10.29173/alr1930

Giroux L., Duchaîne M., Noreau G.-M. and Vézina J. (1997). Le régime juridique applicable aux ouvrages de retenue des eaux au Québec (The legal framework for water impoundment in Quebec). *Cahiers de Droit*, **38**(1), 3–70, https://doi.org/10.7202/043432ar

Government of Alberta. (2010). Facts about water in Alberta. https://open.alberta.ca/dataset/1832cd36-bbeb-4997-ae81-67d3eedfcfe5/resource/18a9d64b-bad8-413a-8c63-77a548ec9d88/download/4888138-2010-facts-about-water-in-alberta-2010-12.pdf (accessed 31 December 2021)

Kennett S. A. (1991). Managing Interjurisdictional Waters in Canada. Canadian Institute of Resources Law, Calgary.

La Forest G. V. (1973). Water Law in Canada – The Atlantic Provinces. Information Canada, Ottawa.

Lelièvre and Angers (eds.). (1856). Questions seigneuriales (Segniorial questions). Côté, Montreal. https://llmc.com/searchResultVolumes2.aspx?ext=true&catalogSet=81086 (accessed 31 December 2021)

Lord G. (1977). Le Droit Québécois de L'eau (Water Law in Quebec). Éditeur officiel du Québec, Quebec.

McKitrick R., Aliakbari E. and Stedman A. (2018). Evaluating the State of Fresh Water in Canada, Fraser Institute. Available at: https://www.fraserinstitute.org/sites/default/files/evaluating-the-state-of-fresh-water-in-canada.pdf (accessed 31 December 2021)

MELCC. (2014). Rapport sur l'état de l'eau et des écosystèmes aquatiques au Québec (Report on the status of water and aquatic ecosystems in Quebec). Ministère de l'Environnement et de la Lutte contre les changements climatiques. https://www.environnement.gouv.qc.ca/eau/rapport-eau/rapport-eau-2014.pdf (accessed 31 December 2021)

MELCC. (2020). Rapport sur l'état des ressources en eau et des écosystèmes aquatiques du Québec (Report on the status of water and aquatic ecosystems in Quebec). Ministère de l'Environnement et de la Lutte contre les changements climatiques. https://www.environnement.gouv.qc.ca/eau/rapport-eau/rapport-eau-2020.pdf (accessed 31 December 2021)

Miner v Gilmour. (1858). E.R. 14, 861.

Ministère de l'environnement. (2003). Synthèse des informations environnementales disponibles en matière agricole au Québec. Direction des politiques du secteur agricole, ministère de l'Environnement, Québec, Envirodoq ENV/2003/0025. Available at: https://www.environnement.gouv.qc.ca/milieu_agri/agricole/synthese-info/synthese-info-enviro-agricole.pdf (accessed 31 December 2021)

Newman D. (2013). Natural Resources Jurisdiction in Canada. LexisNexis, Markham (ON), Canada.

Percy D. (1977). Water rights in Alberta. *Alberta Law Review*, **25**, 142–165.

Percy D. (1986) Water rights law and water shortages in Western Canada. *Canadian Water Resources Journal*, **11**(2), 14–23, https://doi.org/10.4296/cwrj1102014

Percy D. (1987). The Regulation of Ground Water in Alberta. Environmental Law Centre, Edmonton.

Percy D. (1988). The Framework of Water Rights Legislation in Canada. Canadian Institute of Resources Law, Calgary.

Percy D. (1996). Seventy-five years of Alberta water law: maturity, demise & rebirth. *Alberta Law Review*, **35**(1), 221–241, https://doi.org/10.29173/alr1069

Regulation Respecting the Reduction of Pollution from Agricultural Sources. (1997). G.O.Q.2:2607.

Saunders O. J. (1988). Interjurisdictional Issues in Canadian Water Management. Canadian Institute of Resources Law, Calgary.

Saunders O. J. and Wenig M. M. (2007). Whose water? Canadian water management and the challenges of jurisdictional fragmentation. In: Eau Canada: the Future if Canada's Water, K. Bakker (ed.), UBC Press, Vancouver, pp. 119–141.

Schindler D. W. and Donahue W. F. (2006). An impending water crisis in Canada's western prairie provinces. *Proceedings of the National Academy of Sciences*, **103**(19), 7210–7216, https://doi.org/10.1073/pnas.0601568103

Schorr D. (2015). Riparian rights in lower Canada and Canada east: inter-imperial legal influences. In: Imperial Co-Operation and Transfer, 1870–1930: Empires and Encounters, R. Cvetkovski and V. Barth (eds.), Bloomsbury, New York, pp. 107–125.

Statistics Canada. (2014). Irrigation methods and conservation practices on Canadian farms. https://www150.statcan.gc.ca/n1/pub/16-508-x/16-508-x2016001-eng.pdf (accessed 31 December 2021)

doi: 10.2166/9781789062786_0203

Chapter 16

Idaho's Eastern Snake Plain Aquifer: cooperative water policy change for Idaho's groundwater farmers

Katrina Running

Idaho State University, Pocatello, US

ABSTRACT

In 2015, the state of Idaho finalized a cooperative five-year agreement between ground- and surface-water farmers in Idaho's Eastern Snake Plain Aquifer (ESPA). The agreement, which was the result of multi-decadal legal, regulatory, and policy disputes, stipulated a reduction in groundwater withdrawals for the purpose of replenishing the overdrawn aquifer. The state was constrained by its legacy water allocation doctrine called prior appropriation – a framework for water law informally referred to as 'first-in-time, first-in-right' that prioritizes seniority when water rights are contested – becoming particularly fraught with the social-ecological balance shift over time due to in-migration, multi-year drought, and more, leading to a mismatch between supply and demand. The two primary components of the five-year water reduction agreement used a tiered system of usage reductions from 4–20%, with an average of about 13% per groundwater farmer. The requirement also required all farmers to put in water meters on all irrigable land at their own expense. These activities were administered by managers in the eight groundwater districts impacted by the agreement and were reacted to by individual farmers with everything from reluctant acceptance to threats of lawsuits. Impact-wise, our research found farmers' net yields and farm income decreased by around 8% during the first two years of the agreement, but also that by 2020, when the agreement ended, the state's aquifer level measurements actually *surpassed* the initial goal of adding 2 million acre-feet (~2.47 trillion m^3) to the aquifer by about 200 000 acre-feet (~246 696 000 m^3).

Keywords: Aquifer recharge, cooperative water management, drought adaptation, groundwater policy, western United States

16.1 INTRODUCTION

16.1.1 Idaho's water law and background

The state of Idaho, located in the 'mountain west' region of the United States, is experiencing a growing gap between its available supply and demand for water due to a warming climate and changing precipitation and evapotranspiration trends and one of the highest state population growth rates in the U.S. (WPR, 2021). These pressures, combined with increasingly overdrawn groundwater sources, have put a strain on water managers and the state's citizens who need water for a variety of activities including agricultural production and recreational tourism, two of the state's largest and most lucrative industries (du Bray *et al.*, 2018; Elliott *et al.*, 2014; Niles & Hammond-Wagner, 2019). Currently, Idaho's Eastern Snake Plain Aquifer (ESPA) region, which spans most of southern Idaho, produces a sizeable proportion of the potatoes, sugar beets, and wheat grown in the United States, while Idaho's tourism industry brings in about $3.7 billion ($10^9$) annually (UIE, 2016; USDA, 2014). The prominence of these two sectors may be why, as of 2015, Idaho was identified by the United States Geological Survey (USGS) as having the highest per capita statewide water use in the U.S. (Dieter *et al.*, 2017).

Conflict over water and the state's involvement in managing and distributing water existed in Idaho even before it officially became a state. As settlers moved into the area in the late 1800s, among the first rules of any kind established were those relating to natural resource usage and ownership, especially with respect to valuable natural resources such as land, water, ore, and other precious minerals. Initially, the legal framework governing water was based on 'common law', which meant rules derived from English legal doctrine on any issue that had not yet been specifically codified by U.S. federal law (Getzler, 2009). In general, those common law paradigms included strong property rights, including 'riparian rights' for landowners along river corridors. That precedent was quickly rendered insufficient, however, because everyone in the arid west required water to live and work. Moreover, the federal government was pushing development and those moving out west to homestead wanted some assurance that their enormous efforts in tilling previously unplanted land to prepare it for productive farming would benefit themselves and their families (Lenczowski, 1990).

In order to develop a system that was workable and caused minimal barriers to development, stakeholders in the area developed a water allocation system based on the idea of 'first possession' – an established legal concept that was considered a reasonable framework for governing water in the context of the federal government's 1862 Homestead Act designed to incentivize settlers to move west (Harrington, 2012). This policy is typically referred to as 'first in time is first in right' or 'prior appropriation' and has been the foundation of Idaho's water law since. Another key aspect of Idaho's original water management policy was the idea of 'beneficial use' – a concept that became officially enshrined into the state's water management in 1976 when the state adopted an official plan that guaranteed the state sovereignty over its own water resources for 'the benefits of its citizens' (IWRB, 2012).

As the state continued to develop through the 20th century, Idaho continued to build infrastructure to support agriculture including dams and canals to store and move water, and railroads and highways to transport goods. When the technology for groundwater extraction became widespread and affordable, irrigation options expanded from diverted surface water to on-demand groundwater, which was easier to use and more efficient overall as farmers could use every drop they pumped.

Unfortunately, however, this had the unintentional effect of reducing the state's largest 'water bank' (the ESPA) in two ways. One, it reduced the amount of 'incidental recharge' – water that percolated down into the aquifer as a result of flood irrigation practices and the transport of water through unlined canals. And two, it made direct withdrawal from the aquifer possible for the first time. Combined, this meant that not only was less water going in, but far more was coming out, and much more quickly. By the 1980s, state water officials were beginning to notice this decline from their (at that time limited) aquifer monitoring. Since then, steadily increasing temperatures and the continued development of groundwater extraction in the ESPA region led to further losses – an average of about 200 000 acre-feet (AF) per year (or approximately 246 696 000 m^3) from 1952–2015. Figure 16.1 displays the aquifer's storage and spring discharge trends at the Thousand Springs site – a major monitoring point where the ESPA bubbles up to the surface and flows out of cliffs along the river's banks into the Snake River – from 1912–2014.

In the decades that followed, more and more water-related disputes arose, especially between surface water users with the oldest water rights and groundwater users with more junior rights (Slaughter, 2004). The state had to grapple with how to deliver these senior rights holders' full allocations under the 'first in time is first in right' framework without damaging the farming operations relying on

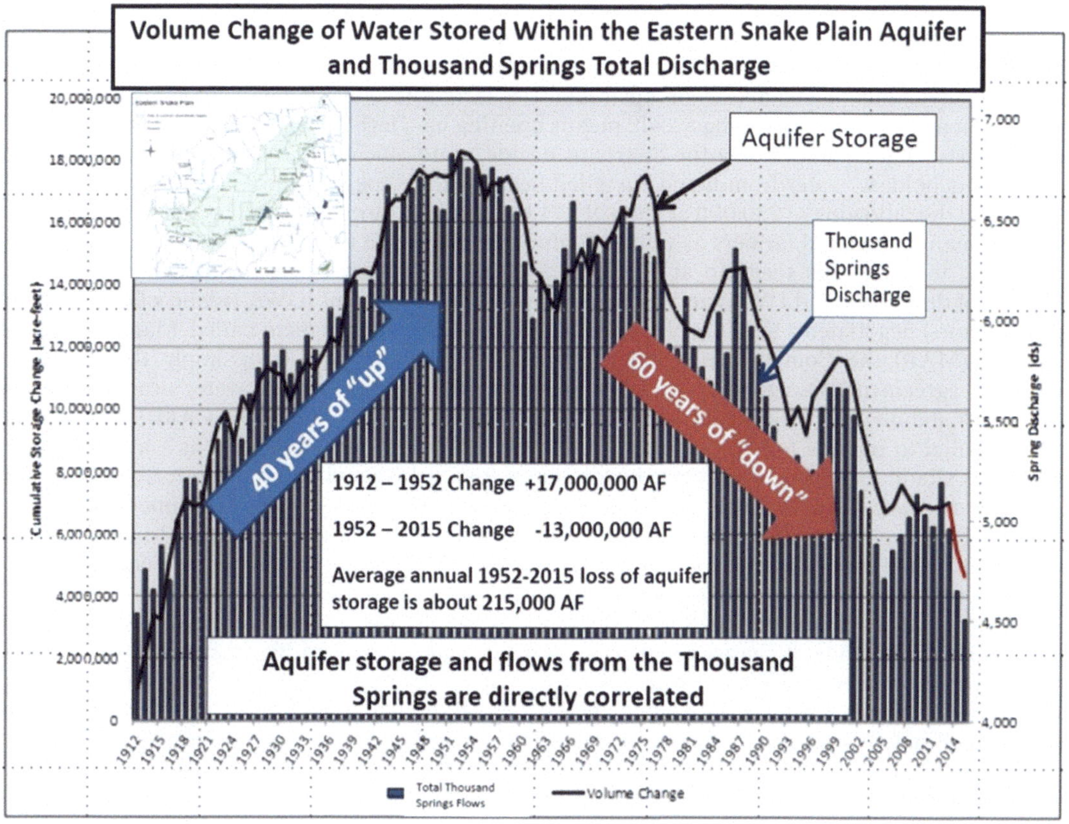

Figure 16.1 Eastern Snake Plain Aquifer level changes, 1912–2014. Source: Idaho Department of Water Resources.

pumped groundwater. In 1994, Idaho made its first big legislative change to how water would be managed towards 'conjunctive management', which meant that surface and groundwater resources were assumed to be hydrologically connected and should thus be managed together. Prior to that, ground- and surface water were considered separate. Groundwater users were able to pump as much water as their water rights allowed without the potential for legal action requiring usage reductions from senior surface rights holders experiencing shortages. Two decades later, in 2014, the state would conduct a full 'adjudication' of all surface water and groundwater rights in the Snake River Basin (Strong & Orr, 2016).

Both the decision to adopt conjunctive management and the adjudication process were controversial among Idaho's farmers. Many were angry about the ongoing legal expenses of frequent changes to water policy. They were also skeptical of the whole ESPA – a fractured and porous basalt aquifer that stretches from Idaho's eastern border with Wyoming to its western one with Oregon and Washington – really being so connected that pumping groundwater actually impacted surface water flows. However, conjunctive management is still in effect, and Idaho's water resources remain monitored and enforced at the state level. The Idaho Department of Water Resources (IDWR) is the main agency in charge of overseeing water management in the state, in coordination with the Idaho Water Resource Board (IWRB), an eight-person board established by the state legislature in 1965 to lead water planning and the financing of water projects in collaboration with the irrigation districts.

Most recently, in the summer of 2015 surface water users ran out of water in the later part of the growing season due to an unusually hot, dry year, while groundwater farmers watered their fields without difficulty. Tensions over this disparity had been festering for years – the Surface Water Coalition (SWC) had placed a 'call' on upstream irrigators whose water rights were mostly junior to theirs about a decade prior (placing a 'call' means opening up a legal dispute that must be adjudicated by the courts). Rather than allow the courts to decide, state officials had stepped in to prevent the junior rights holders' water from being curtailed because they knew a wholesale curtailment would have forced the fallowing of 250 000 acres (101 171 ha) of agricultural land and brought on substantial losses for many individual farmers as well as Idaho's economy as a whole.

Wanting to avoid this scenario at all costs, a group of Idaho's elected leaders along with the directors of the IDWR and IWRB and representatives of the SWC and IGWA (Idaho's Ground Water Appropriators) negotiated a temporary change in Idaho's water management called 'Managed Aquifer Recharge' (MAR), the 'Comprehensive Aquifer Management Plan' (CAMP), or simply 'the 2015 water settlement agreement'. Designed to experiment with temporarily requiring water usage reductions in order to stabilize the ESPA and stave off a decision from the courts, the CAMP was adopted and scheduled to operate for five years – through the 2016–2020 growing seasons – at which point negotiators would reconvene and decide how to move forward depending on the agreement's success.

This temporary, cooperatively negotiated agreement is a real-world experiment into both intentional recharge of underground aquifers and persuading human communities to modify water usage behaviors in service of that goal. And now that the CAMP's term has come to an end, its relative effectiveness can be explored and lessons about what went well and what did not can be analyzed to inform future water conservation efforts. The sections that follow summarize the key aspects of the agreement including how it changed institutional water management in Idaho and how it worked in practice; the process by which it was negotiated and some conjecture about why it was successful after so many failed attempts prior; how it impacted farmers in the region and how they perceived it; how effective it ultimately was; and finally, what can be learned from it that could be of use for stakeholders seeking to develop environmentally and socially sustainable water policy in the arid western U.S. in the face of a steadily warming climate and ongoing social change.

16.2 INSTITUTIONAL CHANGE: THE 2015 WATER SETTLEMENT AGREEMENT

16.2.1 Defining allocation and rules

The CAMP first came into effect on the ground during the 2016 growing season. The agreement requires groundwater irrigators in the eight signatory groundwater districts of the ESPA (shown in Figure 16.2) to reduce total irrigation water consumption by an average of 240 000 acre-feet (AF) per year, or about 296 million m^3 (1 AF = 1233.48 m^3). The CAMP also stipulates that each district must reduce its overall aquifer withdrawals by 12.9%, and that usage reductions are to be calculated based on average use over the previous five years. This disadvantaged farmers who had invested in technology to improve water efficiency before the agreement. It became one of the most unpopular agreements among farmers, many of whom objected to it on the grounds that those past measurements of water usage had been based on the 'power coefficient' method, which is not very precise. The power coefficient method estimates water usage based on a farmer's electricity bill minus an amount calibrated to account for the rest of that farm's energy usage and had been the standard mode of measuring groundwater withdrawal in Idaho since wells began proliferating in the region in the late 1940s. (The terms of the CAMP changed this by requiring water meters on all groundwater wells by 2018.)

The agreement assigned the bulk of the implementation and monitoring power to the individual groundwater districts. This meant that each district decided for itself how to distribute its overall reduction requirement of 12.9% among its own farmers. Indeed, each groundwater district could even decide *how to decide* their allocation assignments: in some districts, managers made these decisions

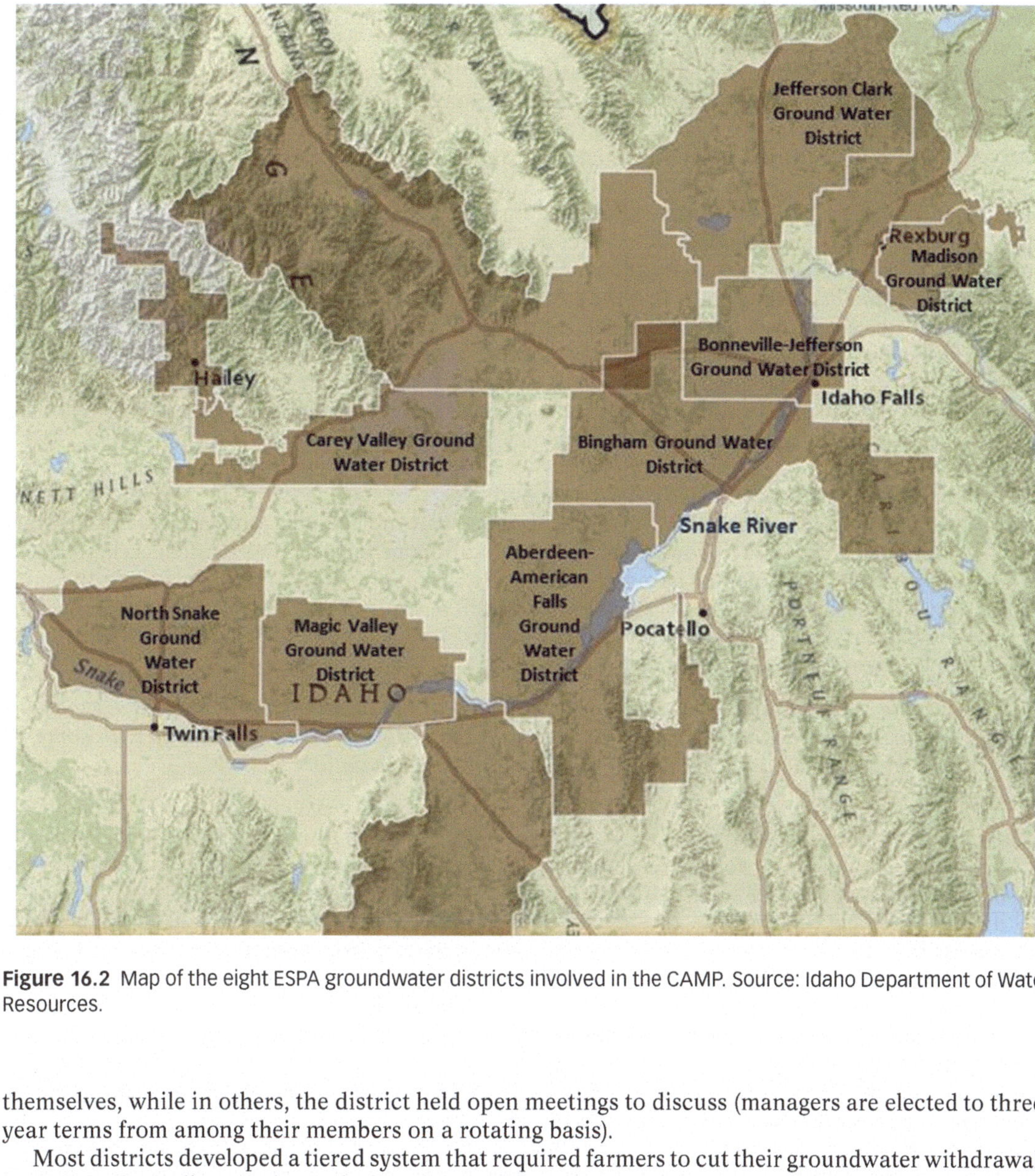

Figure 16.2 Map of the eight ESPA groundwater districts involved in the CAMP. Source: Idaho Department of Water Resources.

themselves, while in others, the district held open meetings to discuss (managers are elected to three-year terms from among their members on a rotating basis).

Most districts developed a tiered system that required farmers to cut their groundwater withdrawals by 4–20% depending on the seniority of their right, though a few districts assigned cuts on relative water usage, compelling farmers with the highest historical use per acre to reduce the most. The most common tier structure included three tiers, with cut-off dates defining senior, mid-range, and junior water rights set strategically to add up to the total amount each district needed to reduce. In these cases, senior water rights holders typically only had to curtail use by about 4%, while the mid-range rights holders were assigned about 13% and the most junior rights holders around 20%. One district assigned all its members the 12.9% average, while another developed a 12-tiered system in order to more precisely distribute the restrictions based on priority dates. And, while municipalities were included in the CAMP, it was fairly easy for them to buy out the portion of their reduction and they were quite minimally affected.

In general, the tiers were designed to pay homage to the prior-appropriation doctrine in spirit by excusing the oldest rights holders from substantial curtailments. This was also sometimes necessary to reach majority support for participating in the agreement *within* districts, as some of the more senior rights holders may have been much more reluctant to agree to participate without it being based on the hope that a decision by the courts would favor them. Complicating matters further is the reality that most farmers in southeastern Idaho do not believe that climate change is caused by humans and thus future predictions based on climate models are correct or valuable in any way. This results in some farmers hoping for – even expecting – temperature averages to go down and/or annual rain and snowfall to increase, causing doubts among many that the CAMP was truly necessary (Running *et al.*, 2017). While the agreement ultimately was adopted, some groundwater district managers reported struggling to persuade farmers who believed their seniority was secure enough for them not to need protection from lawsuits to cooperate (protection from legal judgments was the primary incentive district managers had to motivate consent), and ultimately the precise reasoning for some of the allocation decisions were not known outside of the groundwater district management offices themselves.

16.2.2 The negotiation process and why it succeeded

Despite the substantial challenges, the CAMP was accepted and implemented after years of negotiations and legal wrangling. When asked how the agreement finally came to pass, IDWR officials emphasized the urgency of reaching a solution based on the steady decline of the ESPA's levels since the mid-1900s. Other principal negotiators cited the state's hefty involvement and leadership along with the hydrological data and technical knowledge provided by its state-funded scientists. One IDWR representative explained:

> *I think people realized that, you know what, we need to get to a point where we're actually, instead of just buying water out of the river every year to mitigate an issue, come up with some solutions that would actually change what's going on with the aquifer.*

Conflicts between the SWC and IGWA had been ongoing for decades, and while IGWA had been mitigating the problem by buying water to supply the SWC for all that time, many stakeholders were beginning to feel a long-term solution was necessary in order to avoid immediate curtailment or an ongoing (and likely expensive) legal battle.

Still, the settlement agreement was a tough sell to groundwater users; when asked to explain why they were willing to sign on to the CAMP, groundwater managers from several groundwater districts simply said that the state had left them no choice. During our interviews, managers frequently invoked the metaphor of the state *'having a gun to our heads.'* The settlement agreement had to be agreed because the state legislature would otherwise overturn the prior-appropriation doctrine and reallocate water rights through a centralized water management agency. However, groundwater managers also recognized that the amount of land under threat of curtailment if an agreement was not reached would have put many farmers in their districts out of business.

Yet, many groundwater managers indicated that, despite the dire consequences faced, they almost left the negotiations entirely on several occasions. Several groundwater managers were suspicious of the surface water users' motives, worrying that their ultimate goal was to reduce competition and to supply the crops typically grown in Idaho in order to benefit their own operations. One groundwater district manager defending the groundwater users' interests explained it this way:

> *It was pretty tough to get to, because like I say, they [the SWC] wanted blood. They didn't want any water, they wanted to shut these guys off no matter what. That was the hard line for a long, long time. We went through, oh, it was a long spring. We went to meeting after meeting after meeting and it was to the point there a few times where I wanted to just say, to hell with it, let the state figure it out.*

Another one told us that negotiators on both sides decided to remove the lawyers from some of the negotiating meetings, since the lawyers often pushed their respective clients into refusing to make concessions. That manager told us:

> *Well, it was very contentious and they wouldn't even let the surface water people and the groundwater people in the very beginning be in the same room. They would have all kinds of negotiators, and, well, [state legislators] and then attorneys, and they would go back and forth and it was very contentious. I remember the first few meetings that I attended they wouldn't even let you be in the same room together.*

At that point political officials became the leading compromisers, persuading members of each side that the agreement was important for Idaho and also threatening a total overhaul of the state's water laws as the alternative. Thus, while fear and resistance were widespread, water managers and other government officials were still out to persuade the relevant parties to reach an agreement. They also appealed to farmers' desire for certainty and argued that participating in the joint agreement would suspend all other forms of legal jeopardy in accessing their water for the duration of the agreement. This would mean the farmers included in the agreement would be safe from all legal efforts by irrigators with more senior water rights to cut off their water, thus preventing the possibility of planting a crop and being unable to finish watering it mid-season, or even total curtailment in an especially dry year. One farmer we spoke to acknowledged this, explaining that while he did not like the agreement, it had some benefits:

> *The flip side of that is, at least we know the game we are playing. It seemed like in the pre-settlement, the rules were changing every year. The Director would come out with a new methodology. Right now we know, at least for the term of the settlement agreement, the game we are playing, which is helpful.*

Interestingly, the one significant policy concession the groundwater users won was the development of aquifer recharge sites throughout the state and a commitment by the state to contribute to recharge the aquifer with 250 000 acre-feet per year of recharge water. In Idaho, the IWRB holds rights to surface water for recharge, though those rights are relatively junior and thus only available to users at that position in the priority queue in very wet years or during the winter months. With respect to infrastructure, the existing system of irrigation canals allows recharge while water travels through unlined canals; substantial recharge also occurs in off-canal 'spreading basins' or through injection wells. This commitment was considered a major factor that enabled the agreement's success, with one surface water farmer and member of the SWC summarizing it this way:

> *The legislature said – you guys find a solution. We can find some money. Ultimately the groundwater users could see that recharge and some reductions in use were feasible if there was a big state contribution, too… If we try to address the aquifer every year, and we address it harder in the wet years than we do in the dry years, the aquifer becomes a bank account.*

Groundwater farmers were more likely to accept their own water curtailment requirements if they felt the state – the body that, as they frequently pointed out, had over-allocated the ESPA's resources in the first place – was going to share some of the pain. Another farmer with primarily groundwater rights put it this way:

> *I think that we really are at a crossroads of keeping our valley vibrant in green. The government wanted us to do that. The government was homesteading 160 acres and the deal was if you farmed it and built communities and made the desert blossom, that's how they would give it to you. They didn't give it to you to sit on and vest for 100 years. They wanted people to make it go. So now we've made it go and the government has their point man out there, taking our water from us. I don't think it's ethical. I don't think it's moral. I think it's government run amok and bad science. I don't think there's any proof of damage down there. Unfair. Do I sound like a baby?*

16.2.3 Monitoring, compliance, and enforcement

One of the stipulations of the agreement was that only recharge occurring through managed recharge activities and specifically intended for the CAMP's policy would count towards its annual recharge targets (accidental recharge through unlined canals during the irrigation season would not be considered). The withdrawals were also to be measured by meters on all groundwater wells (made compulsory as part of the 2015 settlement agreement). Thus, final recharge numbers for each year were a sum of all the recharge reports from each of the sites authorized and permitted by the IDWR or the IWRB, while the usage amounts were part of a different calculation designed to accurately measure how much water was being used and whether the mandatory percentage reduction was being met.

It was less clear among state officials and farmers alike how the agreement would be enforced. Given that it was negotiated outside of the courts, it was not legally binding. All eight of the groundwater managers indicated that, because of the differences in climate and soil across the eight districts, they would have strongly resisted a 'one size fits all' agreement in which the state set and enforced curtailment totals.

Instead, the compromise entailed a localized approach in which full permission was granted for the groundwater district managers and members themselves to decide how to implement, monitor, and enforce the CAMP in their districts. This was somewhat illusory, since the groundwater district managers had no specific enforcement power, but many people involved in the negotiations pointed out that nobody else did either. The agreement was thus technically voluntary, and even state officials did not have the power to force farmers to comply (unless they had passed a specific new piece of legislation). The agreement's terms created a formal monitoring process that required periodic updates from all entities involved – information that was then cross-checked through hydrological models and continuous measurement of groundwater levels at 19 strategically chosen 'sentinel wells' across the state.

By providing the data on the aquifer's recharge, this monitoring served as a sort of 'soft enforcement' that worked by appealing to managers' honesty and desire to maintain respect among their peers. Furthermore, as one groundwater district manager recounted:

> *If we cannot get farmers to comply, they're [the lawyers representing both sides of the negotiations] going to figure it out with the state. But the state, I know what they're going to do. They're just going to throw them right back in our district and say 'you guys take care of it.' I know how that works. We've been down this road before. It's true, that's what happens.*

In the end, the groundwater farmers and their district managers along with the legal counsel representing their interests decided that agreeing with the proposed cutbacks was preferable as long as each locale could decide how to meet them. The notion of risking new legislation that could result in even more severe cutbacks or the development of an external enforcement mechanism, or the option of continuing to kick the can down the block without an end in sight all seemed sufficiently unappealing to inspire cooperation and support for the plan.

16.2.4 Effectiveness

In 2020, the IWRB reported initial results of the settlement agreement's impact on the ESPA. The state had been responsible for recharging an average of 250 000 AF of water annually (~29.6 trillion m^3) at both newly developed and already existing sites and the final accounting found they had met and even exceeded their recharge targets; in the winter of 2019–2020, the state recharged almost double its requirement for that year, about 447 432 AF (~53 trillion m^3). Currently, the IWRB estimates that more than 500 000 AF of water is being restored to the aquifer annually, facilitated by the new recharge sites along with renewed awareness about water conservation among farmers and other water users in the ESPA.

Yet, while the agreement succeeded in its short-term goal of reversing the decline of the ESPA's levels, the state has continued to warn that the longer-term sustainability of the ESPA is still in flux and challenges remain (Patton, 2019). Idaho still faces a future in which supply of this valuable resource is likely to fall short of demand, increasing the vulnerability of the agricultural sector (Adger, 2006). And though historically, the state's policy of prior appropriation allowed farmers to manage their water and land almost entirely without government intervention, in a warmer, drier Idaho that is still attracting record in-migrants each year, it is likely that state intervention in water policy is likely to continue to be necessary.

16.3 RESULTS OF THE CAMP AMONG FARMERS

Overall, the conventional wisdom in Idaho is that the 2015 agreement has been a clear success, with aquifer recharge targets met and even exceeded and downstream surface water users generally agreeable about the effects on their operations. Groundwater farmers were less enthusiastic but still generally affirm that the agreement was a success. Based on a survey of the agreement's impacts during the first two years of the policy, groundwater farmers reported groundwater reductions of 11.9% on average along with mean yield losses of 7.9% and income losses of on average 8.6%. Survey responses also did not reveal any farmers who lost their farms or even decided to quit farming due to the agreement. Overall, an average of 33 acres (33.35 ha) per farm were taken out of production in 2016 and 41 acres (16.59 ha) in 2017, and farmers reported spending an average of about $11 173 in 2016 and $13 848 in 2017 to install the required water meters and invest in other technology to help achieve their allotted curtailment. Farmers in our sample farmed an average of 1252 acres (506.67 ha), with a minimum of 0 acres and a maximum farm size of 16 000 acres (6474.97 ha). The most common expenses were putting in required water meters and improving the efficiency of current irrigation systems or switching to a new system, followed by purchasing canal water rights or recharge credits and hiring an outside consultant to perform an efficiency review.

In the survey, the majority of farmers stitched together a suite of adaptation strategies to meet their required irrigation reductions (outlined in Table 16.1). On average, farmers made 9 specific changes out of a possible 27 options we asked about, such as changes to farm management like changing crops grown or crop rotation schedules, leasing land to other farmers, or installing new irrigation technology. Other possible adaptations included changing how their crops were processed and transported to markets and the negotiation of new contracts with large food-processing corporations. Other types of adaptations (e.g., getting a second job or finding other ways to make money from their land such as participating in paid conservation or habitat restoration programs) were also chosen by at least some farmers. This suggests many farmers address risk in a more holistic way than simply employing one-off on-farm management adaptations, and they tend to select a combination of adaptation practices that also reduce temporal, spatial, or community-based risk. As such, it is important for policymakers who seek to incentivize agricultural adaptation to be aware of this diversity of possible choices and how practices they decide to promote may create synergies or incompatibilities with the local farming practices already in use (Agrawal, 2010; Burnham & Ma, 2018; Feola *et al.*, 2015; Taylor, 2015). Offering flexibility, technical support, and local control as much as possible helped this agreement succeed and provides some guidance for water managers in the future attempting to govern water policy change in similarly arid and rural contexts.

16.4 CONCLUSION: STRENGTHS AND WEAKNESSES OF IDAHO'S CAMP EXPERIMENT AND IMPLICATIONS FOR AND FUTURE RESILIENCE

The Eastern Snake Plain is a diverse and productive agricultural basin in the inter-mountain region of the American West with substantial economic value to the region and a prominent role in local culture and lifestyles. Located in a semi-arid high desert climate, more than 74% of this agricultural land is irrigated, with 40% of that irrigated water coming from the groundwater in the ESPA – about

Table 16.1 Typology of adaptations with percent adopting in each category.

Adaptation Type	% Adopting Types	Adaptation Action	% Adopting Actions
Income diversification	34	Took an off-farm job	21
		Started a new business	14
Communal pooling	8	Joined a coop	8
Market exchange	67	Planted higher value crops	18
		Contracted more crops	22
		Contracted fewer crops	14
		Sold land	8
		Put land into CRP* or other government program	14
		Purchased recharge credits	18
		Purchased or rented canal shares	19
Operations management-technology	85	Improved irrigation system efficiency	77
		Switched to more efficient irrigation system	53
		Adopted precision agriculture	21
Operations management- water	88	Dried up corners	48
		Turned off end guns	42
		Irrigated less frequently	59
		Planted less water intensive crops	44
		Used canal water to meet requirements (mitigate)	37
Operations management- crops, soil, and environment	79	Changed crop rotation	53
		Switched to or added dryland acres	25
		Took acres out of production	36
		Changed tillage practices	43
		Reduced livestock herd size	13
Operations management- assets	67	Reduced spending	67
Capacity-building	28	Hired consultant for technical help	28
No adaptation	29	Did not do anything	12
Exit farming	6	Stopped farming	6

*Conservation Reserve Programs (CRP).

one million acre-feet or 404 685.64 ha (IDWR, 2012; USDA NASS, 2014). After rapid aquifer declines in recent decades, however, concerns about decreasing surface water flows in areas hydrologically dependent on the aquifer prompted litigation that resulted in a significant re-envisioning of Idaho water management – first an official change to conjunctive ground- and surface-water management, followed by a comprehensive adjudication and finally the negotiation of the Comprehensive Aquifer Management Plan (CAMP) that, for the first time in Idaho's history, required reductions in irrigation water usage by some groundwater farmers for the purpose of recharging the aquifer.

In Idaho, the foundation water law is the 'first in time is first in right' or 'prior appropriation' doctrine, which says that applied for a water right first has 'priority' over all later applicants, as long as the water is to be used for 'some useful and beneficial purpose'. This principle is still the fundamental tenet of water governance in Idaho, though the collective understanding of what counts as 'beneficial use' has shifted somewhat since the earliest water rights were allocated in the late

1800s – in particular, it is now possible for preservation of fish habitat or other conservation-related goals to qualify rather than only activities with direct economic benefit such as mining, logging, and agriculture. Still, the overall spirit of even the CAMP is to maintain access to water for as many people as possible to protect the state's economic well-being and allow multi-generational farms to continue to produce water-intensive crops because agricultural interests have always been well-represented in the state.

Ultimately, the CAMP was a successful temporary policy experiment that met – and even exceeded – its aquifer recharge goals, though some of this was due to above-average water years, especially during the first few years of the agreement; in both 2016 and 2017, precipitation totals ranged from 115–200% above average depending on the specific location in the state (NWS, 2016, 2017). Moreover, this was achieved without putting any farmers out of business and without reducing individual farm operations' bottom lines beyond what is considered fairly normal year-to-year fluctuation (though many farmers would likely disagree). For these reasons, this water settlement agreement may be a good model that other water managers can learn from to resolve future problems with water scarcity in the face of growing demand.

State negotiators will now decide whether to continue irrigation restrictions, allow each farmers' water rights to be reset to their original terms, enact further restrictions on the same or a different group of farmers, or overhaul Idaho's water allocation policy altogether. In the meantime, the Idaho agreement provides an excellent 'natural experiment' and an opportunity to analyze the process, implications, and impacts of a water usage reduction policy in an arid western state likely to experience continued mismatch between supply and demand for water in the context of both socio-economic and climate-related change in the coming years.

ACKNOWLEDGMENTS

We would like to thank the National Science Foundation (NSF) EPSCoR MILES (Managing Idaho's Landscapes for Ecosystem Services) Program, award number IIA-1301792, and the United States Department of Agriculture's National Institute of Food and Agriculture (USDA NIFA) seed grant, award number 2018-69002-27963, for contributing funding for this research.

REFERENCES

Adger W. N. (2006). Vulnerability. *Global Environmental Change*, **16**(3), 268–281, https://doi.org/10.1016/j.gloenvcha.2006.02.006

Agrawal A. (2010). Local institutions and adaptation to climate change. In: The Social Dimensions of Climate Change: Equity and Vulnerability in A Warming World, R. Mearns and A. Norton (eds.), World Bank Publications, Washington, D.C., USA, pp. 173–197.

Burnham M. and Ma Z. (2018). Multi-scalar pathways to smallholder adaptation. *World Development*, **108**, 249–262, https://doi.org/10.1016/j.worlddev.2017.08.005

du Bray M. V., Burnham M., Running K. and Hillis V. (2018). Adaptive groundwater governance and the challenges of policy implementation in Idaho's eastern snake plain aquifer region. *Water Alternatives*, **11**(3), 533–551.

Dieter C. A., Linsery K. S., Caldwell R. R., Harris M. A., Ivahnenko T. I., Lovelace J. K., Maupin M. A. and Barber N. L. (2017). Estimated use of water in the United States county-level data for 2015. U.S. Geological Survey data release. https://doi.org/10.5066/F7TB15V5 (accessed 5 May 2022)

Elliott J., Deryng D., Müller C., Frieler K., Konzmann M., Gerten D. and Eisner S. (2014). Constraints and potentials of future irrigation water availability on agricultural production under climate change. *Proceedings of the National Academy of Sciences*, **111**(9), 3239–3244, https://doi.org/10.1073/pnas.1222474110

Feola G., Lerner A. M., Jain M., Montefrio M. J. F. and Nicholas K. A. (2015). Researching farmer behaviour in climate change adaptation and sustainable agriculture: lessons learned from five case studies. *Journal of Rural Studies*, **39**, 74–84, https://doi.org/10.1016/j.jrurstud.2015.03.009

Getzler J. (2009). A History of Water Rights as Common Law. Oxford University Press, New York, NY, USA.

Harrington P. R. (2012). The establishment of prior appropriation in Idaho. *Idaho Law Review*, **49**, 23–38.

Idaho Water Resource Board (IWRB). (2012). *Idaho State Water Plan. Idaho Department of Water Resources.* Retrieved October 7, 2021. https://idaho.water.gov.

Lenczowski G. (1990). American Presidents and the Middle East. Duke University Press, Durham, NC, USA.

National Weather Service. (2016). Idaho 2016 Water Year Summary. Sage Winds: Official Blog of the National Weather Service Boise – Serving Southwest Idaho and Southeast Oregon. Retrieved October 20, 2021. https://nws.weather.gov/blog/nwsboise/2016/11/29/idaho-2016-water-year-summary/ (accessed 20 October 2021).

National Weather Service (NWS). (2017). Idaho 2017 Water Year Summary. Sage Winds: Official Blog of the National Weather Service Boise – Serving Southwest Idaho and Southeast Oregon. Retrieved October 20, 2021. https://nws.weather.gov/blog/nwsboise/2017/11/17/idaho-2017-water-year-summary/ (accessed 20 October 2021).

Niles M. and Wagner C. R. H. (2019). The carrot or the stick? Drivers of California farmer support of for varying groundwater management policies. *Environmental Research Communications*, **1**(4), 1–4, https://doi.org/10.1088/2515-7620/ab1778

Patton B. (2019). Idaho Water Resource Board on pace to meet 600,000 acre-foot target for Eastern Snake Plain Aquifer; final report sent to Speaker Bedke. *Idaho Department of Water Resources*, Retrieved October 20, 2021 https://idwr.idaho.gov/IWRB/

Running K., Burke J. and Shipley K. (2017). Perceptions of environmental change and climate concern among Idaho's farmers. *Society & Natural Resources*, **30**(6), 659–673, https://doi.org/10.1080/08941920.2016.1239151

Slaughter R. A. (2004). Institutional History of the Snake River 1850–2004. Climate Impacts Group. University of Washington, Seattle, WA, USA.

Strong C. J. and Orr M. C. (2016). Understanding the 1984 swan falls settlement. *Idaho Law Review*, **52**, 223–270.

Taylor M. (2015). The Political Ecology of Climate Change Adaptation: Livelihoods, Agrarian Change, and the Conflicts of Development. Routledge, London, UK.

United States Department of Agriculture, National Agricultural Statistics Service (USDA NASS). (2014). 2012 *Census of Agriculture*, Washington DC.

University of Idaho Extension (UIE). (2016). Contribution of Agribusiness to Idaho's economic base. In: Economic Contribution of Idaho Agribusiness, P. Watson and L. Ringwood (eds.), University of Idaho, Moscow, ID , pp. 8–11.

World Population Review (WPR). (2021). Fastest Growing States 2021. https://worldpopulationreview.com/state-rankings/fastest-growing-states (accessed 23 June 2021).

doi: 10.2166/9781789062786_0215

Chapter 17

Polycentric governance in Nebraska, U.S., for ground and surface water

Theresa Jedd[1], *Anthony Schutz*[2] *and Mark Burbach*[3]

[1]*Technical University of Munich, School of Governance, Munich, Germany*
[2]*University of Nebraska-Lincoln, College of Law, United States, Lincoln, Nebraska*
[3]*University of Nebraska-Lincoln, School of Natural Resources, Conservation & Survey Division, United States, Lincoln, Nebraska*

ABSTRACT

This chapter describes the locally driven, but centrally coordinated, water governance model in Nebraska, U.S. It offers a snapshot of water resources and the importance of agriculture, then moves to the relevant political institutions in the state, and federal controls related to water quantity. The focus of the chapter is on the Nebraska Department of Natural Resources' (NeDNR) and Natural Resources Districts' (NRDs) management of surface and groundwater, which has some distinct and some overlapping authority. The main area of overlap is in addressing the connection between ground and surface water, particularly in situations when either or both are over appropriated. Integrated management planning is a key tool for basins in crisis, where allocations are fully or over appropriated and there is increased demand or diminished supply. The chapter explains what integrated management planning entails and gives a closer look into instances where it has been implemented. The polycentric model allows for collaborative governance, pushing stakeholders (particularly the agricultural sector) to innovate based on changes in water availability. NRDs can (and do) exercise controls; they do so by using their authority to make institutional changes and sanction violators for over-abstraction. This authority is granted and legitimized by publicly elected boards, an ongoing leadership training network, and a history of locally driven rule-making. However, there are also shortcomings to the model: in particular, it is difficult to address cross-border issues or legal conflicts. Furthermore, there is scant research on its effectiveness in actually preventing groundwater decline. The Nebraska model and its local examples may offer lessons for other basins where water resources have historically been relatively plentiful but are now facing drought stresses and the growing demands of intensive irrigated agricultural production.

Keywords: Drought, federalism, governance, groundwater, integrated water resources management, Nebraska, polycentricity, surface water, United States

17.1 INTRODUCTION

The authority to make decisions about water allocations in Nebraska is distributed across local, regional, state, and federal institutions. At the center of Nebraska's polycentric water governance is the working relationship between sub-state entities called Natural Resources Districts (NRDs) and the state's natural resource agency, the Nebraska Department of Natural Resources (NeDNR).

Generally speaking, surface water is administered by the NeDNR and groundwater is governed by NRDs. While other states use local resource planning units to manage water, the NRD model does not exist in any other state and is unique in its scope of responsibilities for managing the hydrological connections between ground and surface water when maximum allocations have already been issued. This interplay of political authority is an instance of polycentric governance.

Polycentricity is a condition in which decisions are made by multiple distinct but overlapping centers of authority (Ostrom, 2010) and is often applied to environmental governance due to the complex, multi-dimensional, and diverse range of resources subject to overuse or depletion (Carlisle & Gruby, 2017). This framework is fitting for analyzing Nebraska's water governance because of the diversity of actors and levels of decision-making. The need for this mix of authority stems from the fact that there is an abundance of irrigated agricultural production in the state, which has historically thrived. However, in recent decades, some areas of the state now face water stress. Groundwater gives a buffer against variable surface water availability, especially during droughts. Among U.S. states, Nebraska has the most irrigated cropland and pastures (Bleed & Babbitt, 2015) with 8–9 million acres of farmland under irrigation, 32% of the state's total area (U.S. Department of Agriculture, 2019). Most irrigation is supplied by groundwater from the Northern High Plains (Ogallala) Aquifer. Groundwater in the northern region of the aquifer is being withdrawn at a higher rate than it is being recharged. Almost half of the loss in 2000–2009 occurred in the Republican River Basin, which is shared between Colorado, Nebraska, and Kansas. These losses are due to irrigation withdrawal, driven by decreased rainfall and increased surface water evaporation (Peterson *et al.*, 2020).

Nebraska is a state of stark ecological contrasts, which influence the diversity of agricultural production. About a quarter of the state is a Sandhills ecotype with high soil erosion and low precipitation (Dalstrom & Naugle, 2020). These areas are used more for cattle grazing than crop production. In the rest of the state, however, prairie and humus soils are suitable for small grain production (Ibid). The state receives variable annual rainfall with the yearly statewide average ranging from 13.36 inches in 2012 to 35.50 inches in 1915 (Frankson *et al.*, 2017). The average annual rainfall also varies geographically. The eastern portion of the state receives about twice as much rain (35 inches) as the western part (15 inches) (Shulski, 2018). Because of drought, irrigation is a major dimension of agriculture in Nebraska (Ulrich, 2018), and has led to increased yields and an increase in property value. In recent decades, droughts plagued the state from February 2002 to September 2008 and again at a peak the first week of October 2012 with more than 77% of the state in severe drought (National Drought Mitigation Center, 2021). Without groundwater irrigation, the state's agriculture-heavy economy would suffer more than it already does during droughts.

Agriculture dominates Nebraska's water consumption (6100 Mgal/day, U.S.G.S., 2021a), and the state is still experiencing a boom in irrigation. Roughly 300 000 acres were added between 2007 and 2017, while another 440 000 acres were added between 2017 and 2020, reaching a total of 9.03 million acres of irrigated land, mostly from groundwater (Nebraska Association of Resources Districts, 2021). In 2015, usage rates were 5400 Mgal/day for groundwater irrigation and 673 Mgal/day for surface water irrigation (U.S.G.S., 2021a). In contrast, the public consumes 1720 Mgal/day and thermoelectric power generation requires an average of 2900 Mgal/day (U.S.G.S., 2021a).

To date, groundwater supply and access have been relatively stable in Nebraska, even during drought years. The state sits on one of the world's biggest reserves of groundwater, the High Plains Aquifer, which is approximately 174 000 square miles (Johnson *et al.*, 2011). The formation lies beneath the jurisdiction of eight U.S. states and is estimated to hold 2.91 billion (10^9) acre-feet of water. Nebraska holds 65% of the volume (Peck, 2007). Some irrigators, however, do not have easy access to groundwater (Peck, 2007), and in recent years there has been a decline in aquifer volume (U.S.G.S., 2021b). The remainder of the state's needs, then, are supplied by surface water diversions, the two major rivers being the Platte complex (North and South) and the Missouri.

17.2 FEDERAL CONTROLS ON WATER ALLOCATION

The responsibility for governing water quantity resources largely lies with states and sub-state actors in the U.S., who carry out the majority of water delivery and wastewater removal (Stoutenborough & Vedlitz, 2014), though federal responsibilities include providing drought-related economic relief to famers or building new dams and managing reservoirs. Furthermore, some federal obligations explicitly or effectively limit Nebraska's authority.

For example, there are two sources of federal-level resource allocation that apply to the Platte River and Republican River basins, respectively. On the Republican River, Nebraska is a party to an interstate compact with neighboring states, Colorado and Kansas. These interstate agreements are approved by Congress and carry the imprimatur of federal law, displacing state law or actions to the contrary under the U.S. Constitution's Commerce Clause. The Republican River Compact caps the yearly amount of the river's 'virgin water supply' Nebraskans can consume. States are responsible for the Republican River Compact Administration[1]. The Compact limits surface-water diversion and groundwater use that depletes streamflow. It has been one of the key drivers of integrated resource management in the Republican basin (Aiken, 2008).

The second federal restraint occurs through the Endangered Species Act protection of the whooping crane, piping plover, interior least tern, and the pallid sturgeon (Jenkins, 1999). The obligation to protect habitat for these bird and fish species has required Nebraska to reduce its consumption of water in the western two-thirds of the Platte River Basin. Because this habitat is riparian, all activities that deplete streamflow are impacted, including direct diversion of streamflow or indirect diversions through groundwater pumping. This has been a key driver of Nebraska's effort to integrate its water management, effectively limiting consumption during drought. Even within these federal restraints, however, states still wield a great deal of authority over water management. As the Western Governors' Association (2018) claims, states are the 'preeminent authority on water management within their boundaries', with rights to surface water and groundwater management.

17.3 STATE INSTITUTIONS FOR MANAGING WATER ALLOCATIONS

Nebraska is governed by a unicameral legislature seated in Lincoln. The governor presides over most[2] administrative agencies, including the NeDNR by appointing its director. The legislative role in administrative operations is also strong as agencies may not do anything that is not allowed by statute. The NeDNR is a state-level agency, directed by an unelected director who manages a professional staff, carrying out operations under statutory directives and pursuant to regulations adopted under formalized statutory procedures. Its budget is determined through the state appropriation and budgeting process, and its revenue comes from sales and income taxes.

17.3.1 Surface water administration

For present purposes, the NeDNR's main role is administering surface-water rights. These are allocated under 'prior appropriation'; users with historical rights have priority claim to the water. This is different to a riparian rights system which automatically allows property owners along a river, stream, or water body to make diversions. During shortages, riparian systems may allocate water to multiple users equitably, without regard to the date of one's first diversion or any similar vesting of rights. The prior-appropriation system of allocation, predominant in the western U.S, abandons these uncertain outcomes in favor of a system that creates predictable water supplies according to a somewhat strict system of temporal priority. This, in turn, spurs investments in the infrastructure necessary to deliver water to users.

[1]For more information, see the Compact website: http://republicanriver.org/

[2]The Governor has direct control over Code Agencies (e.g., NeDNR) but not Non-code Agencies (e.g., Game and Parks Commission)

Surface appropriations are generally granted for a right to divert water for an authorized beneficial use. If granted, the right's priority dates from the time of application[3]. The system also involves an administrative mechanism for changing or transferring appropriations after they are granted. Transfers are allowed when the original holder files to change the location of use (Nebraska Administrative Code, 2008). The nature of use or the holders of surface-water rights can also be changed through an administrative review process initiated by an appropriator. In addition, appropriations that are not used can be canceled by the NeDNR. In this sense, surface-water rights are often thought of as 'use it or lose it' rights (Ibid).

17.3.2 Groundwater administration

Groundwater appropriation is significantly different, largely due to the NRD institutional structure. State law specifies that NRDs are the 'preferred regulators of activities which may contribute to groundwater depletion' (Neb. Rev. Stat. § 46-702 in Nebraska Association of Natural Resources Districts, 2021). They are local governments with geographic boundaries that correspond with a combination of watershed boundaries and county line administrative units. The boundaries were collaboratively determined in expert hearings on the extent of 'common problem areas'. NRDs' borders resemble a blend of areas that, in former times, were used as Soil Conservation Service districts, Bureau of Reclamation districts, Corps of Engineers drainage districts, Conservation and Survey Division groundwater districts, and Farmers Home Administration water supply districts (Fairchild, 1994). Every watershed in the state has more than one NRD within it. For example, the Republican River Basin has four NRDs with mutually exclusive boundaries (the Upper Republican NRD, the Middle Republican NRD, the Lower Republican NRD, and the Tri-Basin NRD).

An elected Board of Directors governs NRDs. Voting districts comply with a one-person-one-vote legal requirement, so each district has a similar proportion of the district's voters within it. Property and occupation taxes on land's irrigation status are primary sources of revenue. While the publicly elected Boards of Directors govern the district's operations, NRD tasks are carried out with professional staff, including a General Manager who has general administrative oversight. Together, the directors, management and staff work with members of the public and landowners to make decisions about water allocation and drought contingency plans (Nebraska Association of Natural Resources Districts, 2019).

Since Nebraska's water availability varies greatly by location, so do priorities and approaches to management. An area's elevation, climate, groundwater supplies, surface water storage capacity, anthropogenic demand for water, along with underlying environmental, social, economic, cultural, and physical factors influence how susceptible a location is to drought (Hagenlocher *et al.*, 2019). Additionally, the history of cooperation varies between NRDs, with some having engaged in more extensive water-related planning than others.

NRDs manage groundwater against the backdrop of a common-law right to withdraw water from beneath one's property. These correlative rights did not require a permit for their existence, and they could not be lost through the passage of time. As scarcity problems emerged, though, NRDs were given statutory authority to regulate withdrawals. Historically, NRDs did not use this authority very often. In 1980, only one NRD imposed limits on groundwater withdrawal (the Upper Republican). But as connections between ground and surface waters were better understood, NRDs became active water regulators.

17.3.3 Water law and legal conflicts between ground and surface water users

Understanding the development of Nebraska's water law in the context of the conflicts that drove its modification is essential to understanding the present institutional polycentricity. While surface-water rights have generally not existed in Nebraska in the absence of statutory law, groundwater rights were first enunciated in common-law litigation initiated by adjacent landowners. Landowners held the right to withdraw and use water from beneath their properties. As water use began to impact

[3]https://dnr.nebraska.gov/water-admin/faqs#sw-permit

neighbors (due to declining water tables), litigation led to limitations on water use. Generally, these rules allowed landowners to use water on their property for 'reasonable' uses, and, in the event of conflict resulting from a shortage of aquifer supply, the common law requires each landowner to share the burden of reduced use correlatively. These landowner rights still exist to some extent today but, as demonstrated below, they have been largely modified through statutes that empower NRDs to manage water.

Those modifications arose through legislation adopted in the 1970s and 1980s, which allowed NRDs to create Water Management Areas and prescribe controls that would avoid well interference. Well registration and spacing requirements were the main forms of control adopted in this era as groundwater shortages became common in some parts of the state. Within those restraints, the water supply decline was a foreseeable and largely accepted consequence of groundwater pumping (Peck, 2007)[4].

Prior efforts at groundwater regulation proved insufficient to avoid conflicts between groundwater pumping and surface water rights. Drought years placed intense strain on the system, and legal action instigated by surface water users against groundwater irrigators within Nebraska (e.g., *Spear T. Ranch, Inc v. Knaub, 2005*) and by adjacent states (*Kansas v. Nebraska and Colorado*) ensued. In *Spear T.*, the Nebraska Supreme Court first recognized a cause of action against groundwater pumpers that could be pursued by surface water diverters, with liability occurring when the pumping of hydrologically connected groundwater unreasonably interferes with surface water appropriations.

In the case of *Kansas v. Nebraska and Colorado*, Kansas was awarded \$3.7 million for surface water that the U.S. Supreme Court ruled had been illegally diverted from the Republican River by Nebraska irrigators, many of whom were depleting streamflow through groundwater pumping. The Court recognized that groundwater was within the scope of the interstate compact between the litigating states to the extent its consumption impacted streamflow. The conflicts arising from these hydrological connections were a growing point of contention for irrigators.

As a result, Nebraska changed its statutory water law to avoid these conflicts, though it stopped short of providing a clear priority among surface water and groundwater users. Under Legislative Bill 962 (LB962), in the areas where surface water is fully or over appropriated, hydrologically connected groundwater is managed to bring the supply and demand for water into balance. Surface water appropriations are also managed to reduce demands when necessary. To accomplish this, significant changes were made to the institutional structures that, up until that time, had operated independently of one another.

Finally, although concern for water sustainability is found in state statutes, a major driving force of modern groundwater management in the state has been the protection of streamflow for purposes of ensuring compact compliance and meeting the obligations imposed by the Endangered Species Act. This federal authority requires state institutions to abide by limits tied to ecological indicators (e.g., species abundance) while continuing to account for social pressures (e.g., increasing demand for agricultural products), legal constraints (e.g., lawsuits), and meteorological markers (e.g., streamflow).

17.4 INTEGRATING NEBRASKA'S SURFACE WATER AND GROUNDWATER INSTITUTIONS

One of the main challenges of moving toward a more integrated polycentric approach to water management was bridging the gap between the vastly different sets of legal rights (surface water prior appropriation and the correlative rights of groundwater users). Groundwater users vastly outnumber surface water users. The management institutions are also markedly different. The NeDNR agency is tasked by the state to protect water supplies and administer related laws and programs, while the NRDs are governed by boards elected by local residents. The NeDNR is incentivized to engage in basin-wide conservation planning while individual NRDs may face pressure from constituent agricultural producers to develop groundwater resources (Aiken, 2005). To reduce the tensions between surface water protection and groundwater utilization, the NeDNR and NRDs were brought together in integrating management planning.

[4]Current requirements entail purchasing a permit corresponding to the rate of extraction. For <50 gallons per minute the registration fee is \$70 USD, and \$110 USD for >50 gallons per minute (State of Nebraska, 2012).

17.4.1 Integrated management plans and basin-wide planning: accounting for cross-border challenges

The 2004 passage of LB962 directed the NeDNR to declare portions of the state fully appropriated and over appropriated by January 1 of each year. The main driver of the designation is the (in)sufficiency of surface water supplies. The NeDNR evaluates surface water supply as insufficient if the most junior irrigation right holder in the last 20 years has not received 85% of the water needed for a corn crop from May 1 to September 30, which is the irrigation season, or has not received 65% of the water needed for the corn growing period July 1 to August 31 (regulation Title 457 Neb. Admin. Code Chapter 24 in State of Nebraska, 2021). This is called the '65/85' rule.

The Republican Basin and the Platte River Basin are affected by overconsumption (Bleed & Babbitt, 2015). When a basin is declared *over appropriated*, it means that 'existing uses exceed the supply and surface water flows can be expected to drop until either there is no water to use or the cost of using the water is too great to result in beneficial use' (Nebraska Department of Natural Resources, 2005). A *fully appropriated* basin has a cumulative demand for surface and groundwaters that matches available supplies. Basins in this category must be managed carefully in order to avoid entering over appropriation in the case that the uses in a basin overreach the long-term supply (Ibid).

The NeDNR evaluates basins for the presence of hydrologically connected groundwater and projects the impact of pumping on surface water supplies in order to determine whether supply meets usage. This involves an analysis of whether a certain number of years pumping a hydrologically connected well will deplete a river or base flow by a particular percentage. In the case of the Platte River Basin upstream of the Kearney Canal Diversion, the North Platte River Basin, and the South Platte River

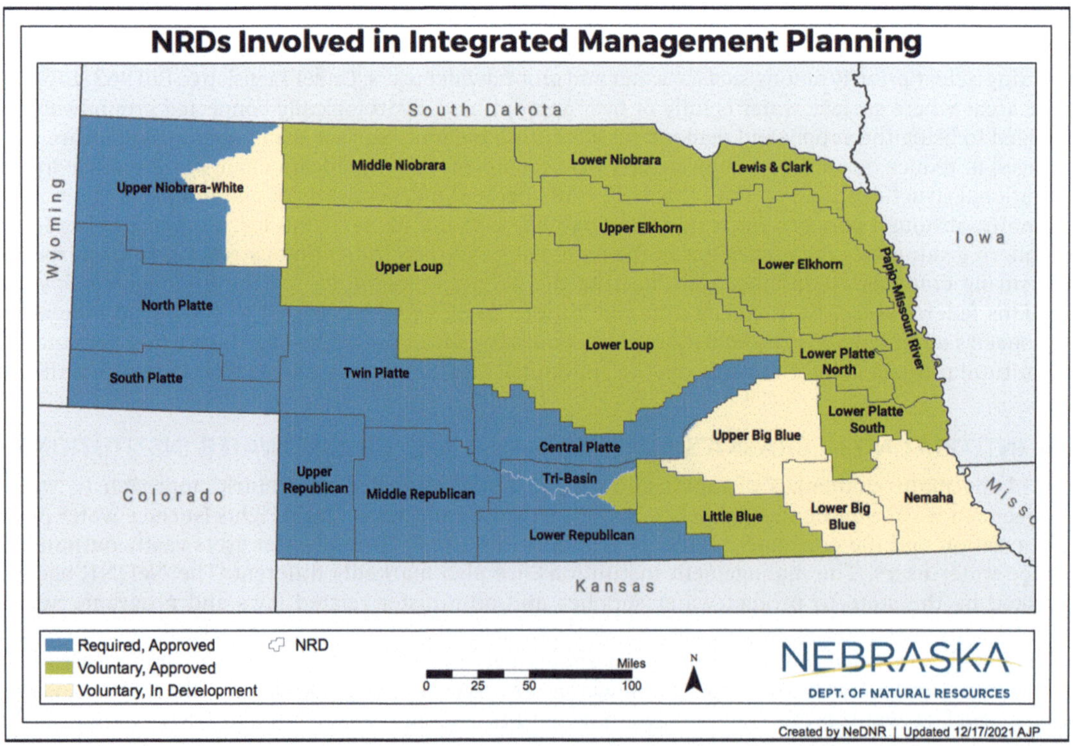

Figure 17.1 Natural Resources Districts with Integrated Management Plans as of December 17, 2021. Source: Nebraska Department of Natural Resources. Reprinted with permission.

Table 17.1 Examples of how NRDs have made institutional changes to allow for various control measures to sanction users and control groundwater over abstraction.*

Natural Resources District and Revision Order Document Name	Examples of Institutional Changes	Types of Sanctioning Allowable for Violations
Upper Republican NRD Order No. 34 Adopting Ground Water Controls (2018)[5]	Amended groundwater rules and regulations for the 2018–2022 period; spacing requirements for industrial and commercial livestock wells; changed criteria for the transfer of irrigated acres	The District can revoke or reduce irrigation allocation permits if compliance is breached. Violators are subject to further sanctioning under District rules and Neb. Rev. Statute.
Upper Big Blue NRD Groundwater Management Rules and Regulations (2020)[6]	New rules for Groundwater Management Area #1 and #2 to supersede previous GMA rules	The District can issue a cease-and-desist order to enforce withdrawal rules and regulations; violators can request an adjudication hearing. The District set new rules for groundwater transfers, and reserves the right to revoke authorization.
Lower Platte South NRD Groundwater Rules and Regulations, Revised (2020)[7]	Establishes a Groundwater Management Area that includes the entire District	Allows for mandatory education requirements to reduce groundwater depletion; authorizes District compliance officers to issue a cease-and-desist order for abstraction violation, followed by subcommittee consideration, and eventually a civil misdemeanor for non-compliance
North Platte NRD Rules & Regulations for Enforcement of the Nebraska Ground Water Management and Protection Act and the Nebraska Chemigation Act (2019)[8]	Establishes procedures for violation of the Nebraska Ground Water Management and Protection Act	The District can issue a cease-and-desist order when acts or activities violate state or District rules; a hearing may follow; penalties may be imposed as determined by the Board and/or future allocation amounts may be accordingly reduced; violations can result in civil penalties from \$1000–\$5000 per day of intentional violation (p. 54)

**For further information see individual Orders and Rules demarcated with footnotes. For an overview of current NRD use of rules and sanctions, see Nebraska Association of Resources Districts (2021), particularly page 8, for a mapped list of management actions and controls used in individual NRDs.*

Basin, the NeDNR issued a '28/40 line', referring to 40 years of pumping resulting in a 28% depletion (State of Nebraska DNR, 2004). This order was renegotiated to the 10%/50 year test – in other words, that 50 years of pumping a hydrologically connected well will deplete a river or base flow by 10% or more (Neb. Admin. Code title 457, ch. 24 in Aiken, 2005). Hydrologically connected areas are the main focus of integrated management since it is the pumping in these areas that impacts streamflow. This is an attempt by the NeDNR to address cross-border problems and link NRD actions together in a coherent way to address stressors in larger hydrologically connected areas. Furthermore, the NeDNR has pushed for repurposing surface water infrastructure to recharge groundwater and maximize the use of surface supplies in lieu of groundwater when the supply of surface water is high.

[5]For specific details see https://www.urnrd.org/sites/default/files/files/20/rulesregsfinal2018.compressed.pdf

[6]https://www.upperbigblue.org/sites/default/files/files/328/rule_5_as_adopted_august_20_2020mod.pdf

[7]Available online: https://www.lpsnrd.org/sites/default/files/files/89/ground_water_rrs_2020_final.pdf

[8]https://www.npnrd.org/assets/site/documents/rules-and-regulations/rules--regs-final-effective–9.11.19.pdf

17.4.2 Procedures for over-appropriated basins

Once a river basin or reach is designated fully or over appropriated, the state and NRDs must co-create Integrated Management Plans (IMPs) within 3–5 years. IMPs stipulate when further water development is no longer possible. IMPs are designed to reduce the strain that groundwater pumping can have on surface water flows, and vice versa, to protect surface water flows that are beneficial for groundwater recharge (Reed & Abdel-Monem, 2015). Figure 17.1 depicts the status of voluntary and required planning.

The IMP process requires setting clear goals and objectives to protect the balance between supply and demand. In drafting plans, each fully or over-appropriated NRD must include a map of the management area (which could be a portion or whole of the NRD extent) and set a minimum of one ground and surface water control and a plan for monitoring availability and use. Once adopted, NRDs and the NeDNR work toward achieving IMP goals by imposing controls. These include, for NRDs, the ability to set pumping volumes, prescribe cropping rotations, limit new irrigated acreage, and build augmentation projects. NeDNR has fewer tools to use, though it can limit new appropriations or prohibit diversions when necessary to achieve IMP goals (e.g., compact compliance and meeting Endangered Species Act obligations). The majority of the controls are set locally within the NRDs.

While the initial push for integrated management was somewhat difficult, today there is a high level of interest. Now, the NRDs routinely use their authority to actively manage groundwater. Those authorities include the ability to allocate the amount of groundwater that can be withdrawn, adopt a rotation system of groundwater use, set well spacing requirements, set flow meters or other measurement devices, reduce the number of existing irrigated acres, limit or prevent expansion of irrigation, and place moratoria on new wells or uses (Neb. Rev. Stat. § 46-73; Peck, 2007). Individual NRDs retain the option to use other rules or regulations in order to implement groundwater management objectives (Neb. Rev. Stat. §46-703 in Schutz, 2015). Table 17.1 below offers examples of how NRDs have used their individual authority to revise rules. The use and combination of these measures varies and NRDs do not all use these authorities in the same way. The differences are most evident in comparisons of sanctioning measures like allocation amount adjustments, flow meter requirements, well moratoria, or required water use reporting.

17.4.3 Implementing controls: North Platte Natural Resources District

As an example of the authority granted under the 2004 NeDNR Order[9], the North Platte NRD used its 2009 IMP to declare the North Platte River valley and Pumpkin Creek areas over appropriated and the remainder of the NRD fully appropriated. This followed years of special attention to the area, particularly after the *Spear T Ranch* legal conflict mentioned above (Aiken, 2005). The resulting allocation controls were set by the Board of Directors for the over-appropriated areas (specifically Pumpkin Creek). This decision was spurred by a multiyear drought. In November 2015, the Board of Directors voted unanimously to reduce the irrigation allocation limit to 70 acre-inches of groundwater per acre for five-year periods (North Platte Natural Resources District, 2021), with a yearly average of 14 acre-inches (North Platte Natural Resources District, 2016). As part of these rules, the 'severely overappropriated'[10] Pumpkin Creek sub-area has a separate limit of 60 acre-inches with a base allocation of 12 acre-inches per irrigated acre. The area, furthermore, is closed to new well permitting (with the exception of replacements for existing wells, or wells for humans and livestock animals), and all irrigation well owners must use flow-meters[11]. The North Platte NRD case illustrates an instance where allocations have been reduced to match declining water availability.

[9]https://dnr.nebraska.gov/sites/dnr.nebraska.gov/files/doc/water-planning/upper-platte/orders/OverappropriatedOrder9-15-04.pdf

[10]https://ngwa.confex.com/ngwa/gw19/webprogramarchives/Paper12669.html

[11]For current information on the Pumpkin Creek Basin Groundwater Management Sub-Area, see the information from the North Platte NRD's Water Management website: https://www.npnrd.org/water-management/pumpkin-creek-basin.html

In addition to reducing allocations, NRDs encourage stakeholder interaction around the topic of decreased water availability, and simulated decision-making. Individual initiatives related to unique challenges, such as drought, are becoming more common. Because of this, some NRDs have held interactive scenario exercises to plan for droughts. The North Platte NRD held a drought tournament in 2016, in which various scenarios were pitched to stakeholders, who had to make decisions about water allocation and resource sharing in order to mitigate potentially devastating impacts. After the tournament, the NRD created a Community Drought Plan, which outlines goals and actions related to education and preparedness for future water shortages[12]

Table 17.1 shows how institutional changes allow for new sanctioning mechanisms. NRD Boards can place moratoria on well withdrawal in stressed basins, but do not have to rely solely on this type of sanction. Not only are these allocations subject to statutory change, they are sometimes re-negotiated on a voluntary basis by a mixture of mechanisms including but not limited to buy downs, short-term leases, and supplementing one source for another – for example diversion of excess surface water into irrigation canals, pits, and reservoirs to make up for reduced groundwater availability.

17.4.4 Advantages of the NRD model

NRDs involve public and private stakeholders in place-based collaborative decision-making. This local involvement builds trust (Sixt *et al.*, 2019). Muñoz-Arriola *et al.* (2021) find that the IMP process has been accepted by local groundwater users because it reduces uncertainties. Collaborative leadership initiatives were built into the institutional design of NRDs. Initially, the head of the State Soil Conservation Service and a state Senator had the vision to obtain additional funding for conservation districts, with the aim of improving water management (Fairchild, 1994). Today, state water leadership networks continue as conduits for scientific, social, and political knowledge about water resources (Burbach *et al.*, 2020).

With their professional expertise, leaders in the NRD Boards reserve the right to decide whether groundwater pumping is allowed in a particular area; a key mechanism to do this is certifying 'irrigated acres'. The North Platte NRD does this by linking allocations to tracts of land and their associated wells, for example. When irrigated acres are separated or sold, they must be re-certified by the district (see North Platte Natural Resources District, 2016) and approved by the Boards.

Integrated management has been incentivized through funding opportunities at the state level; subsequently, all NRDs have developed IMPs (even those not fully or over appropriated). In fact, after the 2017 evaluation of water supply and use, all NRDs in the state either have a required or voluntary plan. In its 2021 evaluation, the state was exempt from carrying out a separate evaluation because all 23 basins are undergoing integrated management planning. The state's position was that there is no need to re-evaluate any of the basin assessments since the 2017 annual report (State of Nebraska, 2021).

17.4.5 Limitations of the NRD model

Though Nebraska's polycentric water governance model is ambitious in its scope to integrate groundwater and surface water management, it is not flawless. Generally, it may suffer from an observed weakness of polycentric systems: when environmental problems exceed the boundaries of a governance system, it is not clear exactly where to turn for the needed guidance or rule-making authority (Morrison *et al.*, 2019). Specifically, Muñoz-Arriola *et al.* (2021) find multiple concerns: that surface water provider participation is limited, the IMP process is not always equitable, irrigators have limited influence especially decisions about the extent of controls on ground water use that impact surface water supplies, and conflicts are not effectively managed. Water quality may be a more appropriate area for NRDs to address: for example, Sixt *et al.* (2019) find that the collaborative

[12]https://www.npnrd.org/drought/climate/weather/drought/drought-planning/

arrangement works well for governing groundwater nitrate levels but say little about its ability to govern water quantity.

Furthermore, there is limited quantitative research available synthesizing the impact of the IMP process on actual ground and surface-water levels. This is either because what is available is anonymized for confidentiality or the information is presented as non-scientific, unqualified demonstration of success. For example, a general trend of net minimal groundwater depletion (roughly 2 m) from 2002 onward in wells within three selected NRDs is presumed to result from a combination of streamflow withdrawal reductions, drought reductions, and also IMP governance resulting in recharge (Muñoz-Arriola *et al.*, 2021). Other sources make the vague claim that, as a whole, groundwater levels throughout the state have benefitted from NRD management, but only when compared to other states above the High Plains (Ogallala) Aquifer. The Nebraska Association of Resources Districts claimed a groundwater increase of more than 85 ft (25.9 m) in areas; meanwhile Texas has seen drops of 234 ft (71.32 m) (Nebraska Association of Resources Districts, 2019).

17.5 CONCLUSIONS

In Nebraska, groundwater and surface water are managed in a somewhat coordinated polycentric model, which offers needed flexibility given the growing demand for irrigation coupled with cyclical changes in water availability. The NRDs' IMPs are a key component for basins in crisis (over appropriation), and the Republican and the Platte River basins offer examples of how Nebraska's water governance model holds up in fully or over appropriated basins. The localized model, together with state and federal planning and legal mechanisms, is designed to mitigate the impacts of diminished supply. It has the advantage of involving public and private stakeholders across levels of governance.

However, there are also shortcomings and limitations. The model of groundwater allocation, in particular, requires a great deal of stakeholder participation and the active oversight of a Board of Directors. When participation and oversight are lacking, lawsuits related to breach of allocations have ensued, pushing the state to revise its statutes, particularly to address the challenge of managing hydrologically connected waters. Furthermore, the disparate legal philosophies and mechanisms used in over-appropriated areas can cause problems: when sanctions are used on groundwater access, irrigators can be subject to the fluctuating availability of surface water. Droughts and annual changes in water availability can create conflicts in hydrologically connected areas. Integrated management planning is now beginning to address these challenges. If NRDs exercise their full potential in designing and implementing these plans, with stakeholder involvement, conflicts may be reduced while increasing the sustainability of water supply in Nebraska.

REFERENCES

Aiken D. (2005). Hydrologically connected ground water, Section 858, and the Spear T Ranch decision. *Nebraska Law Review*, **84**(2005), 962–996. https://digitalcommons.unl.edu/nlr/vol84/iss3/7 (accessed 21 February 2021).

Aiken D. (2008). The republican river: negotiation, arbitration, and a federal water master. *Cornhusker Economics*, **382**. https://digitalcommons.unl.edu/agecon_cornhusker/382 (accessed 21 February 2021).

Bleed A. and Babbitt C. H. (2015). Nebraska's Natural Resources Districts: An Assessment of A Large-Scale Locally Controlled Water Governance Framework. Lincoln, NE. https://digitalcommons.unl.edu/wffdocs/79/ (accessed 18 February 2021).

Burbach M. E., Joeckel R. M. and Matkin G. S. (2020). 2019 Nebraska water leaders academy final report. *Conservation and Survey Division*, **796**, i–48. https://digitalcommons.unl.edu/conservationsurvey/796 (accessed 7 February 2022).

Carlisle K. and Gruby R. L. (2017). Polycentric systems of governance: a theoretical model for the commons. *Policy Studies Journal*, **47**, 927–952. https://doi.org/10.1111/psj.12212

Dalstrom H. A. and Naugle R. C. (2020). 'Nebraska'. *Encyclopedia Britannica*, 3 Dec. 2020, https://www.britannica.com/place/Nebraska-state (accessed 5 March 2021).

Fairchild W. (1994). NRD Oral History Project Interview with Warren Fairchild at the Cornhusker Hotel. March 16, 1994. http://nrdstories.org/wp-content/uploads/Fairchild.pdf (accessed 13 October 2021).

Frankson R., Kunkel K., Stevens L. and Shulski M. (2017). Nebraska State Climate Summary. *NOAA Technical Report NESDIS 149-NE*, p. 4.

Hagenlocher M., Meza I., Anderson C. C., Min A., Renaud F. G., Walz Y., Siebert S. and Sebesvari Z. (2019). Drought vulnerability and risk assessments: state of the art, persistent gaps, and research agenda. *Environmental Research Letters*, **14**(8), 083002, 1–13. https://doi.org/10.1088/1748-9326/ab225d

Jenkins A. (1999). The platte river cooperative agreement: a basin-wide approach to endangered species issue. *Great Plains Research: A Journal of Natural and Social Sciences*, **423** , 95–113. Paper. http://digitalcommons.unl.edu/greatplainsresearch/423 (accessed 15 June 2021).

Johnson B., Thompson C., Giri A. and Van NewKirk S. (2011). Nebraska Irrigation Fact Sheet, Report No. 190. Department of Agricultural Economics. September 2011. Universtity of Nebraska – Lincoln. https://agecon.unl.edu/a9fcd902-4da9-4c3f-9e04-c8b56a9b22c7.pdf (accessed 18 April 2021).

Morrison T. H., Adger W. N., Brown K., Lemos M. C., Huitema D., Phelps J., Evans L., Cohen P., Song A. M., Turner R., Quinn T. and Hughes T. P. (2019). The black box of power in polycentric environmental governance. *Global Environmental Change*, **57**, 101934, 1–8. https://doi.org/10.1016/j.gloenvcha.2019.101934

Muñoz-Arriola F., Abdel-Monem T. and Amaranto A. (2021). Common pool resource management: assessing water resources planning for hydrologically connected surface and groundwater systems. *Hydrology*, **8**(51), 1–15. https://doi.org/10.3390/hydrology8010051

National Drought Mitigation Center. (2021). U.S. Drought Monitor: Time Series. Available online: https://droughtmonitor.unl.edu/DmData/TimeSeries.aspx (accessed 9 December 2021).

Nebraska Administrative Code. (2008). Department of Natural Resources, Rules for Surface Water. Available online: https://dnr.nebraska.gov/sites/dnr.nebraska.gov/files/doc/about/rules/62708Title457FullCurrent.pdf (accessed 2 November 2021).

Nebraska Association of Natural Resources Districts. (2019). 2018 NRD Water Management Activities Summary. Available online: https://www.nrdnet.org/sites/default/files/groundwater_management_summary_2018.pdf (accessed 14 October 2021).

Nebraska Association of Natural Resources Districts. (2021). 2020 NRD Water Management Activities Summary. Available online: https://www.nrdnet.org/sites/default/files/groundwater_management_summary_2020.pdf (accessed 17 April 2021).

Nebraska Department of Natural Resources. (2005). What's the Meaning of LB962's Fully Appropriated Basin Designation?. Lincoln, NE.

North Platte Natural Resources District. (2016). Rules and Regulations for the Enforcement of the Nebraska Ground Water Management and Protection Act. Available online: https://www.npnrd.org/assets/site/documents/rules-and-regulations/gwma-rules-and-regs-11-14-16.pdf (accessed 30 March 2021).

North Platte Natural Resources District. (2021). Water Management: Pumpkin Creek Basin. Available online: https://www.npnrd.org/water-management/pumpkin-creek-basin.html (accessed 15 October 2021).

Ostrom E. (2010). Polycentric systems for coping with collective action and global environmental change. *Global Environmental Change*, **20**(4), 550–557, https://doi.org/10.1016/j.gloenvcha.2010.07.004

Peck J. C. (2007). Groundwater management in the high plains aquifer in the USA: legal problems and innovations. In: Chapter 14 in Groundwater Management in the USA: Opportunities and Threats to Development, M. Giordano and K. G. Vilholth (eds.), CAB International, Wallingford, UK, pp. 296–319.

Peterson S. M., Traylor J. P. and Guira M. (2020). Groundwater availability of the Northern High Plains aquifer in Colorado, Kansas, Nebraska, South Dakota, and Wyoming: U.S. Geological Survey Professional Paper 1864, p. 57. https://doi.org/10.3133/pp1864

Reed C. and Abdel-Monem T. (2015). An Assessment of the Nebraska Integrated Management Planning Process. A report from the Unversity of Nebraska Public Policy Center funded by the Nebraska Department of Natural Resources.

Schutz A. B. (2015). Defining sustainability in nebraska's republican river basin: The LB 1057 task force. *Texas A&M Law Review*, **3**, 771, https://doi.org/10.37419/LR.V3.I4.2

Shulski M. (2018). Nebraska's changing climate – highlights from the 4th National Climate Assessment. University of Nebraska–Lincoln Institute of Agriculture and Natural Resources Crop Watch. Originally published December 6, 2018. Available at: https://cropwatch.unl.edu/2018/nebraska-changing-climate (accessed 22 February 2021).

Sixt G. N., Klerkx L., Aiken J. D. and Griffin T. S. (2019). Nebraska's natural resource district system: collaborative approaches to adaptive groundwater quality governance. *Water Alternatives*, **12**(2), 676–698.

State of Nebraska Department of Natural Resources. (2004). Order Designating Overappropriated River Bains, Subbasins, or reaches, and describing hydrologically connected geographic area. Signed by Director Roger K. Patterson. September 15, 2004. Available online: https://dnr.nebraska.gov/sites/dnr.nebraska.gov/files/doc/water-planning/upper-platte/orders/OverappropriatedOrder9-15-04.pdf (accessed 21 February 2021).

State of Nebraska Department of Natural Resources. (2012). Water Well Registration Instructions: DNR WWR form 5/2012. Available at: https://dnr.nebraska.gov/sites/dnr.nebraska.gov/files/doc/ground-water/contractors/WaterWellRegInstructions.pdf (accessed 22 February 2021).

State of Nebraska Department of Natural Resources (NeDNR). (2021). Annual Review of Availability of Hydrologically Connected Water Supplies. Published by the Nebraska Department of Natural Resources. December 29, 2020. Available at: https://dnr.nebraska.gov/sites/dnr.nebraska.gov/files/doc/water-planning/statewide/FAB/2021AnnualReport/20201229_FAB2021_Final.pdf (accessed 21 February 2021).

Stoutenborough J. W. and Vedlitz A. (2014). Public attitudes toward water management and drought in the United States. *Water Resources Management*, **28**(3), 697–714, https://doi.org/10.1007/s11269-013-0509-7

U.S. Department of Agriculture. (2019). 2018 Irrigation and Water Management Survey: 2017 Census of Agriculture; Volume 3, Special Studies, Part 1. AC-17-22-1. Issued November 2019. Retrieved from https://www.nass.usda.gov/Publications/AgCensus/2017/Online_Resources/Farm_and_Ranch_Irrigation_Survey/fris.pdf

U.S. Geological Survey. (2021a). Water Use Data for Nebraska (Refresh Date June 2018). https://waterdata.usgs.gov/ne/nwis/water_use?format=html_table&rdb_compression=file&wu_area=State+Total&wu_year=ALL&wu_category=ALL&wu_category_nms=--ALL%2BCategories-- (accessed 9 December 2021).

U.S. Geological Survey. (2021b). USGS: High Plains Aquifer Groundwater Levels Continue to Decline. Release date June 16, 2017. https://www.usgs.gov/news/usgs-high-plains-aquifer-groundwater-levels-continue-decline- :~:text=The High Plains aquifer, also,South Dakota, Texas and Wyoming (accessed 18 April 2021).

Ulrich L. (2018). Nebraska's irrigation history: it's complicated. Growing a Healthy Future **7**(2). Published by the University of Nebraska Institute of Agriculture and Natural Resources. Available online: https://ianr.unl.edu/img/magazine/IANR-growing-fall-2018.pdf (accessed 21 February 2021).

Western Governors' Association. (2018). Water Resource Management in the West. Western Governors' Association Policy Resolution 2018-08. Available online: https://westgov.org/images/files/WGA_PR_2018-08_Water_Resource_Management.pdf (accessed 15 June 2020).

doi: 10.2166/9781789062786_0227

Chapter 18

Transboundary water allocation in the Amudarya Basin of Central Asia

Dinara Ziganshina

Scientific Information Center of Interstate Commission for Water Coordination in Central Asia; Associate Professor, Tashkent Institute of Irrigation and Agricultural Mechanization Engineers, Tashkent, Uzbekistan

ABSTRACT

This chapter reviews the transboundary water allocation system in the Amudarya Basin shared by Afghanistan, Kyrgyzstan, Tajikistan, Turkmenistan, and Uzbekistan. The key principles and rules of water allocation in this basin have been set in the Soviet time and validated in 1992 after the countries gained their independence. The treaty-based flexible and specific water allocation formula (percentage of flow) and the operation of joint bodies helped to preserve the resilience of the system during transition. The system would benefit, however, from improved forecasting and an early warning system, annual and long-term planning, coordinated multi-year flow regulation, robust water conservation plans, sound strategies and procedures to deal with droughts and floods. Given the increasing water demand and diminishing water supply in the near future, a more integrated basin allocation planning is required.

Keywords: Amudarya, transboundary waters, Central Asia, water allocation

18.1 INTRODUCTION

The Amudarya, one of the largest rivers in Central Asia, crosses through five states – Afghanistan, Kyrgyzstan, Tajikistan, Turkmenistan, and Uzbekistan – and serves as the source of livelihood for about 27 million people (see Figure 18.1). The Amudarya system is under stress from increasing demand and competition for water for agricultural, industrial and municipal uses due to population growth, economic development and climate change. The Amudarya is among the basins where the water supply is almost fully allocated, and therefore trade-offs among competing interests are the key challenge. The future challenges for transboundary water allocation in the Amudarya relate to increased water demand due to population growth and socio-economic development; climate change implications on river runoff and water demands; increased water demand in Afghanistan (up to 7 km^3/year); and potential changes in the river regulation due to hydropower development. It is estimated that water deficits in the Amudarya may reach up to 9.6–10 km^3/year by 2045 (SIC ICWC, 2017). As such, water allocation arrangements have taken on increasing significance in ensuring that water is managed in a peaceful and sustainable way.

Figure 18.1 Map of the Amudarya basin. Source: Zoi Environmental Network, 2011.

This chapter reviews transboundary water allocation rules and practices in the Amudarya over 25 years from 1991 till 2015 to showcase key achievements and challenges. The first section introduces the overall legal and institutional framework for the water allocation in the basin. The second section details the principles, rules and procedures of the basin water allocation, including in high- and low-water years. The third section reviews water allocation practices in the basin, tracing compliance with rules and procedures. Subsequently, a discussion on strengths and weaknesses of the water allocation system in the basin is presented.

18.2 THE OVERALL LEGAL AND INSTITUTIONAL FRAMEWORK

In 1992, Kazakhstan, Kyrgyzstan, Tajikistan, Turkmenistan and Uzbekistan signed the Agreement on Cooperation in the Field of Joint Management of the Use and Conservation of Water Resources from Interstate Sources (1992 Almaty Agreement), which provides the legal and institutional framework for transboundary water management in the Amudarya and Syrdarya basins. The 1992 Almaty Agreement does not contain a water allocation formula as such, but recognizes that the earlier agreed rules and procedures on water use and allocation from interstate water sources remain in force. The 1992 Almaty Agreement has been supplemented by other regional agreements on joint actions in addressing the Aral Sea crisis (1993), the parallel operation of the energy systems in Central Asia (1999), the use of water and energy resources of the Syrdarya (1998), hyrdometeorological cooperation (1999), the institutional structure of the International Fund for saving the Aral Sea (1999), as well as bilateral arrangements between Uzbekistan and Turkmenistan (1996, 2017, 2021), Tajikistan and Afghanistan (2010, 2014, 2019, 2020), Tajikistan and Uzbekistan (2018) and Kyrgyzstan and Uzbekistan (2018) (CaWater-Info.net, 2021).

The institutional framework for water allocation in the basin was also initiated during Soviet times. On August 27, 1987, the USSR Ministry of Land Reclamation and Water Management established an Amudarya Basin authority for inter-republican allocation of water and operation of water intakes and hydroschemes (named Uprvodkhoz 'Amudarya' and later known as Basin Water Organization or BWO 'Amudarya') (USSR Ministry Decision No 301). The 1992 Almaty Agreement placed BWO 'Amudarya' under the newly established Interstate Commission for Water Coordination in Central Asia (ICWC), made up of heads of national water authorities of five Central Asian countries[1]. The ICWC is mandated to ensure that shared water of the Amudarya and Syrdarya is allocated and used by the riparian countries in a coordinated way. To this end, the ICWC elaborates and approves annual water use limits for each country and reservoirs operation regimes. Moreover, it determines key directions of regional water policy. The Commission takes its decisions on quarterly meetings by consensus. Its decisions on water allocation quotas, rational water use and protection are mandatory for all water consumers and users (1992 Almaty Agreement). The Commission has five executive bodies: BWO 'Amudarya', BWO 'Syrdarya', Scientific Information Centre (SIC ICWC), the Secretariat, and Coordination Metrological Centre.

Acting as an implementation entity, BWO 'Amudarya' is responsible for daily water management and regulation, timely and reliable water allocation among the riparian states according to the agreed limits and provision of sanitary and environmental flow for the Prearalie (former Aral Sea coastal zone) and the Aral Sea. The central office of BWO 'Amudarya' is based in Urgench, Uzbekistan and it has four territorial divisions in Tajikistan (Kurgan-Tyube), Turkmenistan (Turkmenabad) and Uzbekistan (Urgench, Takhiatash).

18.3 WATER ALLOCATION ARRANGEMENTS

18.3.1 Water allocation rules and principles

The water allocation arrangement for the Amudarya was set up in Protocol 566 of the meeting of the Scientific-Technological Council at the USSR Ministry of Land Reclamation and Water Management of September 10, 1987 (hereinafter Protocol 566).

The geographical scope of transboundary water regulation in the Amudarya Basin includes the Amudarya and its major tributaries such as the Vakhsh, the Pyanj, the Kafirnigan (so called 'small Amudarya basin'). Protocol 566 did not allocate water for upstream Afghanistan, but its estimated water withdrawal of 2.10 km^3/year was deducted from the water resources available for allocation for other riparians. Available water resources of the Amudarya are estimated at 74.07 km^3/year, of which 62.1 km^3/year are regulated. The estimation of available water resources excludes diversion of water by Afghanistan (2.10 km^3/year), losses from rivers and reservoirs (3.48 km^3/year) and sanitary releases along the river (3.15 km^3/year).

In terms of hydrological scope, the existing arrangement deals only with surface water allocation, although the Soviet Scheme also estimated exploitable groundwater resources (12.2 km^3) and the

[1]As a consequence of its frustrations with the lack of reform process and perceived neglect of its interests in hydropower development, Kyrgyzstan officially froze its participation in the Interstate Fund for Saving the Aral Sea (IFAS) and its commissions in May 2016.

permissible – without prejudice to surface runoff – groundwater withdrawal (6.6 km^3). It would be desirable to ensure that conjunctive water management and a holistic basin approach is practiced.

The water allocation arrangement in the Amudarya is based on the established (i) limits of water withdrawals for the four riparian countries – Kyrgyzstan, Tajikistan, Turkmenistan, and Uzbekistan, (ii) operation regime of reservoirs, and (iii) water delivery for the Aral Sea and its coastal area (Prearalie).

The water withdrawals limits were set as a percentage of the available runoff: Kyrgyzstan (0.6%), Tajikistan (15.4%), Turkmenistan (35.8%) and Uzbekistan (48.2%). This proportion was determined on the basis of the riparian countries' historical and existing water use, irrigated lands in use, and estimated unit water use (Protocol 566). Given that in arid zones, irrigated agriculture is the most demanding water user, the area of irrigated lands has been taken as the main guiding factor in setting the volume of water withdrawals (1984 Revised Scheme). The future expansion of irrigation was allowed only within the established water limits by reducing specific water use (unit water use) through rationalization, i.e. modernizing irrigation systems and drainage, return water re-use within irrigation schemes, and improving irrigation technique and technology (Protocol 566, para.8).

The operation regimes of the reservoirs have also been driven by the irrigation needs. 'The Revised Master Plan for Comprehensive Use and Conservation of Water Resources in the Amudarya River' approved by Protocol 566 established the irrigation regime of reservoir operation that aligned with the natural hydrological cycle of the Amudarya.

In addition to countries' water limits and the operation regime of reservoirs, Protocol 566 allocated 3.15 km^3 of water annually to provide sanitary flow along the Amudarya with a minimum 100 m^3/s discharge from the Takhiatash waterworks at the lower reaches[2]. It did not differentiate quantities of sanitary flow depending on flow probability, but the 1992 Almaty Agreement requires setting amounts of sanitary flow for every given year, taking into account the water content in interstate sources (Article 4). To maintain the ecologically sustainable status of the Amudarya delta and recharges into the wetland systems on an area of 180 000 ha, an SIC ICWC-led assessment recommended providing 8 km^3 in high-water years, 4.6 km^3 in average-water years and at least 3.1 km^3 in low-water years (Dukhovny & de Schutter, 2003).

In 1993, five Central Asian countries agreed to work towards ensuring water inflows to the Aral Sea required for sustaining its lowered but stable, ecologically acceptable levels and by these means preserving the Sea as an object of nature (1993 Kzyl-Orda Agreement). The ICWC assigned its BWOs to consider the Aral Sea and Prearalie as an *independent water user* and to provide it with water according to the established limits, and called upon all water users to comply with the water withdrawal limits and the established quantities of water to be supplied to the river deltas and the Aral Sea (ICWC, 1993).

18.3.2 Water allocation procedures

Since 1992, the ICWC has made decisions on water withdrawal limits, reservoir operation regimes and water releases for river deltas and the Aral Sea for a hydrological year, separately for growing and non-growing seasons. BWO submits water withdrawal limits and reservoir operation regimes for the approval of the ICWC, based on the riparian countries' demands, the forecasts on flow probability and the actual water-related conditions. After their approval by the ICWC, BWO allocates water among the users and can adjust the agreed limits within ±10%, if required to by changed water availability and water-related conditions. Transboundary water allocation is monitored by national agencies and BWO (for main canals under their jurisdiction). SIC ICWC monitors the water management situation and provides analytical updates to the ICWC members on a regular basis.

---

[2]Sanitary flow is a minimum outflow to meet water quality standards and enable favorable conditions for water uses in the downstream of a reservoir. 'Sanitary rules for the design, construction and operation of reservoirs' (approved by the Chief State Sanitary Officer of the USSR on July 1, 1985 No. 3907–85).

At a national level, water withdrawal limits (water allocation) follow the same pathway from ICWC/BWO via the authorized water ministry, basin organizations, and water users associations to end water users. In Uzbekistan, for example, irrigated agriculture and other sectors issue seasonal requests for water that are collated by the Ministry of Water Management and submitted to the ICWC via BWO. After the ICWC approval, water allocation limits (demands adjusted by actual water availability) go back down to the system through BWO, the Ministry of Water Management and its Basin Irrigation System Authorities.

18.4 ADJUSTING WATER ALLOCATION UNDER DIFFERENT HYDROLOGICAL CONDITIONS

The ICWC can adjust the water withdrawal limits and reservoir operation regimes depending on actual water availability and the current water-related conditions (Article 8 of the 1992 Almaty Agreement). Protocol 566 set basic parameters for adjustments under different flow conditions. It provides that (Protocol 566, para. 7):

> *If actual water availability in the Amudarya river exceeds the estimated one, excess water, i.e. above the established quantities of water withdrawal, must be primarily accumulated by reservoirs, and if flow is very high, a portion of excess water must be delivered to the lower reaches for improvement of sanitary and epidemiological conditions in the area of the Prearalie. If water availability is below the estimated one, water withdrawals of the Republics are to be cut proportionally.*

As mentioned earlier, this provision was supplemented in 1993 by the requirements to secure water for the Aral Sea and the delta. For extreme low-water years, the 1992 Almaty Agreement requires 'a specific decision to deliver water to the areas suffering from severe water scarcity' (Article 4). The ICWC does not have detailed strategies or regulations for water allocation under extreme hydrological conditions; it is guided only by these two rather general provisions.

BWO implements the ICWC-established water withdrawal limits, schedules of reservoir operation, and water supply to the Prearalie. If actual water availability differs from the forecast, the water-related conditions change or if extreme situations occur, BWO adjusts planned values and submits the adjusted (or corrected) limits of water withdrawal and operation regimes of reservoirs to the ICWC members for approval. BWO was permitted to correct the limits of water withdrawals and operation regimes of reservoirs within ±10% of the planned values without prior approval from ICWC (ICWC, 1997).

As a rule, the correction of limits and regimes of water withdrawals is made on a seasonal basis, distinguishing between non-growing and growing seasons. In each case, however, all adjustments should be made by BWO within annual limits. In some years, the Commission allowed the countries to transfer the 'saved' part of their limits to another season (EC IFAS & ICWC, 1994; ICWC, 1998a) or to compensate excessive use in one season during another season (ICWC, 1995, p. 24), but strictly within the total established limits for a hydrological year. These in-seasonal transfers are referred to as 'agreed irregularity' in BWO's reporting.

18.5 MODIFICATION OR REVISION OF THE CURRENT WATER ALLOCATION SYSTEM

The current water allocation structure and principles can be modified only by a collective decision of all parties (Article 14 of the 1992 Almaty Agreement). No specific procedure to initiate such a decision is envisaged.

18.5.1 Water allocation practices over 1991–2015

This section describes the actual water allocation practices in the basin from 1991 until 2015, focusing on the compliance with water limits, flow regulation and water supply to the delta and the Prearalie.

18.5.1.1 Compliance with country water withdrawal limits[3]

Overall compliance with annual water withdrawals limits is generally good for all countries. All riparian countries used less water than entitled according to the country limits approved by the ICWC. Since 1992, on average, Tajikistan has not been using 1.8 km^3 of its annual water withdrawal limit, with wide variations of the non-used quantities of water both during growing and non-growing seasons. In Turkmenistan and Uzbekistan, the average factual water withdrawals have also been 1 km^3 less than envisaged in the annually approved limits. Turkmenistan used over 4 km^3 less water in the dry years of 1999/2000, 2000/01, 2007/08 and 2008/09, and almost 5 km^3 less in 2010/11. Similarly, the water received by Uzbekistan fell below the limit by more than 4 km^3 in the dry years of 1999/2000 and 2000/01, and by more than 6 km^3 in 2007/08 and 2010/11. Since the 2008 growing season, Turkmenistan and Uzbekistan have, on average, been using 2.3 km^3 less water than the allocated limit. Less water use compared to approved limits does not, however, indicate the countries' conscious commitments for water conservation. Rather, it illustrates that limits have been set based on unreliable flow forecasts and that the requested amount of water was unavailable for use.

Less frequently, the riparian countries used more water than approved in the limits. Tajikistan slightly exceeded its limit only during the non-growing season of 1996/97 and the growing season of 2006/07. Turkmenistan overused its limit by more than 1 km^3 in 1995/96 and 2001/02 due to the increased water availability, whereas the limits were initially established on the basis of low flow forecast (78% probability) and, therefore, were cut by 15%. Similarly, Uzbekistan exceeded its limits by more than 2 km^3 in 1995/96, 1997/98 and 1998/99 and primarily during the non-growing season due to excessive actual flow as compared to the forecast.

Even if not used, the countries' water withdrawal limits are secured and fixed. For example, Tajikistan did not use its water limit fully over the last three decades; nevertheless, its limit remains unchanged. The existing agreements have no provisions for a possibility to transfer the non-used share of the allocated limit to another water user-state or to accumulate (or reserve) it for its own future use. The ICWC has addressed this issue partially in its Minutes by mentioning the delivery of non-used water to the Aral Sea[4].

Final documents on the assessment of the proposed Rogun Hydropower Project mention Tajikistan's plans on the future disposal of their water withdrawal limits that were not used in the past. In particular, it says (World Bank, 2014):

> *Tajikistan has not used its full water allocation as set in ICWC annual allocation decisions but has indicated that it intends to do so in the future and this would reduce downstream flows compared to the current pattern. If a dam were built at the Rogun site, Tajikistan has indicated that it would use the difference between its allocation and its actual use to date (averaging 1.2 bcm annually for 2005–11) to fill the resulting reservoir and subsequently for irrigation.*

The existing agreements do not provide for such a possibility, warning the parties of the actions that may 'lead to changes in the agreed water flow' (Article 3, Almaty Agreement). Therefore, the matter may require additional negotiations between Tajikistan and other basin countries to minimize the possible adverse effect.

18.5.1.2 Reservoir operation regimes

Since 1991, the design regime of reservoir operation and flow regulations established in the Soviet regulations has been changed from the irrigation-hydropower mode to the hydropower generation

---

[3] Due to the Kyrgyz Republic's very low share of water withdrawals from the Amudarya, information regarding its use has not been regularly monitored by ICWC.

[4] ICWC instructed BWO 'Amudarya' to prevent the over-utilization of country water limits, and to deliver the non-used volumes of the country limits to the Aral Sea (ICWC, 2010a).

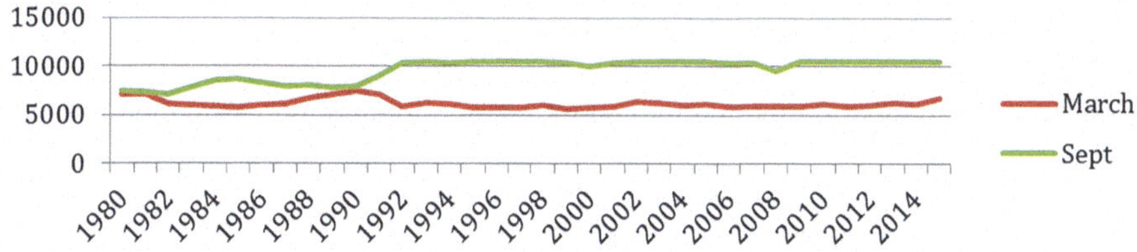

Figure 18.2 Water volume in the Nurek Reservoir by the beginning and end of the growing season, 1980–2015.

mode. This shift led to the situation where water is discharged by the beginning of the growing season and accumulated in summer. For example, the Nurek Reservoir – the largest reservoir located in the Vakhsh – accumulated an average of 7.9 km^3 of water by September (the last month of the growing season) during 1980–1990 and more than 10 km^3 by September during 1992–2015 (Figure 18.2). The

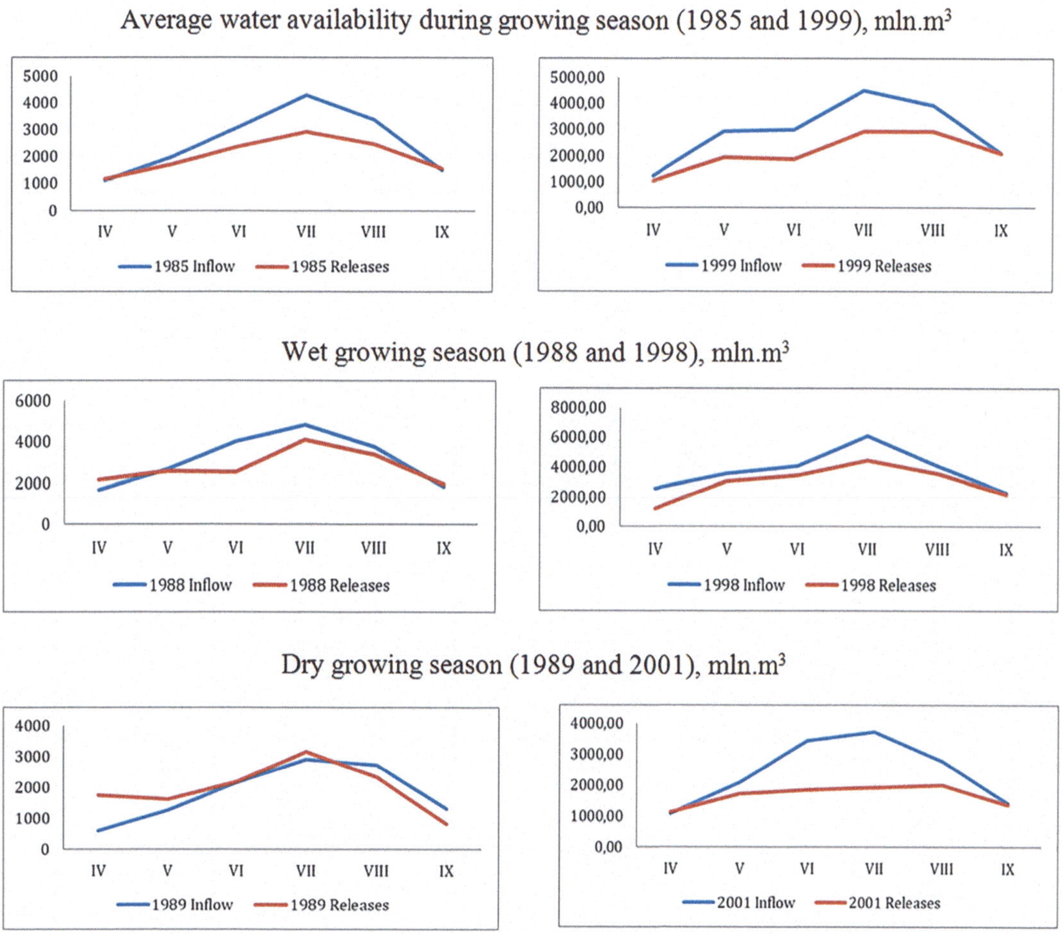

Figure 18.3 Comparison of inflow to and releases from the Nurek Reservoir during the growing season, before and after 1991.

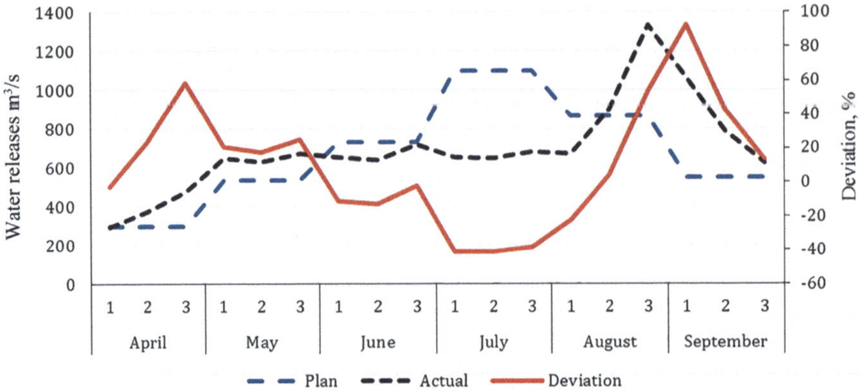

Figure 18.4 Water releases from Nurek over growing season 1997: Plan vs Actual.

water accumulation in the upstream reservoir during summer months worsened the situation in the lower reaches of the Amudarya Basin and jeopardized irrigated agriculture in all countries, including upstream. Comparison of inflows to and releases from the Nurek Reservoir in the growing season under various flow conditions before and after 1991 shows that changes in its operation contributed to water scarcity during low-water years (Figure 18.3). Compared to 1980–1990, average winter water releases have increased by 2 km^3 during 1991–2015. Summer releases remain the same, while inflow to the reservoir has increased by 2.1 km^3.

To accommodate changed requirements of the upstream countries, the ICWC has been approving the operating regimes of the reservoirs, taking into account the requirements of hydropower generation. Still, actual water releases deviated from seasonally agreed water delivery schedules. The actual water releases from the Nurek Reservoir during the growing season corresponded to the agreed schedule only in 2008; in all other years, there were deviations in both directions. For example, in the growing season of 2007, the deviation was an increase of 56.2%, and in the growing season of 2013 the deviation was a decrease of 11.1%. Especially significant fluctuations were observed in ten-day periods. For example, the general deviation during the 1997 growing season was only 1.8%, while releases varied from plus 56% to minus 20% over ten-day periods (Figure 18.4).

18.6 WATER DELIVERY TO THE ARAL SEA AND PREARALIE

The water allocation practice shows that ecosystem needs receive low priority, even though the ICWC repeatedly called upon the countries to assist BWO 'Amudarya' in supplying water to the delta in a timely fashion and in agreed volumes (ICWC, 2010a).

Although over 25 years the Amudarya delta received an average of 8 km^3 of water, which is aligned with the science-based recommendations for wet years, water delivery was highly unreliable and unsustainable throughout the years, seasons, and months, ranging from 29.1 km^3 in 1991/92 to 0.536 km^3 in 2000/01 (Figure 18.5). The delta did not receive a minimum flow of 3.1 km^3 in the dry period from 2006–2009 and had to absorb 20 km^3 excess water in 2009/10.

Protocol 566's requirement on providing sanitary flow along the Amudarya, including a minimum 100 m^3/s discharge from the Takhiatash waterworks at the lower reaches, has not been observed. Figure 18.6 demonstrates high variability in providing sanitary flow across seasons and months.

Water supply to the Amudarya lower reaches during the non-growing season is contained sanitary-environmental flow to irrigation systems, totaling 800 Mm^3, including 150 Mm^3 for the Dashoguz province in Turkmenistan and the Khorezm province in Uzbekistan, as well as 500 Mm^3 for

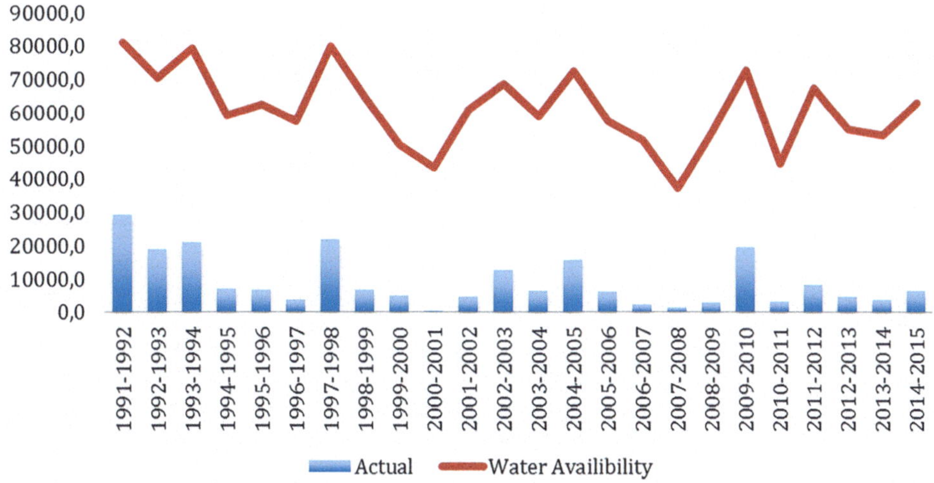

Figure 18.5 Actual water supply to the Amudarya delta, Mm^3. Samanbai station, discharge from Suenli and Kyzketken canals and collecting drains.

Karakalpakstan. Figure 18.7 shows that this flow is also not provided fully, especially in dry years (e.g., 2000/01 and 2008/09).

18.6.1 Water allocation in high-water years

Water allocation in high-water years has been complicated by a number of factors. *First*, the abundance of water in the first years of the ICWC operation (1992–1994) has been exacerbated by increased

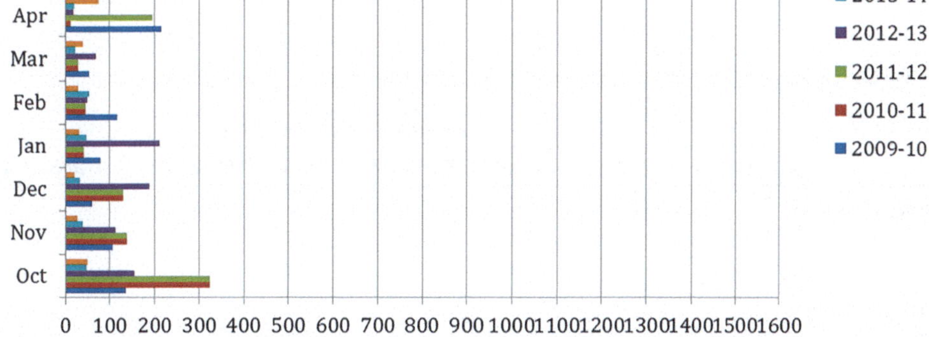

Figure 18.6 Water releases from the Takhiatash waterworks facility, 2009–2015, m^3/s.

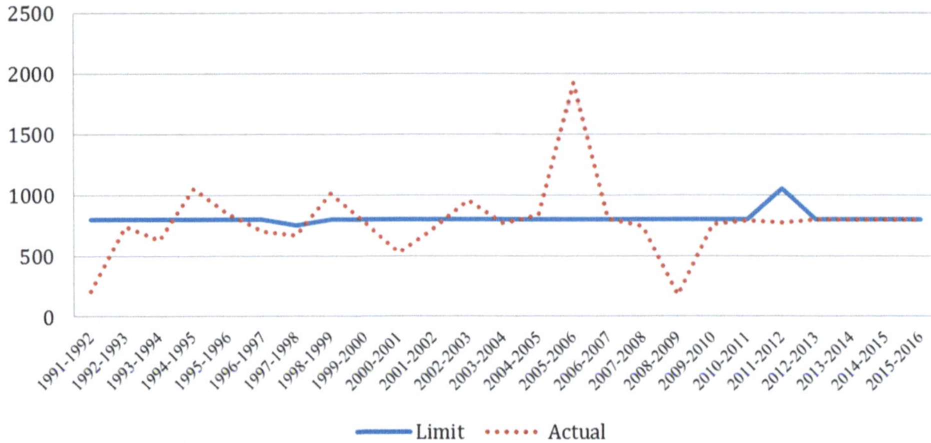

Figure 18.7 Water supply to Amudarya lower reaches as sanitary-environmental flow to irrigation systems, Mm^3.

winter releases from the upstream reservoirs. As a result, BWO 'Amudarya' had been working hard to minimize damage to the riparian countries while dealing with flooding and winter flow routing along the river channel. *Second*, hydromet services failed to provide reliable and timely flow forecasts to be better prepared for the extreme conditions (see e.g., ICWC, 1998b, 2010b). *Third*, in some years, the river channel, after a relatively dry preceding period, was not ready for longer and excessive discharge that caused difficulties in flood routing in the middle and lower reaches. *Fourth*, the reservoir operation regime could not ensure the stability and uniformity of water releases (even excessive compared to the plan) in time. For example, in 1998, despite earlier accumulation compared to the plan, water releases from the Nurek Reservoir in some ten-day periods of May, July, and August exceeded 1500 m^3/s, while in the second ten-day period of July water releases amounted to 1821.1 m^3/s.

Despite these challenges, the ICWC and national water management organizations have consolidated their actions to mitigate high water conditions. Excess flow has been diverted to irrigation systems in Uzbekistan and Turkmenistan to prevent extreme situations in the lower reaches of the Amudarya (See e.g., ICWC, 1998a). Over the course of 25 years, these so-called emergency-environmental releases amounted to 12 km^3, with a maximum of 4.1 km^3 discharged in 2009/10. In some years, such as 1999/2000, excess flow in the Amudarya main channel was transferred to the Zaravshan Basin via the Amu-Bukhara pumping canal (ICWC, 2000a). Finally, as prescribed in Protocol 566, excess water was delivered to the river delta and the Aral Sea.

Table 18.1 Water withdrawals along the river reaches in high-water years (% used from approved limits).

Seasons	Amudarya river reaches			
	Upper (TJ & UZB)	Middle (TM & UZB)	Lower (TM & UZB)	Aral Sea & Delta
Growing seasons				
1998	82	102	103	673
2010	78	90	93	911
Non-growing seasons				
1997/98	93.8	105.3	85.6	
1999/00	69.7	106.8	112.4	

Table 18.2 Water withdrawals along the river reaches in low-water years (% used from approved limits).

Seasons	Amudarya river reaches			
	Upper (TJ & UZB)	Middle (TM & UZB)	Lower (TM & UZB)	Aral Sea & Delta
Growing seasons				
2000	84.2	82.8	48.4	20.5
2001	97.3	91.8	49.5	4.9
2008	92	90.7	45.5	20.8
Non-growing seasons				
2001/02	90.9	116.2	53.6	18.4
2008/09	72.4	81.3	68.3	6.9

Under such extreme water conditions, water management organizations could not provide for uniform and stable water allocation along river reaches (see Table 18.1). As was mentioned above, middle and lower reaches received more water in order to deregulate excess water into irrigation systems or another basin.

Wet years have shown considerable flow losses in the Amudarya and the reservoirs. To address this problem, it was suggested to introduce automation monitoring systems and improve water accounting. Despite Protocol 566's prescription, the riparian countries did not take advantage of high-water years to accumulate excess water in the reservoirs for multi-year flow regulation. As a result, the subsequent low-water years brought great difficulties.

18.6.2 Water allocation in low-water years

The transboundary water management system in the Amudarya has been also tested by low water conditions. In most cases, the water system has not been adequately prepared for water scarcity conditions due to the lack of timely and accurate flow forecasts (ICWC, 2001b) and the absence of multi-year flow regulation, which would allow for more water to be stored in the upper reservoirs (see e.g., ICWC, 2000b, 2001b). In some years, the flow regulation by reservoirs contributed to the aggravation of natural low water conditions.

In dry times, regional and national water management organizations show much higher water use discipline and coordination. Control over river intakes has been strengthened, special joint monitoring groups established, numerous technical meetings at the lower reaches held and water overuse penalized.

Despite all these measures, water users in different river reaches have been shortened differently (Table 18.2). The lower reaches have suffered the most, receiving 45.4% of approved water limits in 2000, 59.4% in 2001, and 47.6% in 2008 in the Dashoguz province of Turkmenistan; 63.8% in 2000, 53.4% in 2001, and 52.8% in 2008 in the Khorezm province of Uzbekistan; and 42.7% in 2000, 43.5% in 2001, and 40.3% in 2008 in the Republic of Karakalpakstan (ICWC, 2000c, 2001a, 2008). In some years, water users in lower reaches and the delta received nearly zero deliveries. For example, in 2001, a critical situation was recorded in the farms in Northern Karakalpakstan, where water availability amounted to only 27.3% (ICWC, 2001a), and in the delta, which was left with almost no water.

BWO also faced difficulties with water allocation in the two driest non-growing seasons – 2001/02 (water availability was 61% of the norm) and 2008/09 (53% of the norm) – when non-uniformity of water allocation had an upward trend further downstream.

To better deal with low-water years in the future, the ICWC took decisions to increase control over water intakes, introduce automation of head structures, estimate channel losses in the middle and lower reaches, and improve accuracy of water measurements at Hydromet's gauging stations and at head intakes (ICWC, 2001a).

18.7 WATER ALLOCATION IN THE LOWER REACHES BETWEEN TURKMENISTAN AND UZBEKISTAN

The most serious problems in water availability and allocation occur in the lower reaches of the Amudarya. As early as in 1995, Turkmenistan and Uzbekistan – two downstream countries – established a bilateral technical group with participation of BWO 'Amudarya' to coordinate their actions on water management (ICWC, 1995). The detailed procedure of its operation was agreed at a 2007 bilateral interagency agreement signed by the heads of the national water authorities. It was decided that water allocation in the Amudarya lower reaches, including the operation regime of the Tuyamuyun Reservoir, will be set every 15 days by a technical group meeting, documented in the form of protocol and approved by the water agencies of Turkmenistan and Uzbekistan. Such water allocation does not go beyond the water limits established by ICWC and suits both countries in the lower reaches. In May 2021, a new agreement established a Joint Uzbek-Turkmen intergovernmental commission on water management issues in order to coordinate actions on the rational use of water resources of transboundary rivers, to ensure the functioning of water facilities located in the border territories of the states of the Parties as well as other issues in the field of water management.

18.8 DISCUSSION: STRENGTHS AND WEAKNESSES

The water allocation system in the Amudarya Basin has its strengths and weaknesses in terms of resilience, equity, efficiency, and environmental considerations.

The 1992, the Almaty Agreement was signed to provide for the continuity and stability of transboundary water management in times of dramatic political, social and economic transformations and to enable a new cooperative platform for the riparian countries to coordinate their future actions. The regime continuity and stability was secured by preserving the Soviet scheme of water allocation, which the countries agreed 'to strictly observe' (Article 2) and 'avoid actions within their respective territories that may affect and cause damage to other Parties, lead to changes in the agreed water flow and to pollution of water sources' (Article 3). But due to changed priorities and needs of the upstream countries under new economic realities, these strict provisions have not always been observed.

The resilience of the system is supported by a treaty-based flexible and specific water allocation formula (percentage of flow) and by the operation of joint bodies mandated to deal with water allocation, taking into account actual water availability and water-related conditions. Despite numerous political, legal, technical, financial and administrative barriers, the ICWC and its executive bodies have been allocating water among the riparian countries for almost 30 years and ensured that the system, which was set up during Soviet times, peacefully adapted to the new realities to the extent possible. Joint bodies helped to increase the resilience to change, serving as a basis for dialogue and continuous interactions.

However, the fixed priority of irrigation, which is taken as the basis for water allocation, and the absence of specific provisions for modifications and revisions of the water allocation system in the basin hindered resilience and raised equity concerns. The water allocation in the Amudarya was based on estimates of irrigation water requirements, taking into account crop water needs and areas under irrigation in the riparian countries. The upstream countries interested in hydropower production have questioned the equity of the water allocation formula that constrains their economic opportunities and future development. The subsequent uncoordinated alteration in the operation regime of reservoirs allowed generation of more hydropower upstream, but also reduced water use efficiency throughout the system. In the absence of formal reallocation mechanisms, the ICWC was forced to accommodate changing requirements of upstream countries.

The proportional division anticipated in the existing regulations serves to guarantee water supplies for each country under any hydrological condition and to ensure equity in sharing any shortfalls. This robust variability approach was not always followed in practice. In particular, during dry years, a higher proportion was abstracted in the upper sections of the river, while the lower reaches, the delta and wetlands received a greater percentage of the total runoff during wet periods only. BWO needs a clearer guidance on countries' priorities in shortening limits and a stronger mandate to ensure proportionality in water allocation.

Environmental considerations receive low priority in the actual water allocation in the basin. The countries agreed to allocate water to the Aral Sea and the delta as an independent water user to sustain its ecological integrity, but water has not been delivered in a sustainable and regular manner in practice. This was the case both due to the general failures of water management as well as engineering mistakes in building water infrastructure in the delta area. Thus, the regulating capacity of the Mezhdurechenskoye Reservoir – which could store excess winter releases for the delta use – was reduced down to 250 Mm^3 from its initial 600 Mm^3 (SIC ICWC, 2015).

The fact that the countries' water limits are fixed and are not canceled or reduced in the case of non-use provides the riparian countries with certain security for the future water supplies, allowing investment in water infrastructure and water use efficiency. The system would benefit from more certainty in annual water use, however. For example, Tajikistan has not been using its annual water shares for several years, but no formal decision on temporary suspension and allocation of its share to another user has been made; neither was its intention to use the entire unused amount of water in the future formalized, leaving the question open for the future.

The ICWC practices exhibit high ad hoc adaptability to changing conditions but lack a long-term coordinated strategy to deal with variability and changes. Water allocation practices under extreme hydrological conditions in the basin have been responsive, rather than preventive, reducing the resilience of the system to climate-induced changes. To mitigate high- and low-water years, the riparian countries and regional organizations have been employing a range of response measures, such as on-the-spot water adjustments, regular technical group meetings, joint monitoring and measurements. However, preparedness to these events has been inadequate due to poor flow forecasts and lack of long-term and annual basin planning. In high-water years, the requirement to accumulate excess water in reservoirs has not been observed due to the absence of multi-year flow regulation and interest in gaining short-term benefits from hydropower production. More precise rules on operational water management in extreme hydrological conditions, with sanctions for non-compliance, would contribute to a more efficient water allocation system. Two general provisions on water allocation in extreme hydrological conditions do not address the full range of problems encountered by water management organizations on the ground. Treaty provisions on coordinated and agreed actions and rational water use and management also do not provide detailed guidelines for action that are needed in high- and low-water situations, which will be more frequent in the future.

Cooperation over water allocation in the basin is focused on short-term solutions and lacks a basin-wide long-term perspective. The ICWC pays more attention to daily operational water management, leaving aside matters related to future development. A call for basin-wide reduction in water use initiated by the ICWC and the IFAS[5] has not been put in action. The 2014 ICWC Action Plan that sought to promote specific regional activities on water saving, integrated water resources management, water accounting, and capacity development has not been fully implemented due to lack of funding and support from the riparian countries and development partners.

Despite the difficulties in implementing treaty obligations, non-compliance procedures have not been developed. A 1992 treaty requirement to elaborate 'a mechanism of economic responsibility and other measures in cases of violation of agreed water use regime and limits' (Article 12 of the 1992 Almaty Agreement) has not been implemented until now.

[5]The IFAS Board by its decision of 12 March 1998 has set the necessity to reduce the annual water withdrawal limits by 1.0–1.5% in all the countries (ICWC, 1999).

18.9 CONCLUSION

The region's arid climate and changing economy, society and ecosystems require transboundary water allocation and management to constantly adapt, as it has for over 30 years. Such adaptation, involving a mix of past practices and present approaches, was not perfect, signaling the need to better prepare for the future. The key areas for improvements include developing joint vision and strategic planning, enhancing legal frameworks and institutions toward a whole-basin approach that would encompass different water users (agriculture, hydropower, drinking supply and ecosystems) and missing basin countries (Kyrgyzstan and Afghanistan), strengthening data, information and capacity, promoting evidence-based decision-making, harvesting the possibilities offered by infrastructure, technology and innovation, enabling multi-sectoral and participatory governance arrangements, paying more prominent attention to water quality and environmental degradation and recognizing multiple facets and values of water (Ibatullin & Ziganshina, 2019). These uneasy tasks can be handled by the riparian countries jointly.

ACKNOWLEDGEMENT

The research leading to this paper has received funding from the Partnerships for Enhanced Engagement in Research (PEER) financed by the United States Agency for International development (USAID) and administered by the U.S. National Academies of Sciences, Engineering, and Medicine (NASEM). The research project on Transboundary Water Management Adaptation in the Amudarya Basin to Climate Change Uncertainties has been implemented by the Scientific Information Center of Interstate Commission for Water Coordination under PEER Cycle 4.

REFERENCES

Agreement between the Republic of Kazakhstan, the Kyrgyz Republic, the Republic of Tajikistan, Turkmenistan, and the Republic of Uzbekistan on Cooperation in the Field of Joint Management of the Use and Conservation of Water Resources of Interstate Sources, Almaty (signed 18 February 1992).

Agreement between the Republic of Kazakhstan, the Kyrgyz Republic, the Republic of Tajikistan, Turkmenistan, and the Republic of Uzbekistan on Joint Actions for Addressing the Problems of the Aral Sea and its Coastal Area, Improving the Environment, and Ensuring the Social and Economic Development of the Aral Sea Region, Kzyl-Orda (signed 26 March 1993).

CaWater-Info.net. Binding and Soft Law Documents in Central Asia. Online: http://cawater-info.net/library/index_e.htm (accessed 14 February 2022).

Decision No 301 of 27 August 1987 of the USSR Ministry for Land Reclamation and Water Resources on the establishment of an Amudarya Basin authority for inter-republican water allocation.

Dukhovny V. and de Schutter J. (2003). South Prearalie – New Perspectives. Tashkent. 2003. Results of the project 'Integrated Water Resources Management for Wetlands Restoration in South Prearalie'. SFP 974357 grant of NATO Science for Peace Program.

EC IFAS & ICWC. (1994). Minutes of Joint Meeting of the Executive Council of the Interstate Committee for Aral Sea Basin and ICWC, 8 June 1994, Khudjand, Tajikistan.

Ibatullin S. and Ziganshina D. (2019). The future of water resources. In: The Aral Sea Basin: Water for Sustainable Development in Central Asia, S. Xenarios, D. Schmidt-Vogt, M. Qadir, B. Janusz-Pawletta and I. Abdullaev (eds.), 1st edn, Routledge, London.

ICWC. (1993). Minutes of the 5th Meeting, 8–9 July, Kzyl-Orda, Kazakhstan.

ICWC. (1995). Results of water use in the Amudarya basin during non-growing season 1994–1995. Materials of the 10th Meeting, 16 February, Shimkent, Kazakhstan. Bulletin No 7, May.

ICWC. (1997). Minutes of the 16th meeting, 22 April, Dushanbe, Tajikistan.

ICWC. (1998a). Results of the non-growing season 1997–1998 and adjustment of operation regimes of the reservoir cascades and water withdrawal limits in the Amudarya basin for the growing season 1998. Materials of the 19th meeting, 15 May, Shimkent, Kazakhstan.

ICWC. (1998b). Materials of the 21st meeting, 23–24 October, Khujand, Tajikistan.

ICWC. (1999). Analysis of water use in the Aral Sea basin. Materials for the 23rd meeting, 11–12 June, Dashoguz, Turkmenistan.
ICWC. (2000a). Performance of water management system in the Amudarya basin during the non-growing season 1999–2000. Materials of the 26th meeting, 29 April, Dushanbe, Tajikistan.
ICWC. (2000b). Minutes of the 28th meeting, 21–22 December, Ashgabad, Turkmenistan.
ICWC. (2000c). Analysis of the low-water year 2000 and measures for 2001 in the Amudarya basin. Materials of 28th meeting, 21–22 December, Ashgabad, Turkmenistan.
ICWC. (2001a). Minutes of the 31st meeting, 23 November, Kurgan-Tyube, Tajikistan.
ICWC. (2001b). Bulletin No 26, April.
ICWC. (2008). Minutes of the 52nd meeting, 5 December, Ashgabad, Turkmenistan.
ICWC. (2010a). Materials of the 54th meeting, 14–15 January, Shimkent, Kazakhstan.
ICWC. (2010b). Materials of the 55th meeting, 3–4 April, Ashgabad, Turkmenistan.
Interagency Agreement between Turkmenistan and Uzbekistan on Sharing Water Resources in the Amudarya Lower Reaches, 26 May 2007.
Protocol No 566 of the meeting of the Scientific-Technological Council at the USSR Ministry of Land Reclamation and Water Management of September 10, 1987, which was approved on December 3, 1987 by the Minister Mr. N.F.Vasiliev (Protocol 566).
Revised Master Plan (Scheme) for Comprehensive Use and Conservation of Water Resources in the Amudarya (1984) approved by Protocol 566 in 1987.
SIC ICWC. (2015). Aral Sea and Prearalie: Summary of SIC ICWC work on monitoring and analysis. Tashkent.
SIC ICWC. (2017). Transboundary Water Management Adaptation in the Amudarya Basin to Climate Change Uncertainties: Key Findings. USAID/PEER Project. http://cawater-info.net/projects/peer-amudarya/key_findings_e.htm (accessed 14 February 2022).
World Bank. (2014). Key Issues for Consideration of the Proposed Rogun Hydropower Project, p. 17. https://www.worldbank.org/content/dam/Worldbank/document/eca/central-asia/World%20Bank%20Note%20-%20Key%20Issues%20for%20Consideration%20on%20Proposed%20Rogun%20Hydropower%20Project_eng.pdf (accessed 14 February 2022).

doi: 10.2166/9781789062786_0243

Chapter 19

Current challenges in the Rio Grande/Río Bravo Basin: old disputes in a new century

Regina M. Buono[1] *and Gabriel Eckstein*[2]

[1]*Non-resident Scholar, Center for Energy Studies, Baker Institute for Public Policy, Rice University; and Principal, Aither, USA*
[2]*Professor of Law and Director of the Energy, Environmental & Natural Resources Systems Law Program, Texas A&M University School of Law, USA*

ABSTRACT

The Rio Grande River traverses 2000 kilometres of the international border between Mexico and the United States. The river and its tributaries are governed by a series of border treaties and institutions, as well as under the domestic laws of each nation. Often lauded for enabling innovative and collaborative governance, in recent years the complicated regime has come under pressure as domestic and international water governance institutions struggle under the strain of climate change, population growth, and other stressors on water supply and demand in the region. This chapter considers three of the major challenges currently facing the Rio Grande River Basin and its riparians: (1) groundwater and ground–surface interactions and related practical and policy implications; (2) engagement with local and regional stakeholders; and (3) Mexico's latest water debt under the 1944 Treaty. It also identifies shortcomings in the regime to address these concerns, as well as innovative responses and solutions that have been crafted at various levels of governance.

Keywords: 1944 Treaty, groundwater, Rio Grande River Basin, stakeholders, transboundary governance

19.1 INTRODUCTION

For over 170 years, the peoples of Mexico and the United States have shared the Rio Grande River Basin (known as the Río Bravo in Mexico). Though the basin traverses over 2000 kilometres of the international border between the two countries, it also ties the two nations together through shared natural resources and wildlife habitats, socio-economic systems, and cultural and historic bonds. Management of the Rio Grande and its tributaries has been governed by a series of border treaties and institutions, the most recent of which is the 1944 *Treaty on the Utilization of Waters of the Colorado and Tijuana Rivers and of the Rio Grande*. Often lauded for enabling innovative and collaborative governance of the three named rivers, in recent years the treaty regime has nonetheless come under intense pressure in the Rio Grande Basin. Domestic and international water governance institutions are struggling under the strain of climate change impacts, population growth, and the attendant impacts to water supply and demand in the region.

This chapter considers some of the major challenges the Rio Grande River Basin faces currently, as the two riparians seek to adjust and adapt long-established systems to accommodate changing conditions and needs. We review the current situation in the region and focus on three areas of importance: (1) groundwater and ground–surface interactions and the related practical and policy implications; (2) engagement with local and regional stakeholders; and (3) Mexico's latest water debt under the 1944 Treaty.

19.2 BACKGROUND

19.2.1 Geology and geography of the region

The Rio Grande River rises in Colorado and flows south and east to the Gulf of Mexico, along the way passing through New Mexico and then forming the border between Texas and the Mexican states of Chihuahua, Coahuila, Nuevo León, and Tamaulipas (Figure 19.1). The streamflow of the river is

Figure 19.1 Map of the Rio Grande River Basin with its principal tributaries (Wikipedia). Source: https://en.wikipedia.org/wiki/Rio_Grande

highly variable but tightly controlled, particularly south of Otowi Bridge in Santa Fe, New Mexico, where diversions and inflows from tributaries are significant.

In the region north of El Paso, releases from Elephant Butte and Caballo reservoirs in New Mexico determine the river flow up to the Mexico–U.S. border (Nava *et al.*, 2016), while confluences with return flows downstream near Fort Quitman, the Rio Conchos, and the Pecos River, again replenish the river downstream. Along its route, the river supports more than two million acres of irrigated agriculture on both sides of the border, and more than six million residents in both countries.

19.2.2 Legal structures and governance at the binational level

Governance of the river at the international level is the responsibility of a binational body created by the *1944 Treaty on the Utilization of Waters of the Colorado and Tijuana Rivers and of the Rio Grande* (1944 Treaty) known as the International Boundary and Water Commission. The Commission is responsible for applying boundary and water treaties between Mexico and the United States and for settling differences that may arise in their interpretation and implementation. It operates through two sections: the Comisión Internacional de Límites y Aguas, based in Ciudad Juárez, Mexico, and the International Boundary and Water Commission, based in El Paso, Texas, in the United States (collectively IBWC/CILA). This is primarily a technocratic organization as each section's leadership team is comprised of a commissioner, who must be a trained engineer; two principal engineers; a legal advisor; and a foreign affairs secretary (Carter *et al.*, 2017).

In addition to creating the IBWC/CILA, Article 4 of the 1944 Treaty allocates the waters of the Rio Grande between Mexico and the United States from Fort Quitman, Texas, to the Gulf of Mexico (Figure 19.2). Under the treaty, Mexico is entitled to all waters reaching the main channel of the Rio Grande from the San Juan and Alamo rivers; half of unallocated flows from the main channel of the Rio Grande below the southernmost international dam; two-thirds of flows reaching the main channel of the Rio Grande from named tributaries in Mexico; and half of all other Rio Grande main channel flows not otherwise allotted by the treaty. The U.S. is allocated all Rio Grande water reaching the main channel of the river from named, smaller tributaries originating in Texas; one-half of unallocated flows in the main channel below the lowest major international storage dam; one-third of the flow reaching the main channel from named tributaries originating in Mexico; and one-half of unallotted flows between Fort Quitman and the lowest major international storage dam. The 1944 Treaty also provides that the one-third of the flow reaching the main channel of the river from the specified Mexican tributaries 'shall not be less, as an average amount in cycles of five consecutive years, than 350 000 acre-feet annually'. It allows Mexico, in the event of 'extraordinary drought or serious accident to the hydraulic systems' on its tributaries, to make up any deficiencies at the end of a 5-year cycle during the following 5-year cycle.

In addition, Article 25 of the 1944 Treaty includes an innovative mechanism – known as the minute system – that authorizes the IBWC to conduct ongoing negotiations to interpret and implement treaty terms. This mechanism has made the 1944 Treaty among the most flexible and adaptive binational treaties globally as it provides a way for the two countries, through their representatives in IBWC/CILA, to adapt management of the border's rivers in response to changing environmental and technical conditions and evolving stakeholder needs. When the 1944 Treaty lacks a clear directive regarding the outcome of an issue, IBWC/CILA Commissioners are able to negotiate an agreement (a 'minute') regarding how to address it. A minute is not an amendment to the treaty, but rather is treated as an interpretation of the 1944 Treaty, and does not require formal approval by the parties' legislatures. If neither government objects to a decision of IBWC/CILA within 30 days of the minute's pronouncement, the minute becomes a binding agreement between the countries.

One other treaty that is especially relevant to the Rio Grande is the *1906 Convention Between the United States and Mexico, Equitable Distribution of the Waters of the Rio Grande*. The treaty governs distribution of the upper portion of the basin before it reaches the Mexico–U.S. border. The agreement requires the U.S. to deliver to Mexico 60 000 acre-feet of water annually in the Rio Grande at a point

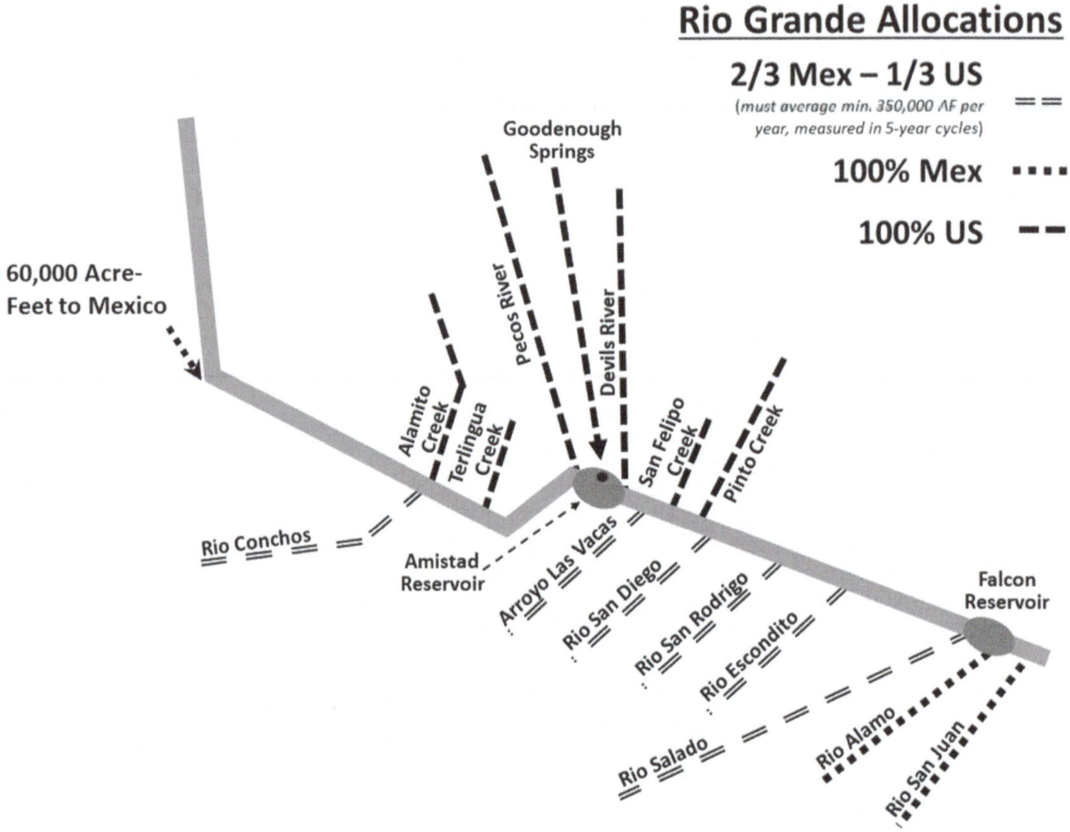

Figure 19.2 Schematic of the lower Rio Grande River basin, its main tributaries and reservoirs, and allocation of water under the 1944 Treaty.

just above the city of Juárez, Mexico. That volume, however, can be proportionally reduced during drought conditions and the U.S. is not required to make up deficits incurred by reduced deliveries. In fact, deliveries to Mexico under this treaty have been reduced in nearly one-third of the years between 1939 and 2015 (Carter *et al.*, 2017). Within Mexico, the water is transported to the Juárez Valley of Chihuahua where it is used primarily for irrigating agriculture.

19.2.3 Legal structures and governance at the national level

19.2.3.1 Surface water

While the 1944 Treaty distributes specific volumes and tributaries from the Rio Grande to both parties, within each country, domestic laws govern the internal allocation and uses of the river's water. Within the United States, the administration of water management and allocation is considerably more devolved and decentralized than in Mexico. In the U.S., inter-state compacts determine the allocation of inter-state surface water to each U.S. riparian state. Thus, the Rio Grande has been apportioned among Colorado, New Mexico, and Texas in accordance with the 1938 Rio Grande Compact. Under Article III of the Compact, while Colorado is entitled to use the river's water within its territory, it is required to deliver a specific volume of water at the Colorado–New Mexico border based on a formula that considers current flows of the Rio Grande and its tributaries at designated gauging stations.

Likewise, New Mexico has similar benefits and obligations regulated by formula under Article IV, however, its obligatory delivery location is not the New Mexico–Texas state line, but rather at a point close to San Marcial in New Mexico, just before the Rio Grande flows into Elephant Butte Reservoir. This delivery point was selected because the U.S. Bureau of Reclamation has managed the reservoir since 1906 and, thereby, controls water flow in the river below the dam as it courses toward the New Mexico–Texas border. Much of the water in the reservoir is reserved for Mexico (60 000 acre-feet noted above) under the Rio Grande Convention of 1906, as well as for the Elephant Butte Irrigation District in New Mexico and the El Paso County Water Improvement District in Texas under the federal 1905 Rio Grande Project Act. Any remaining flows are allotted for use by Texas.

While determinations at the international and national level can impact lower-level allocations, they do so only through a trickle-down process. Thus, water rights and uses by individuals, companies, municipalities, and other users, as well as any reductions in their allocations, are determined exclusively according to each country's domestic laws.

On the U.S. side, after bulk allocations are conducted under the 1906 and 1944 treaties, the three U.S. states of Colorado, New Mexico and Texas allocate the U.S. allotment in accordance with the 1938 Rio Grande Compact. Thereafter, each state can only distribute water in the Rio Grande that has been allotted to it in accordance with the compact. In Texas, according to its Water Code, surface water is owned by the state and held in trust for the public. Individuals and entities who wish to use surface water in Texas must obtain a water right and permit from the Texas Commission on Environmental Quality. Water rights are apportioned based on the doctrine of prior appropriation, which provides that the first person to take a quantity of water from a water source for 'beneficial use' has the right to continue to use that quantity of water for that purpose. The right is perpetual so long as it continues to be used fully in accordance with the permit and can be lost for non-use and other permit violations. Those who likewise take surface water for beneficial use at a later date are junior to those who establish their rights earlier in time, and those rights may be curtailed during low flows.

Both Colorado's and New Mexico's water law regimes, based on their respective state constitutions, are relatively similar to that of Texas in that surface waters are held by the state on behalf of the public, and both embrace the doctrine of prior appropriation (Constitution of the State of Colorado; Constitution of the State of New Mexico). The main difference is that in Colorado, most of the administrative functions of permitting water rights are managed by water courts rather than a state agency (Water Right Determination and Administration Act, 1969); in New Mexico, the State Engineer is charged with administering water rights for the state (New Mexico Statutes, 2016).

In sharp contrast, authority to regulate all water resources in Mexico is explicitly reserved by the federal government under the national constitution (Mexican Constitution, 1917). Surface water allocations are federally granted as concessions or assignments through the National Water Commission (Comisión Nacional del Agua, or CONAGUA) in accordance with the National Waters Law (Ley de Aguas Nacionales, 1992). Concessions are water rights granted for a fee to private parties for terms of 5 to 30 years, but which can be extended upon application. Assignments comprise the transfer of the right to manage, allocate, and charge for water resource uses from CONAGUA to sub-national governmental entities. Mexico uses basin agencies (Organismos de Cuenca) and Basin Councils (Consejos de Cuenca) in order to implement CONAGUA's mandate at the regional and local levels; however, these bodies have little authority or opportunity to engage in decision-making that affects regional or local allocations. For the most part, they function as liaisons between the federal government and local stakeholders (Foster, 2018).

19.2.3.2 Groundwater

With regard to groundwater, national regulations also vary between the two countries, as well as domestically among the individual U.S. states. Like surface water, groundwater in the U.S. is controlled by state law. In Texas, per the state's Water Code, groundwater is privately owned by the overlying landowner as real property. While its use is subject to some regulation by local groundwater conservation districts and the state, the government's ability to regulate groundwater is limited by constitutional provisions prohibiting the taking or overregulation of private property (Constitution of the State of Texas).

In contrast, under New Mexico's Water Code, groundwater is held by the state, but belongs to the public. It is subject to the doctrine of prior appropriation and managed by the state's Water Resources Allocation Program in the Office of the State Engineer. Permits are required for new groundwater appropriations, alterations to existing uses, and drilling of supplemental or replacement wells; however, a water right can be forfeited to the extent that any of the appropriated water is not fully applied to the designated beneficial use within a statutorily-defined time period.

Like in New Mexico, groundwater under Colorado's Water Code is owned by the state on behalf of the public. With the exception of groundwater rights in the Denver Basin, the state employs a modified prior appropriation system based on the type of groundwater at issue. Tributary groundwater (groundwater hydraulically connected to surface streams) is treated as surface water and subject to the state's surface prior appropriation system; non-tributary groundwater rights are subject to prior appropriation in relation to the total amount of recoverable water beneath the overlying land and the aquifer's life expectancy; groundwater rights in designated basins are adjudicated by the Colorado Ground Water Commission under a modified prior appropriation system 'to permit full economic development of designated ground water resources'. Groundwater rights in the Denver Basin aquifer are governed by statutory rules and are appurtenant to ownership of the overlying land.

Just as it does for surface waters, Mexican law vests the Mexican federal government with authority over the management and regulation of groundwater. A permit is required to extract groundwater or discharge wastewater into an aquifer. CONAGUA has jurisdiction nationwide to administer permits and monitor Mexican aquifers; the agency exercises its authority through various sub-agencies organized at the level of hydrologic-administrative regions based on surface hydrogeology. In recent years, Mexico has sought to decentralize water management authority in the country but has not conveyed sufficient resources and authority to local and regional entities to allow them to achieve desired objectives, such as the protection of aquifers and ecosystems and sufficient water quality in drinking water systems (Foster, 2018).

19.3 CURRENT CHALLENGES AT THE BORDER

The treaty regime described above has been lauded by some scholars for enabling innovative and collaborative governance of the named rivers (Mumme, 1993), but circumstances on the border have nonetheless come under pressure in the Rio Grande basin in the decades since 1944 and, in particular, in recent years. Domestic and international water governance institutions are struggling under the strain of climate change impacts, population growth, and the attendant impacts to water supply and demand in the region. The remainder of this chapter focuses on three policy areas where tension is surfacing: (1) groundwater and ground–surface interactions and related practical and policy implications; (2) engagement with local and regional stakeholders; and (3) Mexico's water debt under the 1944 Treaty.

19.3.1 Groundwater

Groundwater has long been treated as the neglected stepchild of the transboundary water regime along the Mexico–US border. Transboundary aquifers are excluded from the existing treaty regime, have rarely been placed on the IBWC/CILA's agenda, and until recently, have only sporadically been studied. Yet, groundwater on the border plays a significant role in agricultural production, economic development, and even the social fabric of the region.

As many as 72 transboundary aquifers and hydrogeological units are surmised to underlay large segments of the 3000-kilometre-long frontier (Sanchez & Rodriguez, 2021). Along Mexico's border with Texas alone, 53 domestic and transboundary hydrogeological formations have been identified, of which nearly 30% have good to moderate aquifer potential (Sanchez *et al.*, 2018). Numerous wells dot the landscape, and millions of people on both sides of the international border rely heavily on these subsurface resources. The Hueco Bolson and Conejos Medanos aquifers, for example, provide all of the freshwater used by Ciudad Juárez's 1.54 million residents, while the Hueco Bolson and Mesilla Bolson aquifers supply approximately one-half of that used by El Paso's 963 000 residents (Far West Texas Water Planning Group, 2021).

The region's transboundary aquifers, however, do not exist in isolation. Many aquifers along the Mexico–Texas border are hydrologically connected to the Rio Grande. For example, the Rio Grande provides significant recharge to the Mesilla Basin aquifer system in the El Paso/Ciudad Juárez area, especially to the upper hydrogeologic unit (the Rio Grande alluvium) (Teeple, 2017). Likewise, groundwater flow in the Allende–Piedras Negras Aquifer, which historically discharged into the Rio Grande, may have recently shifted due to extensive groundwater pumping in the region, which now causes the Rio Grande to seep into the aquifer (Rodriguez *et al.*, 2020). In yet another distinct example, Goodenough Springs (also called Hinojosa Springs) are a series of freshwater springs that now lie submerged below and discharge into the binational Amistad Reservoir, effectively supporting the flow of the Rio Grande. The springs originate in the Edwards Trinity Aquifer emerging from the top of limestone deposits on the extreme western edge of Maverick Basin. Between inundation of the reservoir in 1967–68 and 2005, spring discharge was calculated to decline from a historical mean of 4.03 to 2.03 m^3/s, possibly due to hydrostatic backpressure from Amistad Reservoir (Flores *et al.*, 2021).

In addition, the region's aquifers represent a critical source of freshwater for the border region's distinct environment and ecosystems. For example, groundwater flowing from Cretaceous limestone aquifers into the Rio Grande helps sustain aquatic habitats along the river during dry years, as well as mitigate impairment to water quality (Bennett, 2011). In addition, during low flow conditions, groundwater is estimated to account for as much as two-thirds of the flow in the Rio Grande at the Foster's Weir gage and the point where the river enters the Amistad National Recreation Area (Bennett, 2011). Moreover, flows and aquatic habitats in the river segment above and including Amistad Reservoir have been found to be highly susceptible to groundwater extraction in nearby Terrell County and Val Verde County in Texas (Cutillo *et al.*, n.d.).

Notwithstanding the relevance of groundwater to the Rio Grande and its critical importance to various ecosystems along the frontier, none of the existing treaties between Mexico and the United States address the region's groundwater resources (Eckstein, 2013). The only direct reference to the region's transboundary aquifers is found in Minute 242 from 1973, which limits groundwater withdrawals on both sides of the Sonora–Arizona border near San Luis, Arizona, to specifically enumerated withdrawal targets, and mandates consultation between the parties prior to the development, by either nation, of any groundwater resources along the border that could adversely impact the other country. The two groundwater provisions were intended as temporary measures 'pending the conclusion ... of a comprehensive agreement on groundwater in the border region' (IBWC, 1973, Para. 5). However, internal divisions among the stakeholders, especially on the American side, made agreement unachievable. Five decades later, the temporary provisions of Minute 242 have yet to be realized (Eckstein, 2013).

One other reference to groundwater on the Mexico–U.S. border, albeit very indirect, is found in Minute 289 from 1992, which addresses water quality problems along the border. Hidden among provisions that mostly address the Rio Grande and Colorado rivers, the minute refers to the Integrated Border Environmental Plan, which was adopted by the two countries in the same year. That plan calls for the creation of a water-monitoring program and database to observe both groundwater and surface water quality along the frontier (IBWC, 1992, Para. 4).

Aside from these two instruments, the region's groundwater resources are regulated independently by each country, exclusively under their respective domestic regime.

Missing from the existing treaty regime are mechanisms to address the hydrologic relationship that the Rio Grande has with the various aquifers that flow alongside and below the river. It is easy to imagine a scenario in which extensive groundwater extraction on one or both sides of the river could diminish flows in hydrologically linked segments of the Rio Grande. As noted above, that is already happening with the Rio Grande River losing water to both the Mesilla Basin aquifer system and the Allende–Piedras Negras Aquifer. Under the existing treaty system, however, withdrawals from these aquifers, which are likely exacerbating the leakage, are likely to fall outside the rights and obligations created by the treaty regime. In a similar vein, the contamination of an aquifer (e.g., from sewage overflows, industrial spills, misuse of agricultural products, etc.) that is adjacent to and interrelated with a section of the Rio Grande that receives inflows from the aquifer could cause the pollutants to migrate into the river and affect downstream water uses and users. That scenario also may fall outside the scope of the treaties governing the river.

What is needed are provisions and mechanisms developed through IBWC/CILA that take hydrologically related groundwater into account in the overall management and allocation regime of the Rio Grande. As a first step, this would require compiling the existing information on such groundwater–surface water relationships in the basin, as well as conducting additional research to fill in the significant knowledge gaps that currently exist. At the very least, Mexico and the U.S. should expand their existing system for data and information sharing to include groundwater resources on the frontier.

The two nations also should explore whether groundwater withdrawals can offset allocation rights from the main stem of the Rio Grande or its various tributaries. Likewise, they should consider whether to incorporate a prioritization mechanism into the governance regime to determine when and under what circumstances groundwater or surface water allocations should prevail in times of shortages. In addition, Mexico and the U.S. should assess the impact that shortages in surface water have on groundwater exploitation, even in hydrologically unrelated aquifers.

Lastly, the two nations also should facilitate more opportunities for public participation. The management and governance of transboundary aquifers should not end at the frontier, but rather should be pursued collaboratively by local and regional stakeholders on both sides of the border.

19.3.2 Stakeholder involvement and transparency

IBWC/CILA's role under the 1944 Treaty is primarily carried out with a strong focus on technical issues, which are often analyzed and addressed from an engineering perspective. The organization's current approach to managing the Rio Grande under the 1944 Treaty offers limited stakeholder involvement. This approach has been criticized by scholars, one of whom asserted even nearly 30 years ago that the IBWC is merely 'a social artifact, imperfect at best, and captive to the vicissitudes of time' (Mumme, 1993; Mumme & Collins, 2014).

In response to the criticism, both national sections have established citizens' fora to engage with locals on their side of the border regarding issues in the region. On the U.S. side, the Rio Grande Citizens Forum, established in 1999, addresses the upper stretches of the river to Fort Quitman. The Lower Rio Grande Citizens Forum, established in 2003, offers a venue for residents of the lower reaches of the river (IBWC, 2017). CILA has also created citizens' fora in Ciudad Juárez, Ciudad Acuña, Nuevo Laredo, and Reynosa, Mexico, to facilitate the exchange of information between the Rio Grande Basin border community and CILA (CILA, 2014). The fora in both countries, however, have no formal role in negotiations over the river's management or operations. The U.S. fora have tended to focus primarily on circumscribed, smaller-scale challenges such as saltcedar control, endangered species, and levee remediation, while the Mexican groups appear to have met infrequently, with their agendas determined by CILA.

Stakeholder participation and transparency beyond the citizens' fora vary between the parties, though neither side evidences the extensive stakeholder engagement enjoyed by both nations in managing the Colorado River under the same treaty. In the U.S., stakeholder participation and transparency are, in part, a function of that country's decentralized approach to water management, which requires local participation to operate effectively. Stakeholder engagement in the U.S. also benefits from the greater availability of resources for state-level administrations and agencies that support the development and administration of local and regional water plans, as well as from efforts by private and civil society groups. For instance, in New Mexico, the Nature Conservancy is leading a group of over 90 entities – including federal and state agencies, local governments, health systems, landowner groups, environmental consultancies, irrigation districts, water utilities, and non-governmental organizations (NGOs) – in the creation of the Rio Grande Water Fund (RGWF), which seeks to protect and improve the storage, delivery, and quality of Rio Grande water through landscape-scale forest restoration in the Rio Grande watershed (The Nature Conservancy, 2020). Since it began in 2014, the RGWF has treated 140 000 acres of forest, and continues to generate sustainable funding to support a 20-year program to restore 600 000 acres of forests and Rio Grande headwaters in New Mexico and Colorado. The fund also supports efforts to restore streams and wetlands by installing water sources for cattle and wildlife away from riparian areas and investing in local organizations that reduce wildfire risk through the use of prescribed fire.

In contrast, stakeholder participation and transparency on the Mexican side is largely absent because of the country's centralized approach to water management. Since the vast majority of domestic water-management decisions are made by CONAGUA at the national level, local communities have little to no real opportunity to be involved in meaningful decision-making. Efforts at decentralization proposed by the government over the past two decades have largely been ineffective because they failed to include funding and resources, as well as legal authority to create and enforce laws, for local and regional water-management entities (Foster, 2018). There have been some efforts – including in collaboration with cross-border partners – by environmental and other civic groups in Mexico to implement better conservation and agricultural practices in parts of the region (Borders, 2015). In general, local communities and institutions in Mexico lack adequate information about water availability, how allocation decisions are made, or Mexico's treaty relations with and obligations to the U.S.

The eventual effects of these enduring conditions became apparent in 2020 during Mexico's latest water debt, when violent conflicts erupted between Chihuahua farmers and the Mexican government. The protests were a poignant symptom of the disenfranchisement of local Mexican water stakeholders, and are another example in which decisions on water allocations to farmers, municipalities, and industry in Mexico continue to be made at the highest levels, in Mexico City. Interviews conducted by the media suggested that a lack of transparency by the central government was one of the chief reasons for the protests (Yucatan Times, 2020). Payan (2020) has argued that the conflict in Chihuahua is representative of Mexico's greater governance crisis.

19.3.3 Mexico's recurring water debt

Since the implementation of the 1944 Treaty, Mexico has twice fallen short on its treaty obligations to deliver to the U.S. an average annual 350 000 acre-feet of water down the Rio Conchos and into the Rio Grande. Under the treaty, Mexico is allowed to carry over any incomplete balances of water from one 5-year cycle to the subsequent 5-year cycle in the event of an 'extraordinary drought'. The two countries, however, have disagreed over two critical points. The first is what exactly constitutes an extraordinary drought. The second is whether repayment of a debt from the first 5-year cycle made during the second 5-year cycle must be made concurrently with any debt incurred during the second 5-year cycle, or whether that second shortfall can be postponed and carried over to a third 5-year cycle (Carter *et al.*, 2017).

Between 1944 and the drought that extended from 1994 to 2003, Mexico met its deliveries within each successive 5-year cycle (Carter *et al.*, 2017). Due to intensive expansion of agricultural production in the Rio Conchos basin during the 1980s and 1990s, as well as a number of intermittent droughts, the 1994–2003 drought was especially difficult for farmers in Northern Mexico. As a result, Mexico was unable to deliver the requisite water volumes into the Rio Grande. That water debt was eventually resolved by transferring some of Mexico's water rights in the two international reservoirs to the United States (see IBWC, 2002), as well as by the advent of hurricane-related rains in 2005 (Carter *et al.*, 2017).

The recent water shortfall dispute between Mexico and the United States began during the delivery cycle that began in October 2010 and ended in October 2015. A final accounting for that cycle showed a deficit of 216 250 acre-feet (Carter *et al.*, 2017). That deficit was carried over into the 2016–2020 cycle and threatened to continue into a third 5-year cycle, ratcheting up tension over both the debt and interpretation of the deficit provision as the two countries have long disagreed whether the treaty allows such water debts to be carried over into a third cycle.

The situation led to protests in the summer and fall of 2020 when farmers in Chihuahua learned that their national government planned to pay off the country's water shortfall by increasing Rio Concho flows into the Rio Grande (Mumme, 2020). In violent confrontations that resulted in the death of one female protester, the farmers forcibly took control of three dams on the Rio Conchos to prevent water from being released into the Rio Grande for the United States.

On October 21, 2020, three days before Mexico would have violated its delivery obligations under the 1944 Treaty, IBWC/CILA signed an agreement to resolve the issue. Under Minute 325, Mexico fulfilled its delivery obligations by transferring the entirety of its water in the Amistad and Falcon reservoirs to the United States. The transfer nearly depleted all of Northern Mexico's stored water in the reservoirs, thereby depriving Tamaulipas farmers downstream on the Rio Concho from their winter water supplies. However, by doing so, Mexico abided by the 1944 Treaty and ended the 2016–2020 cycle debt free (Helfgott, 2021). The minute also resolved the long-standing disagreement over Mexico's ability to end two back-to-back cycles, referencing Minute 234 and stating that two subsequent cycles 'may not end in a deficiency'.

In addition, as a long-term measure to improve water management in the basin, Minute 325 officially recognized two pre-existing working groups. The first is the Rio Grande Hydrology Work Group tasked with 'enhance[ing] information exchange, develop[ing] a binational Rio Grande model, and us[ing] the model as a tool to analyze water management scenarios, including scenarios related to future water conservation projects'. The second is the Rio Grande Policy Work Group, which would oversee the Hydrology Work Group and 'consider water management policies in the basin'. The two working groups have been collaborating since 2017 with binational participation to advance modeling capabilities in the basin (IBWC, 2017). Per the minute, the working groups are now tasked with developing a new minute by December 2023 to provide 'increased reliability and predictability in Rio Grande water deliveries to water users in the United States and Mexico' (IBWC 2020, Para. 4).

19.4 CONCLUSIONS

While long-standing and still functionally operational, the 1944 Treaty does suffer from some shortcomings. Most notably, the treaty offers no guidance or direction for dealing with modern circumstances that are beginning to overwhelm the instrument's capacity to generate effective and meaningful responses. Climate change, for example, threatens to make the Rio Grande basin even more arid. Likewise, population growth, as well as economic and agricultural activities, are taxing the regime's existing allocation system and its ability to balance water supply and demand. In addition, the 1944 Treaty provides no references to ecological purposes or to the region's hydraulically linked binational aquifers (Helfgott, 2021). While the treaty could probably continue operating in its current format for a few additional decades, these limitations will only amplify the growing water challenges on the Mexico–U.S. border.

That said, the 1944 Treaty's mechanisms – and, in particular, the minute system – have shown themselves able to facilitate and support innovations in water management. Recent evidence of this includes the work of the two working groups established by Minute 325 currently tasked with the development of a new minute to increase reliability and predictability in Rio Grande water deliveries. Additional lessons may be drawn for the Rio Grande from the experience of minutes developed to advance collaborative governance, stakeholder engagement, conservation, and environmental flows for the Colorado River (Buono *et al.*, 2021). Other initiatives, such as the Permanent Forum of

Binational Waters formed in 2020, will strengthen collaborative efforts across a wider set of scientists, government officials, NGOs, and citizens interested in the sustainability of the Rio Grande Basin. Finally, over the past few years, substantial academic research has been conducted on the region's transboundary aquifers and their relationship to the Rio Grande, contributing to the IBWC/CILA decision to formally to recognize the critical nature of groundwater in the border area and organize its first ever conference focusing on shared aquifers in April 2019. While independent of each other, these three efforts offer hope that human ingenuity and cooperation can move the basin toward better management of shared resources that bind the two nations and achieve long-term sustainability.

REFERENCES

Bennett J. (2011). Trip report for gain loss survey of ground water and spring contributions to flow of the Rio Grande in Texas, February and March of 2011. National Parks Service, U.S. Department of the Interior, June 21, 2011.

Borders G. (2015). In-depth: Climate Change on the Rio Grande, World Wildlife Magazine, Fall 2015. https://www.worldwildlife.org/magazine/issues/fall-2015/articles/climate-change-on-the-rio-grande (accessed 17 November 2021).

Buono R. M., Baggerman J. and Sansom L. C. (2021). Got a minute for the future of the Rio Grande? Considering the prospects for a sustainability minute in the wake of the Colorado's Minute 323. *Water International*, **46**, 543–566. https://doi.org/10.1080/02508060.2021.1915562 (accessed 17 November 2021).

Carter N., Mulligan S. P. and Ribando Seelke C. (2017). U.S.-Mexico Water Sharing: Background and Recent Developments, Congressional Research Services, R43312. https://fas.org/sgp/crs/row/R43312.pdf (accessed 17 November 2021).

CILA [Comisión Internacional de Límites y Aguas Sección Mexicana]. (2014). Foro Ciudadano Documento Marco (Citizen Forum Framework Document). http://www.cila.gob.mx/foros/dm.pdf (accessed 17 November 2021)

Constitution of the State of Colorado.

Constitution of the State of New Mexico.

Constitution of the State of Texas.

Cutillo P., Bennett J. and Braumiller S. (n.d.). Overview Report: groundwater management issues of importance to Big Bend National Park, Rio Grande Wild & Scenic River and Amistad National Recreation Area. Water Resources Division, National Parks Service, U.S. Department of the Interior.

Eckstein G. (2013). Rethinking transboundary ground water resources management: a local approach along the Mexico-U.S. Border. *Georgetown International Environmental Law Review*, **25**(1), 95–128.

Far West Texas Water Planning Group. (2021). Far West Texas Water Plan. Texas Water Development Board. https://www.twdb.texas.gov/waterplanning/rwp/plans/2021/E/RegionE_2021RWP.pdf?d=5589.125000056811 (accessed 17 November 2021).

Flores M. E., Nunu R. R., Wittmeyer G. and Green R. T. (2021). Goodenough Spring Catchment Area Characterization, Amistad Reservoir, Rio Grande Valley. Southwest Research Institute, San Antonio, TX.

Foster J. (2018). Survey of Legal Mechanisms Relating to Groundwater Along the Texas-Mexico Border. Texas A&M University School of Law, Fort Worth, Texas. http://www.law.tamu.edu/prospective/academics/centers-clinics-programs/natural-resources/reports-publications/texas-mexico-groundwater (accessed 17 November 2021).

Helfgott A. (2021). Bilateral Water Management: Water Sharing Between the US and Mexico Along the Border. Wilson Center, Washington DC. https://www.wilsoncenter.org/article/bilateral-water-management-water-sharing-between-us-and-mexico-along-border (accessed 17 November 2021).

IBWC [International Boundary and Water Commission]. (1973). Minute 242: Permanent and Definitive Solution to the International Problem of the Salinity of the Colorado River. Available at: https://www.ibwc.gov/Files/Minutes/Min242.pdf (accessed 17 November 2021).

IBWC [International Boundary and Water Commission]. (1992). Minute 289: Observation of the Quality of the Waters Along the United States and Mexico Border. Available at: https://www.ibwc.gov/Files/Minutes/Min289.pdf (accessed 17 November 2021).

IBWC [International Boundary and Water Commission]. (2002). Minute 308: United States Allocation of Rio Grande Waters During the last Year of the Current Cycle. Available at: https://www.ibwc.gov/Files/Minutes/Min308.pdf (accessed 17 November 2021).

IBWC [International Boundary and Water Commission]. (2017). Lower Rio Grande Citizen Forum. November 8, 2017. https://ibwc.gov/Files/CF_LRG_Minutes_110817.pdf (accessed 17 November 2021).

IBWC [International Boundary and Water Commission]. (2020). Minute 325: Measures to End the Current Rio Grande Water Delivery Cycle Without a Shortfall, to Provide Humanitarian Support for the municipal Water Supply for Mexican Communities, and to Establish Mechanisms for Future Cooperation to Improve the Predictability and Reliability of Rio Grande Water Deliveries to Users in the United States and Mexico. Available at: https://www.ibwc.gov/Files/Minutes/Min325.pdf (accessed 17 November 2021).

Mexico, Ley de Aguas Nacionales. (1992). (as amended).

Mexico, Political Constitution of the United Mexican States of 1917 (as amended).

Mumme S. P. (1993). Innovation and reform in transboundary resource management: a critical look at the international boundary and water commission, United States and Mexico. *Natural Resources Journal*, **33**(1), 93–120.

Mumme S. P. (2020). Beyond the Rio Grande Water Debt, Rice University's Baker Institute for Public Policy Issue Brief. https://www.bakerinstitute.org/research/beyond-rio-grande-water-debt/ (accessed 17 November 2021).

Mumme S. P. and Collins K. (2014). The La Paz agreement 30 years on. *Journal of Environment and Development*, **23**(3), 304, https://doi.org/10.1177/1070496514528801 (accessed 17 November 2021)

Nava L. F., Brown C., Demeter K., Lasserre F., Milanés-Murcia M., Mumme S. P. and Sandoval-Solis S. (2016). Existing opportunities to adapt the Rio Grande/bravo basin water resources allocation framework. *Water*, **8**(291), 1–24.

New Mexico Statutes. (2016). Chapter 72—Water Law, Article 2—State Engineer, Section 72-2-1—Appointment; removal; qualifications; duties; office; private practice prohibited.

Payan T. (2020). Mexico's Water Dispute with the U.S. Is a Symptom of Its Governance Crisis. *World Politics Review*, October 7, 2020. https://www.worldpoliticsreview.com/articles/29112/mexico-s-water-dispute-with-the-u-s-is-a-symptom-of-its-governance-crisis (accessed 17 November 2021).

Rodriguez L., Sanchez R., Zhan H. and Knappett P. S. K. (2020). The transboundary nature of the allende–piedras negras aquifer using a numerical model approach. *Journal of the American Water Resources Association*, **53**(3), 387–408. https://doi.org/10.1111/1752-1688.12843 (accessed 17 November 2021)

Sanchez R. and Rodriguez L. (2021). Transboundary Aquifers between Baja California, Sonora and Chihuahua, Mexico, and California, Arizona and New Mexico, United States: identification and categorization. *Water*, **13**(20), 2878. https://doi.org/10.3390/w13202878 (accessed 5 January 2022)

Sanchez R., Rodriguez L. and Tortajada C. (2018). Transboundary aquifers between Chihuahua, Coahuila, Nuevo Leon and Tamaulipas, Mexico, and Texas, USA: identification and categorization. *Journal of Hydrology: Regional Studies*, **20**, 74–102. https://doi.org/10.1016/j.ejrh.2018.04.004 (accessed 17 November 2021).

Teeple A. P. (2017). Geophysics- and geochemistry-based assessment of the geochemical characteristics and groundwater-flow system of the U.S. part of the Mesilla Basin/Conejos-Médanos aquifer system in Doña Ana County, New Mexico, and El Paso County, Texas, 2010–12. U.S. Geological Survey Scientific Investigations Report 2017–5028. Texas Water Code, Title 2—Water Administration https://doi.org/10.3133/sir20175028 (accessed 17 November 2021).

The Nature Conservancy. (2020). Rio Grande Water Fund: A wildfire and water source protection project. https://www.nature.org/en-us/about-us/where-we-work/united-states/new-mexico/stories-in-new-mexico/new-mexico-rio-grande-water-fund/ (accessed 17 November 2021).

United States and Mexico. (1906). Convention Between the United States and Mexico, Equitable Distribution of the Waters of the Rio Grande.

United States and Mexico. (1944). Treaty on the Utilization of Waters of the Colorado and Tijuana Rivers and of the Rio Grande.

Water Right Determination and Administration Act. (1969). Colo. Rev. Stat. §37-92-101 et seq.

Wikipedia, Map showing the Rio Grande, and its tributaries—within the Rio Grande drainage basin, https://commons.wikimedia.org/wiki/File:Riogranderivermap.png (accessed 17 November 2021).

Yucatan Times. (2020). Water crisis heats up in Chihuahua, Mexico. *Yucatan Times*, October 9, 2020. https://www.theyucatantimes.com/2020/10/115001/ (accessed 17 November 2021).

doi: 10.2166/9781789062786_0255

Chapter 20

Transitioning away from open access: lessons learnt from a comparative analysis of water allocation regimes worldwide

Josselin Rouillard[1], *Christina Babbitt*[2], *Edward Challies*[3] *and Jean-Daniel Rinaudo*[4]

[1]*Ecologic Institute, Berlin, Germany*
[2]*Environment Defense Fund, San Francisco, United States*
[3]*University of Canterbury, Christchurch, New Zealand*
[4]*Brgm, G-EAU, Montpellier, France*

ABSTRACT

The transition from open to regulated access to water resources is a challenging task for water managers who have to address complex environmental, social and economic trade-offs. Water allocation is a powerful tool, yet its implementation is deeply conflictual. This chapter compares the process of transitioning to regulated access in 13 case studies worldwide. It shows the wide diversity of institutional settings and design choices, while exploring why differences occur and considering the advantages and disadvantages of the different approaches deployed in various contexts. It concludes with key takeaways and reflections on the need for ongoing work assessing the environmental, social and economic performance of allocation regimes.

Keywords: Allocations, case studies, governance, institutions, transitioning

20.1 INTRODUCTION

The transition from open to regulated access to water resources, and the setting of water allocations, can be a deeply conflictual and political process that can disrupt perceived historical rights and run against entrenched practices and interests. However, driven by the need to achieve more sustainable water use in over-allocated basins, this process is playing out across the world, amidst intense pressure from climate change, population growth, economic development, and other stressors on water supply and demand. Establishing water resources allocation regimes requires time and much collective effort to succeed, and to our knowledge, none of the cases presented in this book are yet truly successful, that is in adequately balancing environmental, economic and social goals of sharing water.

As Blomquist and Babbitt (Chapter 2) put it, 'there is no reason to expect that transitions from open access to allocations will be easy, quick, or inexpensive, or will be successful upon first attempt'. Furthermore, there is no single model that would fit all situations for an effective, efficient and fair allocation regime. Rather, we should expect allocation regimes to have unique features and require particular reform processes which 'reflect basin characteristics and conditions, uses, preferences and priorities, and the historical, cultural, and political contexts of land and water use'. While we therefore cannot recommend a blueprint for transition, we can critically assess current institutional, social and technological innovations, learn from experience, foster knowledge exchange and promote experimentation through 'institutional bricolage' (Cleaver, 2017).

This book presents reflections on key dimensions of water allocation policies and a unique range of cases of water allocation regimes across the world. Building on the content presented earlier, this concluding chapter aims, first, to compare the process of transitioning to regulated access and key features of allocation regimes in 13 case studies. It shows the wide diversity of institutional settings and design choices, and aims to explain, where possible, why differences occur and point out the advantages and disadvantages of the different approaches deployed in various contexts. This comparison is organised around key features of allocation regimes presented in Chapter 1: the institutional framework, the process of setting extraction limits, (re)allocation rules, and compliance and enforcement. The chapter concludes with key takeaways and reflections on the need for ongoing work assessing the environmental, social and economic performance of allocation regimes.

20.2 ESTABLISHING A FACILITATING INSTITUTIONAL FRAMEWORK

20.2.1 Overview of the main steps of institutional development

The transition from open access to regulated access to water resources necessarily relies on the establishment of an enabling institutional framework which defines the roles and responsibilities of the state and user communities in sharing water resources, the characteristics of water use rights, and links with other policies impacting water use. Although each case reported in the chapters of this book had its own institutional development pathway, some common features appear, which are summarised in Figure 20.1 and in the following paragraphs.

❶ As most abstraction historically drew on surface waters, early institutional structures tended to address conflicts over access to surface water. The riparian doctrine was applied widely across Europe, although appropriation by royal decrees occurred in the middle ages in countries such as France (Chapter 9). Collective institutional arrangements were generally implemented by a local community

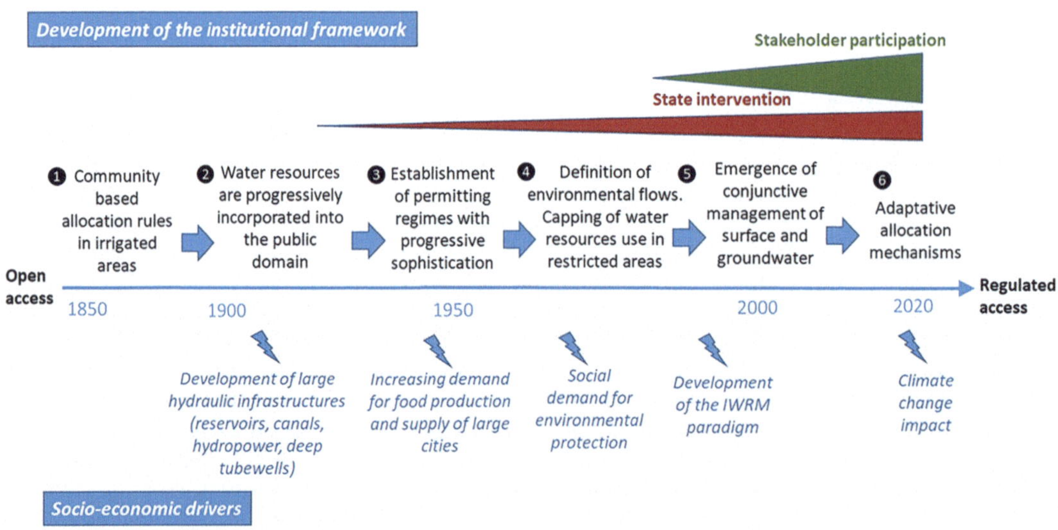

Figure 20.1 Main steps in the transition pathway from open access to regulated use of water resources. (IWRM: integrated water resources management).

or a community of irrigators and oriented towards enabling water use for economic development, as in the case of the allocation systems developed by settlers in Idaho in the late 1800s on the basis of the idea of 'first possession' (Chapter 16), or the *associações de vala* (canal cooperatives) set up by rice farmers in Brazil in the 1920s (Chapter 13). In some places, like Mediterranean Spain (Chapter 8), community-based institutions organising water allocation have been in place since the Arabic period in the early middle ages.

The growth of modern economies through industrialisation (e.g. hydropower development) and large-scale irrigation programmes in the late 19th and 20th centuries had increasingly significant impacts on hydrological regimes, leading many states to incorporate water resources in the public domain ❷ and to increasingly intervene in allocation – in particular through the establishment of permitting regimes regulating access to and use of surface water ❸. This is exemplified in Alberta as early as 1894 (Chapter 15), when the then dominant riparian common law could not accommodate the increasingly intense conflicts over the exploitation of water resources. Later on, in response to growing demand for environment protection, the concept of flows reserved for the environment was progressively incorporated into water legislation and systems of water use restrictions were implemented, progressively taking the form of a cap on water usage ❹.

Groundwater development brought new water management challenges, especially in the second half of the 20th century as surface water became more regulated and technological change facilitated access to groundwater resources. Dropping water tables and heightened tensions over resource availability saw interventions by authorities to control groundwater use and manage access. Controls began with requirements over well registration and locations, sometimes as early as the early 1900s (e.g. New South Wales in 1912; England and Wales in 1945; Turkey in 1960; India in the 1970s). Some major legislative changes were sparked by threats of further litigation between water users (see Nebraska, Chapter 17). This led to the emergence in the late 20th century of truly integrated regulatory frameworks facilitating the conjunctive management of surface water and groundwater systems ❺. Integrated Management Planning in Nebraska, and the recent settlement agreement in Idaho are cases in point. The most recent developments in allocation regimes aim at increasing resilience in a rapidly changing climate, through implementing adaptive allocation mechanisms (see New South Wales, Chapter 12) ❻.

20.2.2 Formalising water use rights

The transition to more regulated control has generally been accompanied by a formalisation of water use rights over the 20th century. The most frequent approach has been to transfer water use rights into the public domain, especially with regard to surface water. Use is managed with permits or concessions formalising individual (or, more rarely, collective) water extraction rights. Those rights, which often have a time-limited validity, can at least in theory be modified, reduced or even cancelled without compensation by the State (for meeting general interest objectives).

Regarding groundwater, although first left to landowners' appropriation, authorities in many countries have ultimately brought groundwater into the public permitting regime, as described in many cases presented in this book (see e.g. Brazil, Colorado in Chapter 19, France). In some places however, the legal framework recognises water as private property (e.g. Texas in Chapter 19, as well as places not covered in the book such as Chile). Those rights can be sold, leased and mortgaged and any reduction in allocation decided by the State would require compensation (buy-back).

Cases exist also where individual use rights are still not formalised through permitting regimes, especially in the case of groundwater (e.g. California). Users may nevertheless decide to formalise individual use rights through adjudication, but this is not systematic and other solutions are being developed to achieve more sustainable management of the resource (see Section 20.2.4).

Finally, hybrid systems of water use rights are also in place. For instance, Spain (Chapter 8) has opted for a dual system, by declaring all water as public, except where users opted to report their historical usage into a Catalogue of Private rights. Those private use rights are thus formalised but not regulated through the permitting regime implemented for public waters. As a trade-off, the law forbids users to modify any characteristics of their private use rights, including location of the extraction point, volume withdrawn, timing, or the purpose of the water withdrawal.

Independently from the ownership issue, water use rights may be defined in very different ways: in the simplest allocation regimes, they consist of an access right, that is an administrative authorisation of the extraction point (well or borehole, pumping station, diversion weir); in that case, users have no limits on extractions. This access right can be complemented by an extraction right, specifying the pumping capacity in flow rate or total volume that can be withdrawn over a specific period of time (irrigation season, low flow period, year). Extraction rights can be further differentiated into entitlements and allocations, an issue examined in Section 20.4.

Extraction rights are generally implemented in water scarce areas, whereas access rights are typically used in less sensitive areas. This is explained by the high transaction costs associated with establishing extraction rights and monitoring actual use, as well as the political costs of establishing a sustainable abstraction cap (see e.g. chapters 2 and 6). Spain prioritises basins with an imbalance between supply and demand as well as those at risk of an imbalance. California identifies high to medium-priority, and critically overdrafted groundwater basins, where Sustainable Groundwater Management Plans must be adopted. Nelson (Chapter 3) supports introducing allocation systems preventatively, in order to avoid 'lock-in to unsustainable use and allow for adjustment at least cost, as needed, over time'. Similarly, Blomquist and Babbitt (Chapter 2) suggest setting an initial cap and controls on water usage, as imperfect as it may be, allowing for adjustments as knowledge on basin conditions improves.

20.2.3 The role of authorities and user communities in allocation decisions

Overall, the cases presented in this book clearly show that the establishment of a water allocation regime results from decisions made across multiple scales. Allocation decisions play out in complex multi-level or polycentric systems of interrelated governing bodies, often acting in partnership with user organisations. The diversity of situations can be characterised by looking at the role of public authorities and the involvement of communities, users and stakeholders in allocation decisions, as depicted in Figure 20.2.

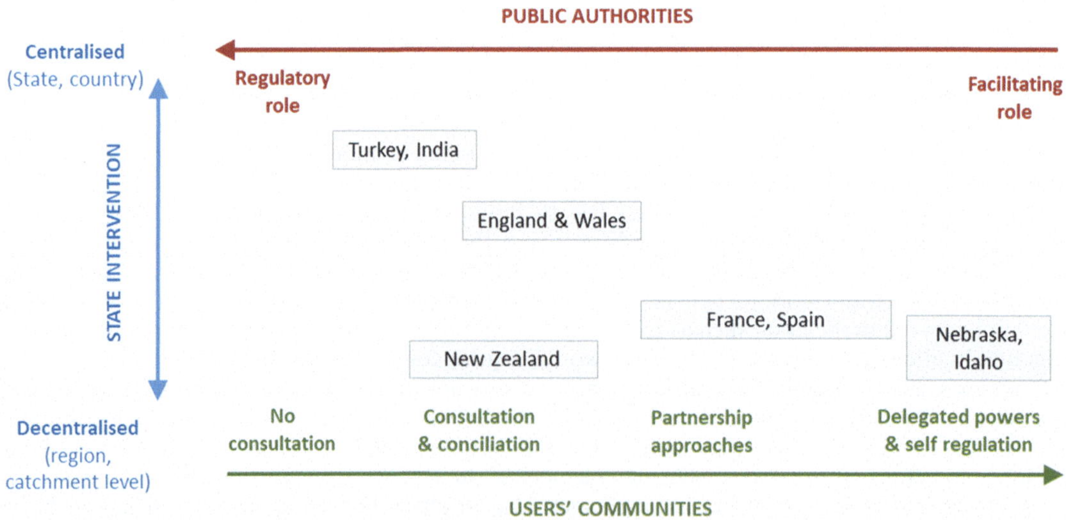

Figure 20.2 Role of public authorities and user communities in allocation. Note: selected examples.

20.2.3.1 The state: from centralised, regulatory interventions to decentralised, facilitative approaches

Concerning public authorities, their role in the transition to regulated access can range from regulatory to facilitative. In regulatory approaches, a centralised state agency makes the key decisions, including defining restriction zones, assessing environmental flows, setting allocation limits and adjusting individual water use permits, based on national procedures. In facilitative approaches, public authorities do not directly participate in allocation decisions, which are devolved to stakeholders; instead, authorities contribute to providing scientific and technical information and to establishing an institutional environment facilitating negotiation and conflict resolution; the State, however, reserves the right to intervene and make allocation decisions if stakeholders do not reach an agreement. As shown in Figure 20.2, most cases covered in this book are hybrid situations, lying on a continuum between those two extreme approaches.

The cases of England and Wales, and New Zealand are illustrative of the more regulatory approach:

- England and Wales illustrate a rather centralised regulatory approach: a national permitting regime defines legitimate users within a wider river basin planning programme under the EU Water Framework Directive; a long-established abstraction licensing regime, originating with the Water Resources Act 1963, is implemented by a central agency; allocation decisions are formalised in Catchment Abstraction Management Strategies and attendant Abstraction Licensing Strategies. Increasingly, this regime is oriented towards environmental sustainability at the catchment scale. Other strongly centralised and regulatory cases are Turkey, India (Chapter 14) and Uzbekistan (in the Amudarya Basin, Chapter 18).
- The case of New Zealand illustrates a more decentralised regulatory approach, where the central government delegates responsibility for water resource management to regional councils. These local government bodies, in consultation with communities and Indigenous groups, create binding statutory plans to deliver on nationally-defined limits and bottom lines, and locally identified objectives, under an overarching policy goal of sustainable management. Allocation decisions are made at the regional or sub-regional level and applied at the scale of catchments or 'freshwater management units' within an integrated management context directed towards ecosystem health and community wellbeing.

At the other extreme of the spectrum, the facilitative approach is illustrated by the US examples (Nebraska, Idaho). In Idaho for instance, in 2015, state authorities brokered a 'Water Settlement Agreement' between competing users of groundwater and surface water from the Eastern Snake Plain Aquifer to restore groundwater levels and sustain surface water flows. Implementation and monitoring of the agreement were devolved to a great extent to the groundwater districts, each of which decided how it would achieve the required reductions. The agreement was mostly achieved because the state threatened to intervene and impose reallocations centrally if users did not compromise.

20.2.3.2 A closer look at user and community involvement

The involvement of communities, users and stakeholders in allocation decisions varies considerably in terms of its form and extent in the cases presented in this book; however, a general trend across jurisdictions has been to make allocation regimes more inclusive, including with regard to Aboriginal values and interests (see chapters 3 and 4). More inclusive and transparent decision-making processes can better take into account local context, build trust, and ultimately craft rules that are more likely to be complied with (Newig *et al.*, 2018).

As public authorities move from centralised regulatory approaches to more decentralised facilitative ones (Figure 20.2), stakeholders are increasingly involved in allocation decisions. In the cases presented in this book, a progression of three tiers can be observed from no consultation and little communication to:

1. Consultation and conciliation, where the focus is on sharing information and creating a degree of shared understanding of the need and basis for capping use and making allocations;

2 Partnership approaches, involving various forms of cooperation or collaboration between users, communities and authorities, where a degree of power sharing is authorised on the part of authorities; and
3 Delegated powers and self-regulation, where user associations and communities regulate access to water.

The first tier of involvement is evident in the cases of India and Turkey. In India, the government has sought to regulate groundwater use since the 1970s, via a national Integrated Watershed Management Programme run by the Ministry of Rural Development. However, this programme was implemented without sufficient adaptation to the needs of specific aquifers and local conditions, and meaningful participation by communities has often been lacking. While there are calls to strengthen participatory processes in decentralised groundwater allocation and management, a techno-managerial paradigm still predominates, which limits scope for participation beyond consultation. Similarly, in Turkey, water users and their representatives are confined to consultative roles in Basin Management Committees and Provincial Water Management Coordination Committees.

The second tier involves communities and users in various forms of partnership alongside authorities in decision-making. For example, the cooperative approach adopted among French environmental authorities, river basin councils, and users involves co-defining management targets and allocations at river basin and catchment levels. Other examples include the collaborative 'catchment-based partnerships' that have proliferated in England and Wales since 2013.

In the third tier, water authorities may opt to delegate decision-making and management powers to communities or groups of water users, with requirements to self-regulate and self-monitor. In the US examples where the State has a facilitative role (see Section 20.2.3.1 above), the Natural Resource Districts in Nebraska exercise delegated authority to make institutional changes and sanction violators for over-abstraction (Chapter 17). In Brazil and France, for example, authorities have issued collective permits to groups of users who allocate to individual members and monitor use. Other instances of power delegation in decision-making can be seen in the allocation and monitoring responsibilities devolved to groundwater districts in Idaho (Chapter 16), and to irrigation collectives in New Zealand (Chapter 11). Forms of self-regulation may be efficient and effective in certain circumstances, but may fail in others. They can be captured by specific interests (Lopez-Gunn & Cortina, 2006) and fail to achieve environmentally sustainable management of the resource (see e.g. Chapter 2, Rouillard *et al.*, 2021). Hence, a key question is how to create institutions that build on the synergies between state and community control, rather than seeing them as antagonistic.

20.2.4 Establishing a wider supportive policy framework

Water allocation regimes seldom work in isolation in the transition from open access to more sustainably governed water resources. In particular, controls on water abstraction will need to be accompanied by a coherent and integrated policy framework that provides incentives and compensatory mechanisms to soften transitions to regulated access (Figure 20.3). Even a well-designed allocation regime can be undermined by perverse incentives in other sectors, such as subsidies that encourage over-consumption of water resources.

In the cases presented in this book, subsidies of various kinds are widely used to promote behaviours consistent with water conservation or to compensate for the economic impact of reduced allocations. In the Turkey, India and EU examples, subsidies are targeted at practices and technologies to prevent water losses and leakage, as well as to promote agri-environmental practices consistent with water conservation.

Subsidies need not be in the form of direct payments to water users. For example, in the US state of Idaho, the State committed to making a significant contribution to aquifer recharge to offset the costs of required reductions in groundwater withdrawals. Similarly, in France where most of the reductions required to achieve sustainable extraction caps affect the agricultural sector, farmers can access

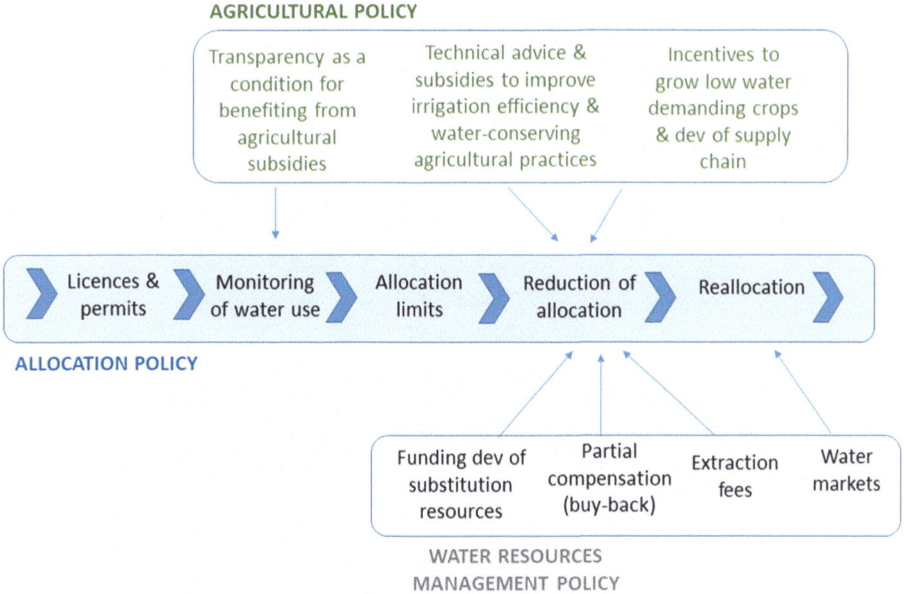

Figure 20.3 Elements of an integrated policy framework for the transition to regulated access with water allocations.

financial support to build storage schemes that can capture flood flows and winter runoff. With other societal groups opposing these solutions, 'contracts' are emerging to deploy an integrated approach to meeting allocation limits by making support for storage schemes conditional on adopting less water-demanding crops, increasing water use efficiency, and building resilience in farming systems through sustainable soil management, agro-ecological practices, and organic farming. This case highlights that policy and institutional arrangements should bring private-sector investments and community activity into line with objectives for water allocation and management.

Aside from subsidies, various other economic instruments may be deployed to complement allocation regimes, as outlined by Perez-Blanco (Chapter 6). Water charges or tariffs can incentivise water saving and efficiency gains on the part of water users, although, in many cases presented in this book, they are not set at sufficiently high levels to have an incentive effect. Various forms of pecuniary payments may be used to encourage temporary reductions in water withdrawals (see Brazil, Chapter 13), markets may facilitate transfer of water permits and increase efficiency of water use at basin or aquifer scales (see New South Wales, Chapter 12), and financial sanctions for non-compliance with allocations are an important tool available to regulators wherever education and engagement cannot secure compliance (see Section 20.5).

20.3 SETTING THE ALLOCATION CAP

As the cases presented in this book show, the process of setting an overall cap on allocations involves several steps and types of assessments (see Chapter 1 and Figure 20.4). All of these assessments are fraught with major technical difficulties and scientific uncertainties that complicate negotiations over the setting of the allocation cap. Below, we focus on three challenges which cases presented in the book have commonly identified: integrating environmental needs; addressing temporal variability of water resources; and accounting for connectivity between water sources.

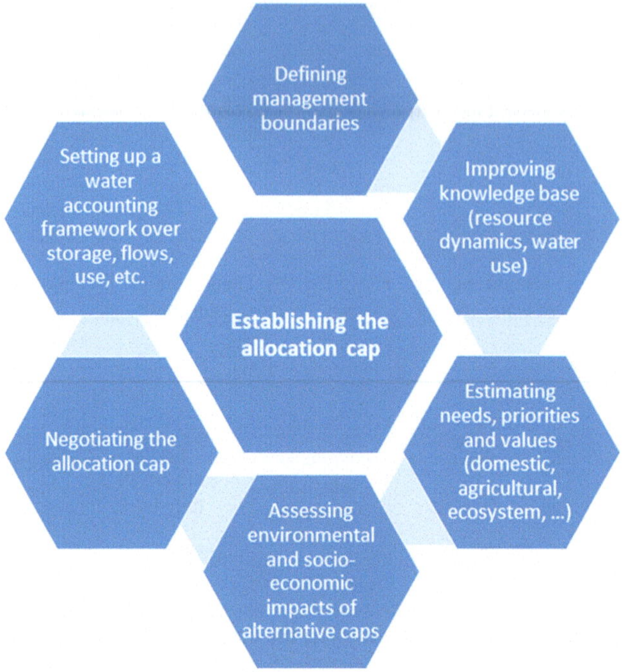

Figure 20.4 Key steps and challenges in setting an allocation cap.

20.3.1 Integrating environmental needs

In the majority of cases in the book, the main approach to integrating environmental needs in allocation limits is to set minimum flows for rivers, sometimes complemented by minimum aquifer levels where groundwater discharge plays a role in sustaining base flows during the low flow period. Minimum river flows and groundwater levels are used as thresholds for implementing temporary restrictions on water use (e.g. when rivers approach or drop below these levels). They may also be taken into account when establishing volumetric allocations to water users (see e.g. chapters 9 and 12).

In the chapter on environmental flows, Stein *et al.* (Chapter 5) highlight that, ideally, flow requirements should be set based on their functional role in maintaining resilient and healthy aquatic ecosystems. This means not only ensuring that minimum flows are not breached during the low flow period, but also maintaining the full range of flows needed to sustain ecosystems (e.g. including high and variable flows). According to the authors, greater use of functional flows may offer greater flexibility to allocate water among environmental and agricultural uses in times of need. However, no cases report using such an approach, although Spain does set various flow requirements (Chapter 8).

In the cases presented in this book, flow requirements are usually set according to water quality standards (i.e. maintaining certain flows for their dilution effect), sediment or temperature characteristics in rivers. In Turkey (Chapter 10), minimum flows are based on arbitrary criteria regarding hydrological characteristics using the Tennant method. In theory, European countries are required under the EU Water Framework Directive to maintain near natural hydrological and morphological character to maintain healthy aquatic ecosystems. In practice, flow requirements are still based on the needs of specific (protected) species rather than on maintaining whole ecosystem functions. Ecosystems and key protected species are taken into account in the US under the Endangered Species Act (see Chapter 17).

The task of determining the 'natural' character of a river or groundwater body and establishing the degree to which the natural flow of a river can be modified without hindering water-dependent ecosystems is still fraught with scientific challenges. This has exacerbated disputes and tensions over setting the 'right' minimum river flows, groundwater levels and volumetric allocations for environmental needs – and their implementation, especially where it has involved reallocating water from agriculture to the environment. Authors in the book advise appropriate attention be paid to local conditions, characteristics, and contexts when setting allocation limits and establishing a transparent process that allows for integration of not only physical considerations, but also social and historical ones (see chapters 2 and 5).

20.3.2 Addressing the temporal variability of the resource in the allocation cap

Available water resources are, by nature, highly variable between months, seasons, years and decades. However, the temporal dynamics of water resources are varied within river basins, as one may find:

- Rivers largely modified by reservoirs, where water storage can, in theory, support delivery of specific quantities of water downstream across the year;
- Rivers in which flows are not supported by reservoir storage and are thus likely to present a more variable flow pattern;
- Groundwater bodies, where there is often limited understanding and knowledge of water storage capacities, water levels, and aquifer structure and dynamics. Some will be connected to surface water bodies and terrestrial ecosystems while others will be shared with other river basins;
- Water transfers to or from other river basins, and unconventional water such as treated wastewater, desalinated water and intercepted rainwater.

When transitioning to regulated access, two general strategies appear to be applied in the cases studied in this book. The first strategy is usually to manage scarcity and drought conditions by restricting use when specific flows or groundwater level limits are reached. Particularly suitable for rivers with more variable flows (i.e. not supported by large storage capacities), these temporary limits on the authorised licensed flow rates are highly dynamic, allowing full use of the resource in times of abundance and providing a form of 'safety net' in emergency situations. However, as scarcity conditions worsen (due to aridification or higher levels of consumptive water use in the system), short-term, emergency limits become more problematic. A race to the bottom may start with agricultural users behaving strategically to appropriate more water before restrictions commence, resulting in even more frequent restrictions (e.g. Chapter 9). Short-term, emergency limits do not address structural over-allocation problems, leaving holders of more junior rights with more uncertain security of supply.

The second strategy involves adopting a collective volumetric cap (further specified in individual caps for authorised users, see Section 20.4.1). Volumetric caps limit withdrawals for specific timescales (e.g. monthly, seasonal, annual and interannual). Methodologies for defining volumetric caps differ between cases and include statistical and political exercises to define the acceptable security of supply to the authorised users. The volumetric approach is particularly suitable for surface water and groundwater systems where storage capacity negates the impact of rainfall variability between years. However, this approach has created tensions where the cap prevents increasing storage of abundant winter flows (Chapter 9).

20.3.3 Accounting for connectivity between water resource types

Recognising interactions between water sources – in particular between surface water and groundwater systems – is of increasing concern to allocation regimes in order to avoid unintended impacts of regulation of one source on another unregulated source (chapters 3 and 5). The cases collected in this book reflect a gradual process of increasing integration of allocation systems across surface water and groundwater. Several cases in the book have, in particular, moved to more integrated accounting across water resource types. In New South Wales for instance, Water Sharing Plans have rules linking the management of surface water and groundwater where there is strong connectivity between the two resources. A proportion of groundwater recharge is assigned to the environment, recognising thereby that groundwater levels and discharge into surface ecosystems are still impacted even if extraction is less than recharge.

Cases of active conjunctive management remain rare in the cases outlined in this book. In Idaho (Chapter 16), conjunctive management of surface water and groundwater was adopted in 1994 in order to tackle dwindling groundwater resources driven by a reduction in incidental groundwater recharge and increased rates of groundwater withdrawal when farmers switched away from diversion-based irrigation to groundwater-based irrigation. Recent political agreement strengthened conjunctive management rules by diverting flood waters (and therefore water not appropriated by anyone) to recharge groundwater through the existing system of irrigation canals and dedicated spreading basins and injection wells. In Nebraska, surface water infrastructure has been repurposed to support groundwater recharge outside the irrigation season and to maximise the use of surface water in lieu of groundwater when surface supplies are high.

Other resource connectivity issues reported in the cases involve the increasing utilisation of previously unused treated wastewater, which has implications on downstream users who previously benefited from wastewater discharges. Changes in use patterns, especially higher consumptive uses such as agriculture, need to be accounted for in water balances as they can pose significant challenges in fully and over-allocated basins where several uses rely on wastewater-fed downstream flows. In Spain (Chapter 8), modification of concessions is required when a single user plans to reuse their wastewater, and reuse by another user is subject to a separate concession.

20.4 ALLOCATION AND REALLOCATION RULES

When transitioning from an open to a formally regulated system, decisions must be made on how to share the allocable pool through establishment of allocation and reallocation rules between users. We identify five steps, each with their own challenges and solutions (Figure 20.5).

20.4.1 Defining authorised users at initial implementation of the allocation cap

Historically, various legal principles and norms applied to prioritise uses in times of scarcity, such as the riparian doctrine or the rule of prior appropriation (see e.g. Chapter 15 for a discussion on the two regimes in Canada). In the cases presented in this book, authorities and user associations typically recognise historical users as the legitimate users, and their historical water use has generally been recognised ('grandfathered') as entitlements. This illustrates the difficulty of sharing water more equitably, at least when first formalising water use rights.

It is worth noting that several cases describe provisions for exemptions from registration and permitting. This may be based on the intended water use (e.g. drinking water for personal use or for animals), the source of water (e.g. spring water, rainwater), or the amount of water. For instance, Spain exempts users from the need to obtain a permit for withdrawals of less than 7000 m^3 per year, but registration is still required. Several authors discuss fairness and equity issues in relation to exemptions, as well as the risk of opening the door to overexploitation. As a result, some legislatures are progressively removing exemptions. In England and Wales, for instance, the rights of irrigators and the Crown to take unlimited amounts of water were removed in 2017, but the law still exempts abstraction of less than 20 m^3 per day (equivalent to $\sim$7000 m^3/year).

20.4.2 Adjusting individual allocations to the allocation cap

Where the sum of individual entitlements exceeds extraction limits, rules are needed to ramp down, or claw back, allocations to match extraction limits. However, modifying entitlements runs against entrenched views on historical water use rights and can face major legal constraints. The

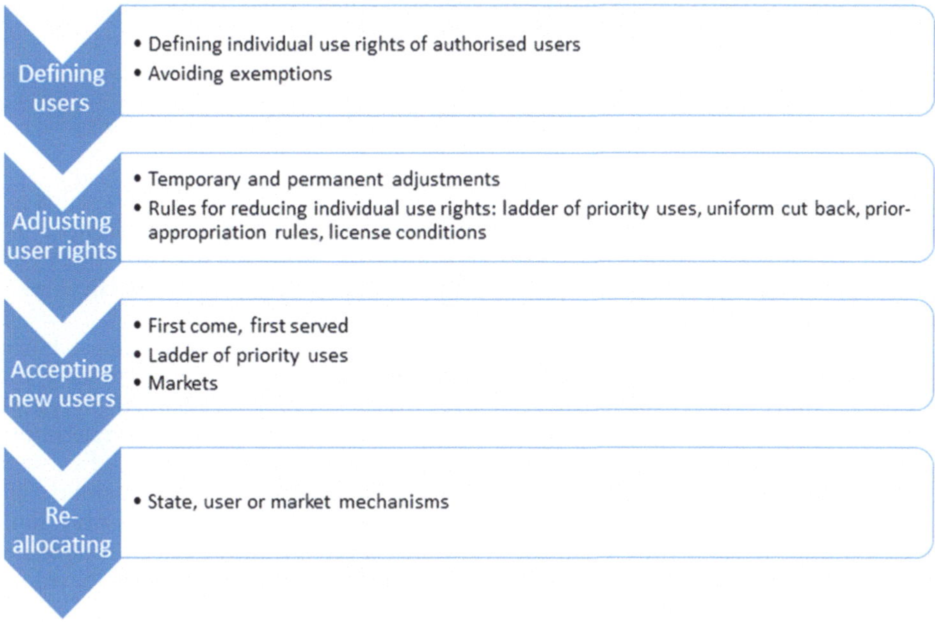

Figure 20.5 Types of allocation and reallocation rules observed in the case studies.

cases presented in this book describe two overarching strategies that have been adopted in reducing allocation.

20.4.2.1 Temporary reductions

The first strategy involves maintaining entitlements, for example at their historical levels, but reducing them temporarily to meet extraction limits. In this system, entitlements act as a right to a share of the available resource; hence, the user is authorised to withdraw only up to their allocation, not their full entitlement. This approach is the most widely implemented, and is used in particular during temporary drought restrictions. In Spain, it is used where institutional rigidity and the threat of significant compensation claims have prevented major permanent reductions of existing concessions in over-allocated basins. It has also been used in other contexts such as Idaho, where voluntary cuts were agreed upon until groundwater reserves were replenished, at which point allocations would return to the original specification of the entitlement. Some cases formally 'unbundle' long-term entitlements and annual allocations. New South Wales for instance presents a sophisticated accounting framework combining extraction limits, entitlements and allocations (called Available Water Determinations), where allocations are tied to resource availability throughout the year (see Chapter 12).

20.4.2.2 Permanent reductions

In some cases, the choice has been to permanently ramp down entitlements to reduce over-allocation and increase the likelihood that a given user is able to use their full entitlement in any year (in other words, to increase the security of supply to authorised users). This approach has been implemented at great social and political costs in New South Wales (Chapter 12). In France, several agricultural user organisations have established a multi-annual process to reduce individual volumetric allocations to a sustainable extraction cap. This approach was facilitated by the designation of water as a 'common good to the Nation' and development of a strong narrative against any private appropriation of water resources. Authorities in England and Wales have taken a similar approach and facilitated the process by removing from the statutes all compensation measures for revoking and amending licences.

20.4.2.3 Rules used to reduce use rights

Rules to reduce use rights (temporarily or permanently) are described in several cases covered in the book:

- A first approach and the easiest one, is to remove unused or underused water rights (e.g. from valid permits). In the US, legal systems have historically accommodated the need to exploit water for 'reasonable' and 'beneficial' use, recognising therefore that use patterns can affect the legitimacy of a use right. However, this approach risks encouraging full use of licences by users who fear losing them otherwise.
- A second approach is to specify a ladder of priority uses. Many countries have established such prioritisations to implement drought restrictions. These typically prioritise drinking water over agricultural water uses. This approach can also be used to implement permanent reductions to entitlements (e.g. preserving entitlements for drinking water over those for agricultural uses) (e.g. Canada, France).
- A third approach exists in prior-appropriation systems which prioritise 'senior' rights (i.e. longer-standing, based on the date of permit issuance or well construction) over 'junior' (i.e. more recently assigned) rights. Junior rights are more severely affected by cuts than senior rights when cuts are implemented. In some cases, statutes have superposed a priority-based system to prior-appropriation to protect essential uses (e.g. Alberta in Canada).
- A fourth approach involves applying uniform reductions across entitlement owners (e.g. % of entitlement). This implies that all users are equally affected by cuts. It is coherent with some aspects of legal principles recognising the 'riparian rights' of surface water users and the 'correlative' rights of groundwater users to use a common water resource. It appears to be implemented more frequently for temporary cuts than for permanent reductions in use rights.
- Fifth, some countries, such as Spain, require that licensed users increase their efficient use of resources according to efficiency targets set in their concessions. This is implemented where the same use can be maintained with a smaller quantity of water. In agriculture, for example, this may lead to investments in more efficient irrigation. In order to avoid the 'saved' water being redirected to other consumptive purposes, potentially resulting in increased net water consumption, the saved water is subtracted from the licence, and no new entitlements are issued with the saved water.
- Sixth, in some places (e.g. New Zealand, Chapter 11), licence conditions provide a specified security of supply. Those with a lower security of supply will be limited in their use before users with a higher specified priority. The advantage of this system is that users can seek to obtain the most appropriate level of supply security when applying for a permit.

20.4.3 Accepting new users

Another challenge is how to prioritise between existing and prospective users when the resource is considered fully allocated. Issuing additional water use rights would impact environmental flows or reduce other users' rights or security of tenure. The authorisation process therefore requires consideration of the degree of flow or volumetric commitment, impacts on downstream flows, ecosystems and protected areas, and other users' security of supply. In most cases covered in the book, where a river or groundwater basin is considered over-appropriated and 'closed', new users are only accepted when water is 'freed up' when a legitimate user surrenders, loses or sells their water use rights.

In the case where market transactions are allowed, the water use rights may be acquired by buying the water use right itself, or indirectly by buying the land to which it is attached. Few cases allow market transaction of the water use right itself (e.g. New South Wales, Spain, England and Wales) although this, in theory, leads to a more economically optimal allocation of water use rights (see Chapter 6).

In most cases, new users must buy (or rent) the land to which the water use right is attached, or apply for an authorisation for instance in the form of an administrative permit. Most authorisations are issued with a 'first come, first served' approach, wherein prospective users are prioritised according to the date of their original request (e.g. on a waiting list). This approach is applied in Brazil and New Zealand. This approach does not permit optimisation of the economic value of the allocable water as in the case of markets. A prioritisation that favours pre-defined water uses is however sometimes applied by the authorities or organisation responsible for issuing authorisations to use water. For instance, in Spain, the same priority ladder that is applied during droughts is implemented to prioritise the issuance of permits (e.g. for drinking water before agriculture). In France, rules have been defined by agricultural user organisations to prioritise higher-value agricultural production systems over lower-value ones, or young farmers (to encourage farm renewal and investments).

20.4.4 Facilitating state, user or market reallocation

In most systems presented in this book, the state reallocates water use rights. For instance, in the Uzbekistan part of the Amudarya Basin (Chapter 18), users apply to the government for yearly allocations. Similarly, in Turkey (Chapter 10), water reallocations are decided by state authorities. Where some form of devolution to users or communities exists (see Section 20.2.3), non-state actors may themselves manage reallocation of water. In Brazil and France, for instance, user associations have powers to reallocate among irrigators in their management areas.

Several authors (e.g. chapters 2 and 6) support the use of economic instruments to overcome institutional rigidity and enable rapid response and user adjustment to changing conditions. Cases in the book include incentive programmes, such as buy-backs of entitlements and short-term leases (e.g. Nebraska, New South Wales), pecuniary payments for fallowing land (e.g. California, Brazil), water markets (user-to-user trades, New South Wales) and water banks (user-to-authority trade) (e.g. Spain and England & Wales).

20.5 COMPLIANCE AND ENFORCEMENT

Enforcing allocation rules is a key challenge reported in most of the case studies covered in this book (e.g. Spain, France, New South Wales). Monitoring of water use and enforcing compliance with water use rights is particularly challenging in agricultural basins where thousands of individual extraction points may exist. Compliance issues reported in the chapters of this book as elsewhere in the literature (Schmidt *et al.*, 2020) include illegal extraction points, unlicensed use (e.g. domestic borehole used for irrigation), non-compliance with licence specification (in terms of volume, timing or place of extraction), failure to report use data, or tampering with water monitoring devices (e.g. meters). Compliance can be improved using five main mechanisms (Figure 20.6): (i) using modern technologies for water use monitoring; (ii) involving users and other societal groups in monitoring and enforcement activities; (iii) developing a graduated, progressive approach of enforcement; (iv) increasing transparency of water use information and compliance; and (v) encouraging the development of social norms of compliance and collective responsibility.

20.5.1 Technology

Many advanced cases initially relied on estimates of water use, but have moved to compulsory installation of water metering (e.g. Idaho, Spain). Modern monitoring technologies may also allow better control of water use in agriculture, for instance through the use of satellite imagery to identify irrigated areas, telemetry for real-time monitoring of water extraction, and smartphone applications allowing frequent water use reporting.

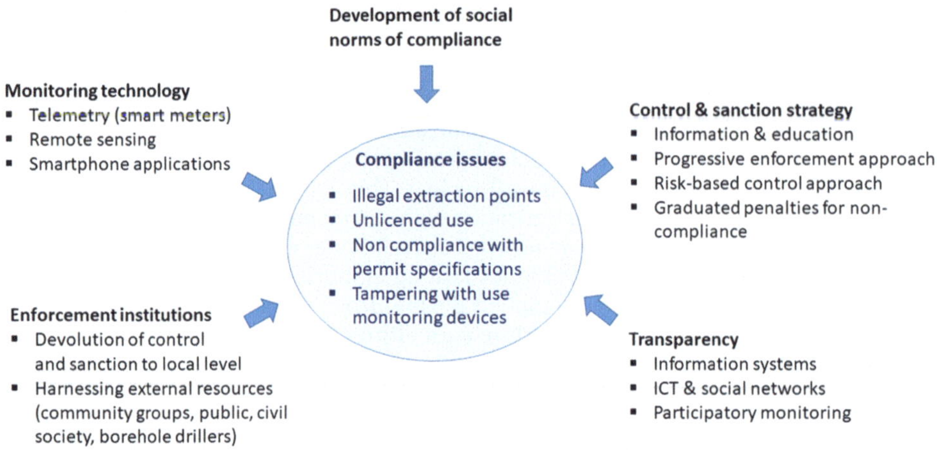

Figure 20.6 Five main mechanisms facilitating compliance with allocation regimes.

Several cases in the book report significant effort in improving surveillance and monitoring technologies. The installation of meters, along with record-keeping and reporting, have been made compulsory in several cases (e.g. Spain, Idaho, France, New Zealand, England and Wales), sometimes accompanied by telemetering and cross-checked with remote sensing (Spain).

20.5.2 Institutions

Enforcement institutions can be adapted to encourage compliance with allocation rules, in particular by (partly or fully) devolving control and sanctions to water user communities themselves, or to local institutions with greater legitimacy than government agencies. Another strategy involves enrolling other social groups, such as the public, experts, community groups, or civil society groups in compliance monitoring. Such an approach can harness human resources and capacity from within the community to augment or supplement insufficient government resources for compliance monitoring, but it requires careful management and trust-building among involved stakeholders.

Some countries have taken this general approach, enabling greater degrees of local control over allocations to enhance compliance with allocation limits. Natural Resource Districts in Nebraska, for example, are empowered to implement cease and desist orders and fines of up to $5000 per day for non-compliance. Brazil and France use collective licences granted to agricultural user associations to enhance compliance. These associations are tasked with registering all abstraction points in the course of establishing allocations, and subsequently reporting meter readings. Despite limited formal powers to sanction non-complying users, the process in France has resulted in identification of multiple unrecorded abstraction points, mainly thanks to peer pressure and self-policing within the agricultural community. The threat of further state intervention in the case of non-compliance played an important role. Similarly, in Idaho, communities of users and local organisations are given powers to enforce allocations, although compliance with recently agreed cut-backs (which were not legally binding) was driven by the threat of further state intervention.

20.5.3 Enforcement strategy

The effectiveness of enforcement activities can also be enhanced by implementing a graduated, progressive approach of enforcement relying on (i) education, information, advice and prevention activities; (ii) notice (requiring improvements) and warnings; (iii) administrative sanctions; and (iv) court action. Under such an approach, most enforcement action is concerned with directing compliance, while activities involving administrative remedies and criminal proceedings are far less common. A risk-based approach can also help to optimise the effectiveness of limited government resources, by prioritising regulatory activities and deploying resources based on an assessment of the risks that users pose to regulatory objectives, for example at sites of intense pumping, or in areas with significant water-dependent ecosystems.

This is illustrated with the case of England and Wales, where the Environment Agency applies a graduated approach, starting from advice and guidance, followed by warnings, enforcement notices, civil sanctions and criminal proceedings. Similar approaches are implemented in France, New Zealand and Australia (Holley *et al.*, 2020; Montginoul *et al.*, 2020).

20.5.4 Transparency

Transparency of water monitoring is another factor likely to enhance compliance. Transparency can be fostered using open digital information systems, employing information and communication technologies, and mobilising social networks or participatory monitoring initiatives. The benefits of transparency in monitoring and compliance, using reliable, trusted and readily accessible information about water sources and water allocations, are highlighted by Nelson (Chapter 3).

Examples of the use of such technologies are found in several chapters of the book. In England and Wales, environmental authorities have focused on digitising individual licence management via an online user interface that allows users to access their licence details, and to submit abstraction records online. In addition, water abstraction licensing was streamlined with other environmental permit requirements to ensure consistency. In New Zealand and Brazil, authorities and agricultural user associations responsible for agricultural allocations offer similar services, sometimes developing computer applications and single interfaces to provide real-time monitoring and reporting to farmers. This can make the state of the water resource and water takes transparent to all users in order to support preventative measures to avoid the imposition of use restrictions. In India, Aleska *et al.* (Chapter 14) present a participatory approach to mapping, measuring and monitoring groundwater dynamics and use.

20.5.5 Social norms

Last but not least, compliance is highly dependent on social norms that emerge within each particular cultural context and that determine whether and to what extent water users comply with allocations set by agencies. Engaging non-government actors (community leaders or 'champions', who influence others through their good example) to promote broader social norms will provide the true 'glue' that cements and holds cooperative compliance behaviours together (Holley *et al.*, 2020). Consolidating a social norm of collective responsibility and compliance can be more effective than using 'enforcement sticks', in particular in the absence of sufficient enforcement staff or resources on the ground.

20.6 CONCLUDING REMARKS

The intention with this book was to present a range of water allocation regimes regulating agricultural water use in overexploited basins and contributing more widely to the transition away from open access to surface and ground water resources. More specifically, it aimed to fill a perceived gap in the scientific literature by providing a simple, concrete and more 'operational' description of such water allocation regimes. Hence, we hope this edited volume will be a source of information and inspiration to practitioners and scientists alike in their work striving to reform water allocation regimes in the transition towards more sustainable water use.

Based on our reading of the material presented in this book, and without striving for an exhaustive synthesis, we would like to briefly highlight some key takeaways that scholars and practitioners working on water allocation reforms may want to consider (see also the principles set out by Blomquist and Babbitt in Chapter 2):

- First, any reform needs to be implemented on the basis of trusted, transparent water availability and use data. This serves to build a common understanding of the challenges and the degree of transformation required in the exploitation of water resources. Systems must be in place to monitor conditions with triggers to enable action based on data that are accessible, transparent and reliable.
- Second, careful attention must be given to the governance of the water allocation process. Authorities and entities in charge of the process should have the authority and credibility to manage water resources and set allocation rules. At the same time, local control (with state oversight) in setting and enforcing allocation and reallocation rules may be an effective way to secure commitment to the reform process. In any case, emphasis during the reform process should be on securing stakeholder participation and buy-in to the allocation reform. Users need to understand the urgency of the situation and why allocations are needed.
- Third, setting a cap on water withdrawals and (re)allocating available water essentially entails questions of fairness and social justice. Users want to be treated fairly and need to know what the rules are so they can plan accordingly. This entails attention to both procedural and distributive justice.
- Fourth, capping and allocating water resources should not be seen as a rigid and fixed process. Water resources are dynamic in nature and allocations should respond to this variability to ensure environmental effectiveness. At the same time, greater user acceptance is likely if users have some flexibility on how to use their allocations within the limit required to keep the system environmentally effective. This may imply allowing users to carry-over some amount when it will not negatively impact effectiveness. It also calls for robust transfer and trading schemes which reallocate water between uses according to needs.
- Sixth, attention must also be given to the wider policy framework to ensure sufficient coordination across governmental agencies and avoid, for example, sectoral subsidies that run against the goals of the water allocation regime.
- Finally, although not always possible, it is preferable not to wait until the situation is dire, as this will result in more rigidities and limit adaptive capacity.

Beyond these recommendations, we would like to conclude on the urgent need for further work assessing the performance of water allocation systems. In line with the three pillars of sustainability, and the overarching need to adapt to climate change, allocation systems should support outcomes that are environmentally effective, socially equitable, and economically efficient, while being resilient in the long term. Some of the cases presented in this book provide some evidence of such sustainability performance. For instance, some positive signs of environmental recovery are visible in for example Nebraska and Idaho, where groundwater levels have been partially restored. Many cases also undeniably demonstrate increased participation of communities and users, and the increasing consideration of human and Aboriginal/Indigenous rights. And reallocation rules in several cases also allow for redistributing allocations to more junior permit holders, new users or higher value uses.

At the same time, many of these cases are still in the early phases of the transition process, and are constrained by opposition, lack of participation, and institutional rigidities. A robust assessment framework is needed to analyse the performance of water allocation systems in a long-term perspective, and adequately consider their capacity to work well under stress by maintaining effective, fair and economically optimal outcomes over time, taking into account the impacts of climate change. As water resources will increasingly face pressures from climate change, population growth, economic development, and other stressors on water supply and demand, scholarship on water allocation regimes must continue to identify, describe and assess innovations in water allocation systems and promote their diffusion across contexts to support efforts to transition towards sustainability.

REFERENCES

Cleaver F. (2017). Development Through Bricolage: Rethinking Institutions for Natural Resource Management. Routledge, Abingdon, UK.

Schmidt G., De Stefano L., Bea M., Carmody E., van Dyk G., Fernández-Lop A., Fuentelsaz F., Hatcher C., Hernández E., O'Donnell E. and Rouillard J. J. (2020). How to tackle illegal water abstractions? Taking stock and lessons learned from international experiences. Fundacion Botin, Madrid, 37pp.

Holley C., Mutongwizo T., Pucci S., Castilla-Rho J. and Sinclair D. (2020). Groundwater regulation, compliance and enforcement: insights on regulators, regulated actors and frameworks in New South Wales, Australia. In: Sustainable Groundwater Management, J. D. Rinaudo, C. Holley, M. Montginoul, and S. Barnett (eds.) Springer, Cham, Switzerland, pp. 411–433.

Lopez-Gunn E. and Cortina L. M. (2006). Is self-regulation a myth? Case study on Spanish groundwater user associations and the role of higher-level authorities. *Hydrogeology Journal*, **14**(3), 361–379, https://doi.org/10.1007/s10040-005-0014-z

Montginoul M., Rinaudo J.-D. and Alcouffe C. (2020). Compliance and enforcement: the Achilles heel of French water policy. Sustainable Groundwater Management, J. D. Rinaudo, C. Holley, M. Montginoul, and S. Barnett (eds.) Springer, Cham, Switzerland, pp. 435–459.

Newig J., Challies E., Jager N. W., Kochskaemper E. and Adzersen A. (2018). The environmental performance of participatory and collaborative governance: a framework of causal mechanisms. *Policy Studies Journal*, **46**(2), 269–297, https://doi.org/10.1111/psj.12209

Rouillard J., Babbitt C., Pulido-Velazquez M. and Rinaudo J. D. (2021). Transitioning out of Open Access: a closer look at Institutions for Management of Groundwater Rights in France, California, and Spain. *Water Resources Research*, **57**(4), e2020WR028951, https://doi.org/10.1029/2020WR028951

Index

E

F

G

H

I

L

M

N

O

P

R

S

T

U

W

IWA Publishing's authorised EU representative for General Product Safety Regulations is Diane D'Arras, 15 rue Duret, 75116 Paris, France, e-mail: safety@iwap.co.uk.

Printed and bound by CPI Group (UK) Ltd, Croydon, CR0 4YY
12/05/2026
02108877-0006